MAGNESIUM: CURRENT STATUS AND NEW DEVELOPMENTS

Magnesium: Current Status and New Developments

Theoretical, Biological and Medical Aspects

edited by

Theophile Theophanides
and
Jane Anastassopoulou
National Technical University of Athens, Greece

KLUWER ACADEMIC PUBLISHERS
DORDRECHT / BOSTON / LONDON

A C.I.P. Catalogue record for this book is available from the Library of Congress.

ISBN 0-7923-4821-4

Published by Kluwer Academic Publishers,
P.O. Box 17, 3300 AA Dordrecht, The Netherlands.

Sold and distributed in the U.S.A. and Canada
by Kluwer Academic Publishers,
101 Philip Drive, Norwell, MA 02061, U.S.A.

In all other countries, sold and distributed
by Kluwer Academic Publishers,
P.O. Box 322, 3300 AH Dordrecht, The Netherlands.

Printed on acid-free paper

Printed in the Netherlands

TABLE OF CONTENTS

III. TOXICOLOGY

IV. AGRICULTURE

V. METABOLISM

VII. CARDIOVASCULAR DISEASES

X. MISCELLANEOUS

PREFACE

This book presents the lectures and posters of some of the main leaders in the field of magnesium research and medical applications delivered at the 8th International Symposium on Magnesium, which took place on 5-9 October, 1997 in Heraklion, Crete under the chairmanship of Professor Theophilos Theophanides.

The meeting was sponsored by the National Technical University of Athens and Ministry of Industry, Energy and Technology.

The aim of this meeting was to promote research and applications of magnesium and interface between medical doctors, clinicians and scientists in order to understand the mechanisms responsible for magnesium involvement in the pathogenesis of diseases, its biological significance, metabolism and many other utilizations which are associated with membranes and cells. The success of this interface is due to the contribution of its participants who came from all continents and to their high scientific level.

The topics, which were presented, and the questions, which were asked, concerned mechanisms of mode of action of free magnesium cations, Mg^{2+}, hydrated cations, Mg^{2+}. 6 H_2O, and magnesium linked cations $Mg^{2+}.LxH_2O$, where L: ligand and x = 1...6 H_2O.

We would like to express our utmost gratitude to the sponsers and to extend our deep appreciation and thanks to all those who helped and encouraged the scientific and material organization of this meeting. We wish to thank all the members of the scientific committee and the organizing committee.

Jane Anastassopoulou Theophilos Theophanides

Athens, July 15, 1997

MAGNESIUM: AN INTRODUCTION TO ITS SUPRAMOLECULAR CHEMISTRY AND BIOLOGY

T. THEOPHANIDES[1], J. ANASTASSOPOULOU[1], G. DE MUNNO[2], J. DURLACH[3] AND M.BARA[3]
[1]National Technical University of Athens, Chemical Engineering Department, Radiation Chemistry and Biospectroscopy, Zografou Campus, Zografou 15780, Athens, Greece, [2]Dipartimento di Chimica, Università della Calabria, Arcavacata di Rende, Consenza, Italy, 3. S.D.R.M. Hopital Saint Vincent de Paul, Avenue Denfert-Rochereau, Paris, France

1. INTRODUCTION

Despite the major advances of magnesium research during the past 20 years, there is still much to be learned about the mechanism of action of magnesium cations in the many chemical and biological processes in which the water molecules are intimately involved. Very little systematic study has been conducted of magnesium chemicals that may inhibit or prevent cancer. The effect of counter anion also has not been studied to its basic considerations and magnesium ions in cardiovascular cells. The supramolecular chemistry of magnesium in biological systems is the starting point of description of the chemistry and biochemistry of magnesium. Magnesium ions are present in all cells and in all reactions in which ATP is a substrate (Mg^{2+}.ATP)[1]. In this substrate complex, Mg^{2+} is chelated between the phosphates β and γ and is capable of approaching and bind in a reversible fashion to other sites of specific proteins. However, its precise function in the specific cleavage of the terminal phosphoryl group of the ATP molecule is still unknown [2]. In addition, it intervenes in all the nucleic acid syntheses, in nucleotides or lipids and glucides and in the activation of muscular contraction. This cleavage involves hydrolysis of ATP to ADP and Pi and Mg^{2+} is known to be essential for this de-phosphorylation. The hydrolysis of ATP at the Pγ position has not been determined clearly. It proceeds through the attack of nucleophiles, such as OH^- and H_2O on the Pγ

T. Theophanides and J. Anastassopoulou (eds.), Magnesium: Current Status and New Developments, 1-5.

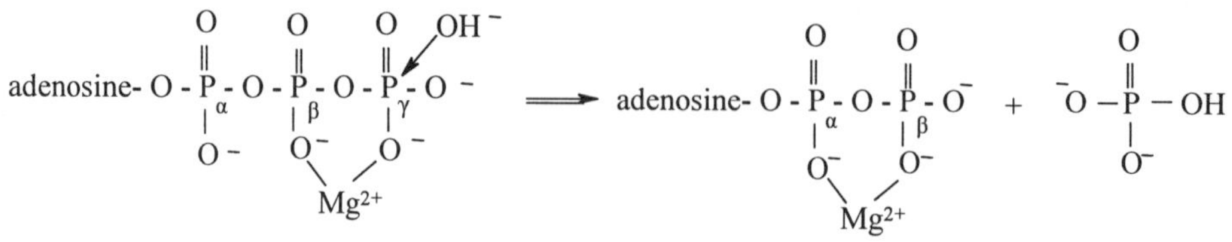

atom. A proposed mechanism of hydrolysis of ATP has been published [2], in which the nucleophile attacks the Pγ atom, which forms a P_β-O-Pγ chelate. This chelate leads to the formation of $HPO_4^{=}$ by breaking the Pγ-O bond and forming a Pα-O-Pγ chelate. Mg^{2+} is a structure construction or stabilizing ion, particularly in all biological systems, where water is the biological molecule by excellence to which it is attached avidly [3].

In the cells there is also antagonism of Ca^{2+} against Mg^{2+}, particularly in the kidney-stone formation, where, if the ratio of Mg/Ca >0.7 then this is normal. However, if this ratio is < 0.7 then "calculus" type stone-forming is taking place. In these cases oral Mg^{2+} supplementation prevents the kidney-stone formation [4].

2. THE BEHAVIOR OF Mg^{2+} IONS IN WATER SOLUTIONS AND IN BIOLOGICAL SYSTEMS

Mg^{2+} is the major divalent cation present in the cell. Mg^{2+} in its relationship with the other alkali and alkaline-earth metal ions, such as Li+, Na+, K+ and Be^{2+}, Mg^{2+}, Ca^{2+} will be used in nature in all living processes as a competitor, however not always on equal terms. For instance, in biology the concentration of the above ions is very unequal and therefore this competition is unfair. The following parameters are very important in the competition for biological sites. The charge on the metal, the size and the number of water molecules coordinated to the metal play a major role in the above competition. Selectivity for organic anion sites is based on the stability constants of the complexes formed. The effect of the ion and its hydration characteristics plays an important role in the formation of coordinated supermolecules which stabilize a complex·

Mg^{2+} ions exist in aqueous solutions not as naked ions (Mg^{2+}), but as aquated cations $Mg^{2+}x.H_2O$, where the water molecules are coordinated water molecules [3], and can be dissociated as follows:

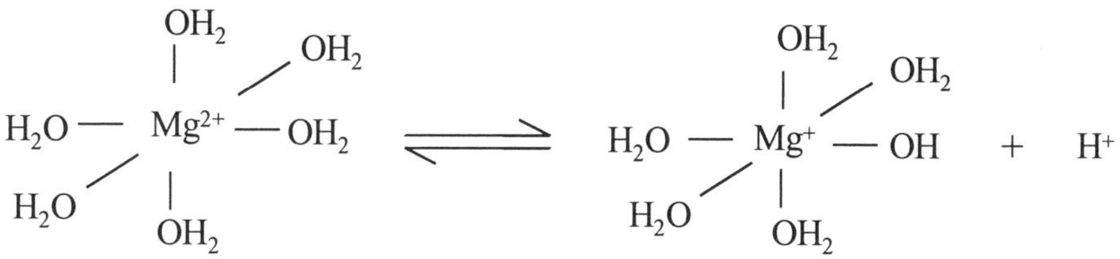

each magnesium atom in $Mg^{2+}xH_2O$ is coordinated to six water molecules in an octahedral environment. This has been found in the crystal structure of $[Mg(H_2O)_6(1\text{-}Mecyt)_6]\ (ClO_4)_2.H_2O$[5]. The structure of this supermolecule is illustrated in Figure 1.

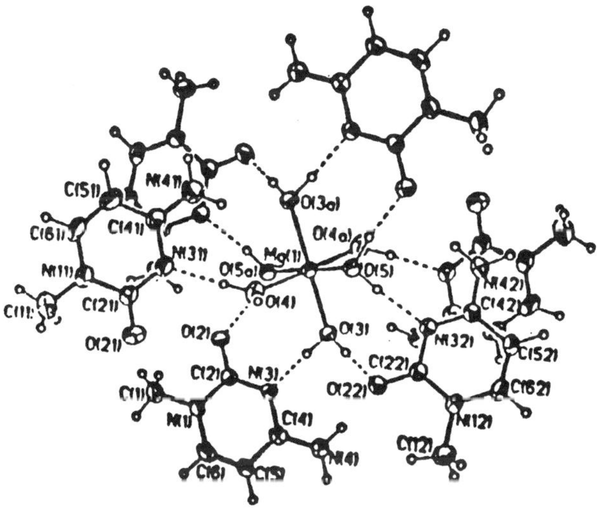

Figure 1. The supramolecular complex cation $[Mg(H_2O)_6(1\text{-}Mecyt)_2]^{2+}$

The crystal structure and atomic labeling of the $[Mg(H_2O)_6(1\text{-}Mecyt)_6]^{2+}$ ion with interatomic distances and hydrogen bonds are also shown. The equilibrium dissociation process of the aquated ion $Mg^{2+}.6H_2O$ may be influenced by the hydrogen bonds of the supermolecule and the process of breaking the O-H bond to form H^+ protons and hydroxyl groups which depend on the rate of water exchange with the metal. This point of course is concerned with the thermodynamics or equilibrium properties of Mg^{2+}.
In the following reaction:

$$Mg^{2+} + H_2O \rightleftharpoons Mg^{2+}(H_2O)$$

The exchange is slower than in the alkali or alkaline-earth cations and these aquated magnesium ions could act as slow regulators in the processes in which it is involved. Consequently $Mg^{2+}.xH_2O$ ions are structure stabilizers [6,7] and its counter anions.

In nucleic acids each nucleotide consists of a negatively charged phosphate group, a ribose sugar and an aromatic base. The negative changes are neutralized by strong cation coordination or by weak cation binding. A detailed analysis of the binding sites and their role in double helix formation permits us to learn more about the three-dimensional structure of polynucleotides [8].

Magnesium-water octahedron fits loosely on the pockets formed by the polynucleotide and can be or not coordinated on one site. Another magnesium site in the helical polynucleotide occurs at another position of the polynucleotide bound to a negative phosphate. Sometimes magnesium is found directly coordinated to the phosphate and hydrogen bonding through water molecules to the phosphates or other sites. Some of these possibilities have been found in the structures obtained by us very recently [9]. The binding constants for Mg^{2+} are here of the order of 10^3.

In nucleotide systems one can imagine both specific and general binding to be of great importance with $Mg^{2+}x.H_2O$ [10].

References

1. Szent-Gyoergyi, A. (1956) in O.H. Gaebler (ed.), *Enzymes: Units of biological structure and function*, Academic Press, New York, p. 393.
2. Yokshikawa, K., Shinohara, Y., Terada, H., Kato, S. (1987) Why is Mg^{2+} necessary for specific cleavage of the terminal phosphoryl group of ATP?, *Biophs. Chem.*, **27**, 251-254.
3. Anastassopoulou, J., Andreopoulos, A.G., Theophanides, T. (1996), The interactions of magnesium salts with water in biological systems, in M J Halpern & J Durlach, (eds.), *Current Research in Magnesium*, John Libbey, London, Chap 33, p. 137-140.
4. Williams, R.J.P., Magnesium: *An introduction to its biochemistry*, (1993), in N.J. Birch (ed.), Academic Press, New York, p. 15
5. Gaday, M.A., De Munno, G., Mrdaglia, M., Anastassopoulou, J., Theophanides, T. (1997) First Isolation of Supramolecular Assemblies containing Nucleic Bases and Magnesium (II) Hexahydrate Ions, *Angewante Chemie, Int. Ed. Engl.*, **36**, 511.
6. Theophanides, T. (1996), Biological Implications of Magnesium Salt at the Molecular Level, *Magnesium Res.*, **9**, 259.
7. Durlach, J., Bara, M., Theophanides, T. (1996), A hint on pharmacological and toxicological Differences Between Magnesium Chloride and Magnesium Sulphate, or of Scallops and Men, *Magnesium Res.*, **9**, 217-219.

8. Teeter, M.M., Quigley, G.I., Rich, A. (1980), Metal ions and transfer RNA, in T.G. Spiro, ed., *Nucleic acid-metal ion interactions*, John Willey and Sons, New York, Chapt. 4., pp. 145-177.
9. Armentano, D., De Munno, G., Regina, M., Anastassopoulou, J., Theophanides, T.(1997), Supramolecular Mg(II) complexes of cytosine and 1-methylcytosine, in T.Theophanides, J. Anastassopoulou, eds., *Magnesium: Current status and new developments*, Kluwer Academic Publishers, The Netherlands, this volume.
10. Theophanides, T., Angiboust, J-F., Polissiou, M, Anastassopoulou, J. Manfait, M. (1990), Possible role of water structure in biological magnesium systems, *Magnesium Res.*, **3**, 5-13.

NEURODEGENERATIVE DISEASES AND MAGNESIUM STATUS

JEAN DURLACH
SDRM, Hopital Saint-Vincent-de-Paul, 74-82 Avenue Denfert-Rochereau, F-75014 Paris, France

1. INTRODUCTION

Neurodegenerative diseases (NDD) are disorders of the central nervous system (CNS) due to the selective and premature decay of a functionally related population of neurons, occurring without obvious causes nor inflammatory stigma, but with slow and ineluctable evolution and a few pathological hallmarks of neuronal abiotrophy.
These have been called the "gravestones" of neurons e.g. neurofibrillary tangles, amyloid plaques, spheroids, Bunina and Lewy bodies.
Dementia of the Alzheimer type (D), Parkinson's disease (P) and amyotrophic lateral sclerosis (ALS) are the most frequent age-related human neurodegenerative diseases.
They are major social problems and require growing medical attention. 2020 project 16% of the total population to be elderly. Research in the field of neurodegererative disease is in the forefront of basic and clinical neuroscience [1].
It is therefore logical to take into consideration every theory with an overall approach of neurodegenerative diseases. This forthcoming environmental proposal mainly relies on data concerning the Western Pacific Amyotrophic Lateral Sclerosis Parkinsonism-Dementia complex (WP ALS-PD) considered as "a prototypal human neurodegenerative diseases [2] and on extrapolation from it to the bulk of neurodegenerative disease (NDD) [3]. Studies of WP ALS-PD are intended to reveal clues on the etiology and pathogenesis of this and of clinically related neurodegenerative disease [4]. NDD would be due to an accelerated aging process in certain populations of neurons due to the noxious synergy of 1. Increased environmental slow deleterious factors (such as slow toxins) and of 2. Decreased environmental protective factors (Mg deficient intake particularly).
The aim of this study is: To recall clinical, morphological, epidemiological and physiological data concerning WP ALS-PD. To make a critical appraisal of the overall theory of neurodegenerative diseases and of its theoretical and practical consequences.

2. WESTERN PACIFIC AMYOTROPHIC LATERAL SCLEROSIS-PARKINSONISM-DEMENTIA COMPLEX (WP ALS-PD): CLINICAL, MORPHOLOGICAL, EPIDEMIOLOGICAL AND PHYSIOLOGICAL DATA

T. Theophanides and J. Anastassopoulou (eds.), Magnesium: Current Status and New Developments, 7-16.

Guamanian neurodegenerative disease first entered the neurological literature at the century when a high incidence of so-called "hereditary paralysis" was reported. Later it was realized that three apparently unlike conditions occurred at extraordinary high rates in Guam area: motoneuron disease (ALS), Parkinson's disease (P) and Alzheimer's-like dementia (D) [5]. Guam ALS is characterized by progressive limb weakness, spasticity, bulbar dysfunction, fasciculation and muscle atrophy. Patients with Guam PD exhibit early loss of fine movements, decreased blinking, bradykinesia, resting tremor progressive dementia, Parkinsonian rigidity, lower and upper-motor neuron signs (or both), incontinence of urine and feces dysphagia, anarthria and eventually flexion-contractures of all joints. 5% of Guam ALS patients subsequently develop PD whereas 38% of PD display ALS [6]. Morphologically abiotropic lesions are characterized by widespread neurofibrillary tangles in association with prominent neuronal degeneration [7] and are found in upper motor neurons, in lower motor neurons, the substantia nigra, the hippocampus and the medial basal forebrain [3].

Several other foci of endemic ALS-PD were found in Asia and Oceania in three Western Pacific population groups. These included the Chamorro people in Mariana Islands (Guam and Rota), The Auyu and Jakai people of West New Guinea and the Japanese residents of the Kii peninsula (Honshu island) [4,8,9].

Early epidemiologic studies suggested the possibility of a genetic etiology [10.11], but subsequent familial aggregation analyses failed to reveal findings characteristic of any simple mendelian inheritance patterns [12, 13 and WP ALS-PD occurred in disparate ethnic groups [4, 8, 9]. No slow virus has been implicated [5, 14]. Heritable and infections etiologies have been, virtually ruled out [4, 5, 15].

The post-Second World War decline of the occurrence of WP ALS-PD in all three high incidence disease foci [4-6] coupled with the absence of demonstrable heritable or transmissible factors [4, 5, 15] had led to focus the search for the cause of this degenerative disease on nontransmissible environmental factors that are disappearing as the susceptible population groups acculturate to modern way. Epidemiologic study has shown that preference for traditional Chamorro food is the only of 23 tested variables significantly associated with an increased risk for PD [9]. An early suggestion incriminated the toxic seed of the false sago palm (*Cycas circinalis L*) which was used in traditional food and medicine. Laboratory investigation of cycak seed revealed the presence of various toxins and particularly of an "unusual" non protein aminoacid: L-BMAA (β-N-methylamino-L-alanine), an excitotoxic aminoacid which activates NMDA receptors at high concentrations and non NMDA and metabotropic receptors at low concentrations [6, 16]. This was the first hint of a **"slow toxin"** that is to say a deleterious factor whose neurodegererative effects will be expressed only many years after [5].

This bioorganic neurotoxin presents some structural similarity to another "unusual" excitotoxic aminoacid: L-BOAA (β-N-oxalyl-amino-L-alanine) which also displays preferential affinity for the non-NMDA receptors. L-BOAA is an exogenous neurotoxin present in the grass pea (*Lathyrus sativus*) whose excessive consumption may cause lathyrism: a central motor-system disease [16]. It seems important to note that the excitotoxicity of both L-BMAA and L-BOAA mainly concerns non-NMDA receptors.

Paradoxically large consumption of cycad seed [16] or of grass pea [17] may be perfectly well tolerated. This might be the result of the methods of washing and fermentation, which successfully remove the noxious principles.

The neurotoxicity of these aminoacids highly varies with experimental models failing to induce an experimental model akin to WP ALS-PD [5, 6] or displaying many of the motor-system and behavioral changes of WP ALS-PD [6, 16. It may also be due to the presence of physiological levels of bicarbonate [2] or of various toxic cofactors: bio-organic such as cycasin [16] or inorganic such as pollutant metals e. g. aluminum or manganese [4]. That this last mechanism together with the lack of protective factors (e.g. calcium and magnesium deficiencies) should be cofactors of bio-organic neurotoxins, is only hypothetical at the present time. But environmental heavy metal exposure *per se* may constitute the fundamental cause for WP ALS-PD.

In the three foci of endemic neurodegenerative complex excessive amounts of toxic metals including aluminum (Al) and manganese (Mn) were present in garden soil and drinking water. Metal analysis of the CNS tissues from necropsied cases showed high levels of Al, Mn (and Ca) [8, 18]. However metal neurotoxicity varies greatly depending on various factors: species (rabbits and cats are more vulnerable than other mammals such as rats, mice and primates), nature of the metal salts and of the food intake [19]. It was hypothesized that calcium and magnesium dietary deficiencies might increase the neurotoxicity of metals. Secondary hyperparathyroidism due to low Ca/Mg could increase the absorption and deposition of metals such as Al, Mn and Ca in central nervous system and develop later into the neurodegeneration characteristic of ALS-PD [18-21]. When speculating on the environmental characteristics of ALS-PD complex, it seems necessary to study animal models in a condition more similar to the environment of endemic areas of the neurodegenerative disorder where people in a Ca/Mg deficient state have been chronically exposed to neurotoxins since early in life. But these studies practically concern low Ca-Mg diet with Al load only [19-24] and not low Ca-Mg diet simultaneously with all exogenous noxious factors: the toxins from cycad seed and all the possible toxic metals not only Al, but also Mn.

The well studied model with unbalanced diet associating low Ca and Mg intake and high dietary content of aluminum may be considered as a neurodegenerative disease model corresponding to WP ALS-PD. This is particularly true of studies with **chronic** Al load in rabbits [20]. The neuropathologic characteristics of this endemic human neurodegenerative disease were the appearance of neurofibrillary tangles (NFT) in AMMON's horn, temporal cortex, midbrain and pons. Al analysis by PIXE (particle induced X-ray emission spectrometry) revealed a high level of Al signals with positive correlation between Al content and NFT frequency [22]. Al intoxication combined with Ca/Mg deficiency is able to induce abnormalities in cytoskeletal protein such as neurofilaments and/or tau protein and may cause the formation of NFT and degeneration of neurons [20, 22]. As in the human necropsied central nervous system tissues, increased concentration of Al, Mn and Ca were found in CNS tissues of rats with low Ca-Mg dietary intake plus Al, while mg concentration was either decreased [24] or within the normal range [19]. Combined Al intoxication with Ca-Mg deficiencies is a reasonable model to investigate the pathogenesis of neurodegenerative diseases and eventually to screen their treatments.

These models may also be considered as models of **magnesium deficit,** but they do not concern simple magnesium **deficiency** reversible with more oral physiological magnesium supplementation. Magnesium deficiency cannot result in neurodegenerative disease. Combined Al intoxication with Ca-Mg deficiencies is not reversible through physiological oral magnesium supplementation. It therefore constitutes a type of experimental **magnesium depletion** model, instrumental in the investigation of the pathogenesis of magnesium depletion and in the screening of its still unknown possible treatment [25]. Even now, data from these models of magnesium depletion may contribute to our knowledge of the differences between **the mechanisms of magnesium deficiency and those of magnesium depletion**. As a rule, no changes have been found in brain magnesium concentrations during the course of magnesium deficiency in adult rats [25, 26], but in the experimental and clinical type of magnesium depletion related to pollutant metal load combined with low Mg and Ca dietary intake, magnesium concentrations were decreased in various structures of the nervous central systems [25, 26]. It is interesting to discuss the respective noxious importance of the two parameters (Ca and Mg deficiencies) which can decrease the protection against neurotoxins.

Ca deficiency plus Al load seems less deleterious than Ca/Mg deficiencies plus Al load (21, 24, 27) and magnesium deficiency plus Al load appears more noxious than Ca/Mg deficiencies plus Al load (19). Al and Ca concentrations in CNS tissue were higher. There was a negative correlation between Al and Mg levels while a positive correlation between Al and Ca was observed. The Ca/Mg ratio in the CNS tissue showed a strong positive correlation with Al content. Neuronal morphometric alterations were observed only in the low magnesium groups. It is speculated that magnesium depletion by increasing the Ca/Mg ratio in the CNS tissues further accelerated the uptake of Al into the CNS, which promoted the neurodegenerative processes [19]. Lesser noxiousness of magnesium deficiency when combined with Ca deficiency might be mainly ascribed to a lower increase in cellular calcium.

The physiopathologic mechanisms of neurodegenerative and neurotoxic insults associate several intricated and interactive factors: e. g. excitotoxicity, depolarization, decreased energy metabolism and cationic gradient, increased Na and Ca cellular influx, oxidant and arachidonic cascades [1,28-32]. Therefore among the three models of acquired reversible magnesium depletion [26] one appears of special interest. Although these three models associate various types of stress able to induce excitotoxicity with magnesium deficiency in adult rats, magnesium depletion due to systemic kainic acid (KA) plus magnesium deficiency [33] appears as the model the most closely linked to the main basic mechanisms of age-related neurologic disorders [25]. It associates low dietary magnesium intake (the most noxious decrease in neuroprotective factors) [19] to the most deleterious excitotoxicity in elderly: kinic acid, an agonist of non NMDA receptor such as L- BMAA and L-BOAA [2, 4, 16]. NMDA receptor number actually decreases in aged central nervous system unlike KA receptor number [34]. Aged animals were compromised in response to excessive activation of KA receptor as they were fourfold more sensitive than adult rats [1,34,35]. This reversible model of magnesium depletion mimicking some main mechanisms of environmental degenerative disease may be useful for screening various treatments of not only magnesium depletion but also of possibly linked psychiatric and geriatric disorders [25,26,33]. Further neurochemical and morphologic studies are desirable for better evaluation of the significance of this model as a screening tool.

Relying on these data on endemic WP ALS-PD complex a general environmental theory of the bulk of the age-related neurodegenerative diseases has been hypothetized: NDD would be due to an accelerating aging process caused by the sum of increased environmental slow deleterious factors and of decreased environmental protective factors (a low magnesium intake particularly)

3. CRITICAL APPRAISAL OF THE ENVIRONMENTAL GENERAL THEORY OF NEURODEGENERATIVE DISEASES: THEORETICAL AND PRACTICAL CONSEQUENCES

The value of this theory versus endemic WP ALS-PD, sporadic corresponding neurodegenerative diseases and other neurodegenerative diseases will be discussed.

3.1 ENDEMIC WP ALS-PD COMPLEX

First of all, postulated nutritional deficiencies of Ca and Mg have only been inferred from indirect therefore questionable data: low concentrations of Ca and Mg in garden soils and in river water [8,18-21] and not from direct evaluation of Ca and Mg dietary intake. Recent evidence has besides cast doubt on the premises of previous investigators that Ca and Mg contents of the Guamanian water supply are low [36].

Status of Ca, Mg and heavy metals was directly evaluated in 12 patients with WP ALS-PD complex and 12 Chamorro control subjects [36]. Patients had normal values for serum total and ionized calcium, 25-OH-D, and 24h calciuria. No patient had reduced Mg and P values, although a minority of patients and controls had low urinary magnesium concentrations. Median blood serum and 24 h urine collections for heavy metals (Al, Mn, Pb, Cd, Cu, Fe, Hg and Zn) were statistically similar in the patient and control groups, except for a slight elevation in blood, but not urine Pb in the patient group. Concentrations of heavy metals in hair and nails were similar in the two groups. There is no evidence in support of a major role of Ca/Mg deficiencies combined with heavy metal loads in the development of WP ALS-PD complex [36].

Value of the environmental pathogenesis of the endemic prototypal neurodegenerative complex itself is therefore not firmly established.

3.2.NON FOCUSED ALS, P AND D

It is hazardous to extrapolate from physiopathological data concerning endemic WP ALS-PD to **sporadic** cases of Amyotrophic Lateral Sclerosis (ALS), Parkinson's disease (P) and Dementia of the Alzheimer's type (D). It has never been observed that patients with ALS or P or D had low Ca and Mg dietary intake. In Alzheimer's disease investigation of magnesium status disagrees with the hypothesis of magnesium deficiency, but stresses the possible role of some types of magnesium depletion which might for instance be due to alterations of albumin [30,37-40]. In Parkinson's disease, another type of magnesium depletion has been found: grain magnesium concentration was significantly lower in cortex, white matter, basal ganglia and brain stem although calcium concentration was unchanged in all anatomic subregions. Aluminum concentration was high in the substantia nigra, caudate nucleous and globus pallidus [41]. In amyotrophic lateral sclerosis the non focused sporadic cases showed onset at later age and with shorter duration of the illness than the endemic focused cases [22]. The pathogenic importance of cumulative toxic metal loads-of Al and Mg particularly-seems uncertain in ALS, P and D. These diseases and the neurological patterns of pollutant metal chronic toxicity present some similarities and differences [22,36,40-44], but even analogous patterns do not demonstrate an identical aetiopathogenesis [45].

3.3.OTHER NEURODEGENERATIVE DISEASES

More criticism can be expressed concerning extrapolation from environmental data on WP ALS-PD complex to other degenerative diseases presenting no common symptomatology.

In multiple sclerosis some stigma of dysregulation of magnesium metabolism may be observed: extra-cellular deficit in ionized free magnesium [46], decreased magnesium concentration in the central nervous system, especially in white matter (which included demyelinated plaques). The calcium content of white matter has also been found decreased although Ca/Mg ratio was increased [47]. Significance of this type of magnesium depletion is still unknown. The most hypothetical environmental proposal concerns the physiopathology of slow transmissible spongiform encephalopathies: in animals bovine spongiform encephalopathy and in humans the corresponding Creutzfeld Jacob disease. This theory relies on some epidemiological data, a few therapeutic effects of parenteral magnesium therapy and on extrapolations from various biological data on magnesium. Environmental neurotoxins would consist of high doses of lipophilic organophosphates. Intraneuronal magnesium depletion would be allegedly due to low magnesium intake plus organophosphate neurotoxicity able to initiate the aberrant synthesis of mutant prion protein [48-50].

Further research is still necessary in support of this attractive hypothesis.

3.4.THEORETICAL AND PRACTICAL CONSEQUENCES

These studies arouse several remarks.

Little can be acknowledged for sure at the physiopathological level concerning the environmental general theory of neurodegenerative diseases. These studies have however permitted **to develop several experimental models** of magnesium depletion which constitute promising tests for study of the physiopathology of different types of acquired magnesium depletions and for screening potentially efficient drugs both in these magnesium depletions and in the related type of neurodegenerative diseases.

Because of the basic differences between the two types of magnesium deficit: magnesium deficiency and magnesium depletion [44] it may be deduced that **usual oral magnesium therapy through simple physiological oral magnesium supplementation** will never have **any curative effects** in case of neurodegenerative diseases [25]. Magnesium depletion requires more or less specific correction of its causal dysregulation, the aim of the treatment being to directly cure the implicated pathogenic and pathophysiological mechanisms of the neurodegenerative diseases e.g. through metal chelating agents, glutamate receptor antagonists, free radical scavengers, calcium antagonists. Moreover pharmacological use of magnesium salts whether parenteral magnesium or high oral magnesium synaptic acetylcholine in the central nervous system. Magnesium exerts a powerful synaptic effect by inhibiting acetylcholine release and increasing the number of synaptic vesicles in the presynaptic terminal, and possibly by stimulation of acetylcholinesterase [51]. Magnesium supplementation Alzheimer's disease seems contraindicated. Similarly in multiple sclerosis, magnesium therapy could also be risky. Experimental allergic encephalomyelitis is regarded as the most suitable model for multiple sclerosis in humans [52, 53] and as an experimental model of cell mediated immunity, which is strongly implicated in the aetiological process of multiple sclerosis.

4. CONCLUSION

From these very interesting studies mainly based on endemic WP ALS-PD data, it appears difficult to infer a general environmental scheme of the bulk of neurodegenerative diseases.

But they have generated various experimental acquired models of magnesium depletion whose pathogenic mechanisms are linked to those of neurodegenerative diseases, particularly when excitotoxicity and magnesium deficiency were combined.

Today, magnesium supplementation carried out with the usual magnesium salts either at physiological or at pharmacological doses appears of little help as a curative treatment of neurodegenerative diseases. It might even be contraindicated in such cases as Alzheimer's disease and multiple sclerosis. But the experimental models of magnesium depletion related to neurodegenerative diseases should constitute promising tools for the screening of effective treatments.

References

1. Dawson Jr., R., Geal, M.F., Bondy, S.C., Di Monte, D.A. and Isom, G.E (1995) Excitotoxins, aging and environmental neurotoxins: implications for understanding human neurodegenerative diseases. *Toxicol. Appl. Pharmacol.* **134**, 1-17.
2. Allen, C.N., Spencer, P.S. and Carpenter, D.O. (1993) β-N-methylamino-L-alanine in the presence of bicarbonate is an agonist at non-N-methyl-D-aspartate-type receptors, *Neuroscience* 54, 567-574.
3. Calne, D.B., Eisen, A., McGeer, E. and Spencer, P. (1986) Alzheimer's disease, Parkinson's disease and motoneuron disease: abiotropic interactions between ageing and environment? *Lancet II*, 1067-1070.

4. Spencer, P.S., Kisby, G.E. and Ludolph, A.C. (1991) Slow toxins, biologic markers and long-latency neurodegenerative disease in the Western Pacific region, *Neurology* 41 (suppl. 2), 62-66).
5. Lewin, R. (1987) Environmental hypothesis for brain diseases strengthened by new data, *Science* **237**, 483-484.
6. Spencer, P.S., Nunn, P.B., Hugon, Jlk Ludolph, A.C., Ross, S.M., Roy, D.N. and Robertson, R.C. (1987) Guam amyotrophic lateral sclerosis – Parkinsonism – Dementia linked to a plant excitant neurotoxin, *Science* **237**, 517-522.
7. Hirano, A., Malamud, N. and Kurland, L.T. (1961) Parkinsonism – dementia complex an endemic disease on the island of Guam II. Pathological features, *Brain* 84, 662-679.
8. Gajdusek, D.C. and Salazar, A.M. (1982) Amyotrophic lateral sclerosis and parkinsonian syndromes in high incidence among te Auyu and Jakai people of the Westh New Guinea, *Neurology* **32**, 107-126.
9. Reed, D., Labarthe, D., Chen, K.M. and Stallones, R. (1987) A cohort study of amyotrophic lateral sclerosis and parkinsonism – dementia on Guam and Rota, *Am. J. Epidemiol.* 125, 92-100.
10. Plato, C.C., Reed, D.M., Elizan, T.S. and Kurland, L.T. (1987) Amyotrophic lateral sclerosis/parkinsonsm – dementia complex of Guam, *Am. J. Hum. Genet.* **19**, 617-632.
11. Plato, C.C., Cruz, M.T. and Kurland, L.T. (1969) Amyotrophic lateral sclerosis/parkinsonism-dementia complex of Guam, *Am. J. Hum. Genet*, **21**, 133-141.
12. Plato, C.C., Garutto, R.M., Fox, K.M. and Gajdusek, D.C. (1986) Amyotrophic lateral sclerosis and parkinsonism-dementia on Guam -a 25 year prospective case- control study, *Am. J. Epidemiol.* **124**, 643-656.
13. Zhang, Z.K., Anderson, D.W. and Mantel, N. (1990) Geographic patterns of parkinsonism-dementia complez on Guam, *Arch. Neurol.* **47**, 1069-1074.
14. Gibbs Jr., C.J. and GAjdusek, D.X. (1982) An update on long term in nino and in vitro studies designed to identify a virus as the cause of amyotrophic lateral sclerosis, parkinsonism-dementia and Parkinson's disease, in L.P. Roland (ed.) *Human Motor Neuron Diseases*, New – York, Raven Press, 343-353.
15. Kurland, L. T. and Molagaard, L.A. (1982) Guamanian ALS: hereditary or acquired?, in L.P. Roland (ed.) *Human Motor Meuron Diseases*, New York, Raven Press, 165-177.
16. Spencer, P.S., Allen, C.N., Kisby, G.E. and Ludolph, A.C. (1994) On the etiology and pathogenesis of chemically induced neurodegenerative disorders, *Neurobiol. Aging* 15, 265-267.
17. Haque, A., Hossain, M., Wouters, G. and Lambein, F. (1996) Epidemiological study of lathyrism in northwestern districts of Bangladesh, *Neuroepidemiology* **15**, 83-91.
18. Yase, Y. (1972) The pathogenesis of amyotrophic lateral sclerosis, *Lancet II*, 292-296.
19. Mitani, K. (1992) Relationship between neurological diseases due to aluminium load, especially amyotrophic lateral sclerosis and magnesium status, *Magnes. Res.* **5**, 203-213.
20. Kihira, T., Yoshida, S., Uebayashi, Y., Wakayama, I. and Yase, Y. (1994) Experimental model of motor neuron disease: oral aluminium neurotoxicity, *Biomed. Res.* **15**, 27-36.
21. Yasui, M., Ota, K. and Garruto, R.M, (1995) Effects of calcium –deficient diets in the central nervous system and bones of rats, *Neurotoxicology* **16**, 511-517.
22. Kihira, T., Yoshida, S., Mitani, K., Yasui, M. and Yase, Y. (1993) ALS in the Kii Peninsula of Japan, with special reference to neurofibrillary tangles and aluminium, *Neuropathol.* **13**, 125-136.
23. Kihira, T., Yoshida, S., Komoto, J., Wakayama, I. and Yase, Y. (1995) Aluminum-induced model of motor neuron degeneration: subperineurial infection of aluminum in rabbits, *Neurotoxicol.* **16**, 413-424.
24. Yasui, M., Ota, K. and Yoshida, M. (1997) Effects of low Ca and Mg dietary intake on the central nervous system of rats and calcium-magnesium related disorders in the amyotrophic lateral sclerosis focus in the Kii peninsula of Japon, *Magnes. Res.* **10**, (in press).
25. Durlach, J. and Bac, P. (1997) Mechanisms of action on the nervous system in magnesium deficiency and dementia, in M. Yasui, M.J. Strong, K. Ota and M.A. Verity (eds.), *Mineral and Metal Neurotoxicology*, CRC Press, Boca.
26. Durlach, J., Bac, P., Durlach, V., Bara, M. and Guiet-Bara, A. (1997) Neurotic, neuromuscular and autonomic nervous forms of magnesium imbalance, *Magnes. Res.*, 10 (in press).
27. Yasui, M., Yase, Y., Ota, K. and GArruto, R.M. (1991) Evaluation of Mg, Ca and Al metabolism in rats and monkeys maintained on Ca deficient diets, *Neurotoxicology* 12, 603-614.

28. Landfield, P. W., Thibault, O., Mazzanti, M.L., Porter, N.M. and Kerr. D.S. (1992) Mechanisms of neuronal death in brain aging and Alzheimer's disease: role of endocrine-mediated calcium dyshomeostasis, *J. Neurobiol.* **23**, 1247-1260.
29. Rayssiguier, Y., Durlach, J., Gueux, E, Rock, E, and Mazur, A. (1993) Magnesium and ageing. I. Experimental data: importance of oxydative damage, *Magnes. Res.* **6**, 379-394.
30. Durlach, J., Durlach, V., Bac, P., Rayssiguier, Y., Bara, M. and Guiet-Bara, A. (1993) Magnesium and ageing. II. Clinical data. Aetiological mechanisms and pathophysiological consequences of the magnesium deficit in elderly patients, *Magnes. Res.* **6**, 379-394.
31. Couratier, P., Sindou, P., Tabaraud, F., Diop, A.G., Spencer, P.S. and Hugon, J. (1995) Modulation of tau neuronal expression induced by NMDA, non NMDA and metabotropic glutamate receptor agonists, *Neurodegeneration* **4**, 33-41.
32. Schulte, S., Muller, W.E. and Friedberg, K.D. (1995) In vitro and in vivo effects of lead on specific 3H-PN 200-110 binding to dihydropyridine receptors in the frontal cortex of the mouse brain, *Toxicology* **97**, 113-121.
33. Bac, P., Herrenknecht, C., Pages, N., Dupont, C. and Durlach, J. (1996) Recersible model of magnesium depletion induced by systemic kainic acid injection in magnesium deficient rats. I. Comparative study of various magnesium salts, *Magnes. Res.* **9**, 281-292.
34. Tamaru, M.., Yoneda, Y., Ogita, K., Shimizu, J. and Nagata, Y. (1991) Age related decreases of the N-methyl-D-aspartate receptor complez in the rat cerebral cortex and hippocampus, *Brain Res.* **542**, 83-90.
35. Dawson Jr., R. and Wallace, D.R. (1992) Kainic acid-induced seizures in aged rats: neurochemical correlates, *Brain Res. Bull.* **29**, 459-468.
36. Ahlskog, J.E., Waring, S.C., Kurland, L.T., Petersen, R.C., Moyer, T.P., Harmsen, W.S., Maraganore, D.M., O'Brien, P.C., Esteban-Santillan, C. and Bush, V. (1995) Guamanian neurodegenerative disease: inbestigation of the calcium metabolism/heavy metal hypothesis, *Neurology* **45**, 1340-1344.
37. Glick, J.L. (1990) Dementias: the role of magnesium deficiency and an hypothesis concerning the pathogenesiw of Alzheimer's disease, *Med. Hypotheses* **31**, 211-225.
38. Durlach, J. (1990) Magnesium depletion and Alzheimer's disease, *Magnes. Res.* 3, 217-218.
39. Borella, P., Giardino, A., Neri, M. and Andermarker, E. (1990) Mg and K status in elderly subjects with and without dementia of the Alzheimer type, *Magnes. Res.* **3**, 283-289
40. Jolly, D.H., Poitrinal, P., Millart, H., Kariger, E., Blanchard, F., Collery, P. and Choisy, H. (1993) Blood, zinc, copper, magnesium, calcium, aluminum and manganese concentrations in patients with and without Alzheimer type dementia, *Trace Elements in Med.* **10**, 192-195.
41. Yasui, M., Kihira, T. and Ota, K. (1992) Calcium, magnesium and aluminum concentrations in Parkinson's disease, *Neurotoxicology* **13**, 593-600.
42. Calne, D.B., Chu, N.S., Huang, C.C., Lu, C.S. and Olanow, W. (1994) Manganism and idiopathic Parkinsonism: similarities and differences, *Neurology* **44**, 1583-1586.
43. Lai, J.C.K., Chan, A.W.K., Minski, M..J. and Lim, L. (1997) Relationship between manganese and other minerals, in M. Yasui, M.J. Strong, K. Ota and M.A. Verity (eds.), CRC Press, Boca Raton, 297-303.
44. Wolters, E.C. (1997) Manganese-induced Parkinsonism, in M. Yasui, M.J. Strong, K. Ota and M.A. Verity (eds.), CRC Press, Boca Raton, 319-325.
45. Durlach, J. (1995) Editorial Policy of Magnesium Research: general considerations on the quality criteria for biomedical papers and some complementary guidelines for the contributors to Magnesium Research, *Magnes. Res.* **8**, 191-206.
46. Altura, B.T., Bertschat, F., Jeremias, A., Ising, H. and Altura, B.M. (1994) Comparative findings on serum ionized Mg^{2+}, *Scand. J. Clin. Lab. Invest.* 54 (supl.217), 7781.
47. Yasui, M. and Ota, K. (1992) Experimental and clinical studies on dysregulation of magnesium metabolism and the aetiopathogenesis of multiple sclerosis, *Magnes. Res.* **5**, 295-302.
48. Purdey, M. (1994) Are organophosphate pesticides involved in the causation of bovine spongiform encephalopathy (BSE)? Hypothesis based upon a literature review and limited trials on BSE cattle, *J. Nutr. Med.* **4**, 43-82.
49. Purdey, M. (1996) The UK epidemic of BSE: slow virus or chronic pesticide-initiated modification of the prion protein? Part 2: An epidemiological perspective, *Med. Hypotheses* **46**, 445-454.

50. Purdey, M. (1996) The UK epidemic of BSE: slow virus or chronic pesticide-initiated modification of the prion protein? Part 2: An epidemiological perspective, *Med. Hypotheses* **46**, 445-454.
51. Janka, Z. (1992) Significance of magnesium in functioning of the nervous system, in T. Fazekas, B. Selmeczi and P. Stefanovits (eds.), *Magnesium in Biological Systems*, Akademia Kiado publ., Budapest, 212-221.
52. Lassmann, H., Bruller, C., Bradl, M. and Linington, C. (1988) Experimental allergic encephalomyelitis: the balance between encephalitogenic T lymphocytes and demyelinating antibodies determines size and structure of memyelinated lesions, *Acta Neuropathol.* **75**, 566-576.
53. Chambron, J., Namer, I.J., Steibel, J., Gounot, D. and Armspach, J.-P. (1994) Etude par imagerie de resonanve magnetique du role de la barriere hemato-encephalique dans la pathogenese de l' encephalomyelite allergique experimentale: applications a la sclerose en plaques, Bull., *Acad. Natle. Med.* [178], 1647-1665.
54. McCreary, P.A., Battifora, H.A., Laing, G.H. and Hass, G.M. (1996) Protective effect of magnesium deficiency on experimental allergic encephalomyelitis in the rat, *Proc. Soc. Exper Biol. Med.* **121**, 1130-1133.
55. Hass, G.M., McCreary, P.A., Laing, G.H. and Galt, R.M. (1980) Lympho-proliferative and immunologic aspects of magnesium deficiency, in M. Cantin and M.S. Seelig (eds.), *Magnesium in Health and Diseade*, Spectrum Publ., New-York, 185-200.

SECRETAGOGUE-EVOKED MAGNESIUM AND CALCIUM TRANSPORT IN MOUSE PANCREATIC ACINAR CELLS

MARIA M. GEADA, F.CH.* MOOREN, *W. DOMSCHKE AND J. SINGH.
*Department of Applied Biology, University of Central Lancashire, Preston, PR1 2HE, England, UK. and * Department of Medicine, University of Münster, D-48129, Münster, Germany*

1. SUMMARY

This study investigates the effect of 10 pM cholecystokinin-octapeptide (CCK-8) on intracellular free magnesium $[Mg^{2+}]_i$ and calcium concentrations $[Ca^{2+}]_i$ in magfura-2-AM and fura-2-AM loaded single pancreatic acinar cells, respectively employing the technique of microspectrofluorimetry. CCK-8 evoked an initial sharp rise in $[Mg^{2+}]_i$ (Mg^{2+} release) followed by a decrease to a new steady-state level (Mg^{2+} efflux). On removal of CCK-8 $[Mg^{2+}]_i$ returned to the pre-stimulated basal level (Mg^{2+} reuptake). In contrast, CCK-8 gave rise to Ca^{2+} oscillations. When acinar cells were co-loaded with BAPTA (10 μM) and either magfura-2-AM or fura-2-AM, CCK-8 evoked only the normal decrease in $[Mg^{2+}]_i$ and a slight $[Ca^{2+}]_i$ elevation (10 nM). This results suggest that the initial rise in $[Mg^{2+}]_i$ seen with CCK-8 in normal conditions may be due to Ca^{2+} interfering with the magfura-2-AM signal. Both thapsigargin (0.5 μM) and ionomycin (5 μM) evoked a marked decrease in $[Mg^{2+}]_i$ in magfura-2-AM loaded acinar cells and an elevation in $[Ca^{2+}]_i$ from fura-2-AM loaded acinar cells. The results indicate that cytosolic Ca^{2+} is associated with secretagogue-induced decrease in $[Mg^{2+}]_i$. When acinar cells were pre-treated with either forskolin (10 μM), N^{-w} Nitro-L-arginine (N-LA; 2 mM), 8-Bromo guanosine cyclic mono-phosphate (8 Br. Cyclic GMP; 100 μM) or staurosporine (1 μM), CCK-8 elicited only a small decrease in $[Mg^{2+}]_i$ compared to a much larger response with CCK-8 alone. In contrast, genestein (10 μM) and 12-O-tetradeconyl phorbol 13-α acetate (TPA; 1 μM) augmented the decrease in $[Mg^{2+}]_i$ evoked by CCK-8. The results indicate that Ca^{2+} mobilising secretagogues can stimulate Mg^{2+} efflux which is also mediated by a number of intracellular mediators.

2. INTRODUCTION

T. Theophanides and J. Anastassopoulou (eds.), Magnesium: Current Status and New Developments, 17-27.

Magnesium (Mg^{2+}) is necessary for many cell functions, such as activating enzymes, stabilising the structure of membranes and ribosomes, and modifying ion channel and co-transport activities[1]. In secretory epithelia, fluid secretion is regulated by $[Ca^{2+}]_i$ and cytosolic sodium $[Na^+]_i$. $[Mg^{2+}]$ plays a key role in these processes as an essential co-factor for the activation of Ca^{2+} and Na^{+i} pumps. Mg^{2+} has also been found to evoke physiological dose-dependent activation of K^+ channels in mouse parotid acinar cells, and the $Na^+/K^+/Cl^-$ cotransporter[2,3] in red blood cells. Thus, intracellular regulation of Mg^{2+} is critical for many cellular functions encompassing those of fluid and electrolyte secretion, enzyme activity, cell growth and proliferation[4]. Despite such physiological significance, factors regulating $[Mg^{2+}]$ are still poorly understood.

There are several different signalling pathways involved in pancreatic enzyme secretion. One pathway involves the activation of phospholipase C resulting in the hydrolysis of membrane bound phosphotidylinositol,4,5-bisphosphate(PIP_2) which leads to the formation of inositol 1,4,5-trisphosphate(IP_3), inositol 1,3,4,5-tetrakisphosphate (IP_4) and diacylgycerol (DG). IP_3 and IP_4 act to mobilise intracellular Ca^{2+} while DG stimulates the phospholipd Ca^{2+} dependent enzyme protein kinase C[6]. An increase in $[Ca^{2+}]_i$ is thought to activate nitric oxide synthase (NOS) activity, which is responsible for the production of nitric oxide (NO). This being a physiological activator of guanylate cyclase, which converts guanosine triphosphate into cyclic GMP[7,8]. which is involved in the regulation of several important cellular processes, including the control of epithelia cell growth, differentiation and electrolyte secretion[9]. Another signalling pathway, also present in various tissues including pancreatic acinar cells, involves tyrosine kinases, being activated by insulin, epidermal growth factor and fibroblast growth factors[110]. These agonists bind to membrane receptors and cause the activation of phospholipase $C\gamma$[11]. Information of the action of these signalling pathways on Ca^{2+} mobilisation is widely available, however, information about the action of these on Mg^{2+} mobilisation is currently lacking. The present study was designed to investigate the effects of the different signalling pathways involved in the control of intracellular Mg^{2+} homeostasis in mouse pancreatic acinar cells.

3. METHODS

Adult white mice weighing between 20-24 g of either sex were killed by severance to the vertebral column. The pancreas was rapidly removed and minced for 5 min in modified Krebs-Ringer-Hepes (KRH) solution (2 ml per pancreas) comprised of (mM): NaCl, 140; KCl, 4.7; $MgCl_2$, 1.2; HEPES, 10; Glucose, 10; KH_2PO_4, 1.2; $CaCl_2$, 1.28., and pH 7.4. Pancreatic segments were acutely digested with 200 U ml^{-1} of pure collagenase at 37°C for 10 min with continuos shaking (120 cycles/min) employing established methods[12,13]. Cells were kept as a concentrated suspension for up to 4 hours without any noticeable changes in viability (<95%, as assessed by trypan blue extrusion). Two millilitre aliquots of acini were loaded in 5 μM of either magfura-2-

AM or fura-2-AM for 20 min at room temperature. At the end of loading, cells were washed twice by centrifugation and incubated for 30 min to allow complete de-esterification of the dye. Loaded cells were plated onto glass cover slips coated with Cell-TaK (Biopolymers(1 μl/cm^2) and kept at an ambient room temperature for further 1 hour before use.

4. MEASUREMENT OF $[Mg^{2+}]_i$ AND $[Ca^{2+}]_i$

Plated cells were mounted in a home built chamber (internal volume 0.28 ml) placed on the stage of a Nikon Diaphot-TMD inverted microscope equipped with a Fluor X 100 oil immersion objective. Acini were continuously perfused at a flow rate of 3 ml min^{-1}. The superfusing solution contained 0.05% BSA to prevent adhesion of peptide hormones to the internal surfaces of tubing and was equilibrated with air at 37°C from multiple reservoirs. Single acinar cells were optically masked using a pinhole diaphragm stopped down to an optical diameter of 8-10 μM, the size of a single acinar cell without background fluorescence. Epifluoresecence microscopy was used to monitor changes in magfura-2-AM or fura-2-AM fluorescence signals using a dual excitation wavelength spectrofluorimeter (Delta-scan Photon Technologies International (PTI)). Excitation from a 450-W xenon lamp was provided by two monochromaters preset at 340 nm and 380 nm. The excitation light from the two monochromaters was alternated using a rotating chopper mirror and was delivered to the sample, via optic fibres, through a dichroic mirror (DM 400). Emitted light intensities were collected at an emission wavelength of 510±10 nm by a bandpass filter set within the microscope and was analysed with an integrated PTI software. Each signal was averaged and recorded at 1 sec intervals. Autofluorescence determined in single cells were found to contribute to less than 3-5% of the emitted signal and was therefore not subtracted[13]..

5. STATISTICAL ANALYSIS AND REAGENTS

All data are expressed as the mean ± S.E.M. Statistical significance was tested using the Student's t-test. Only values of $P<0.05$ were considered significant. Magfura-2-AM , fura -2-AM and BAPTA were from Molecular probes, Eugene, USA. Cell-TaK was obtained from collaborative Research Incorp., Bedford, USA. All other reagents were from Sigma Chemicals (München, Germany).

6. RESULTS

Changes in $[Mg^{2+}]_i$ and its relationship with $[Ca^{2+}]_i$

In this study changes in the ratio 340/380 nm, employing the technique of microspectrofluorimetry, are referred to as changes in $[Mg^{2+}]_i$. The resting levels of

$[Mg^{2+}]_i$ in this series of experiments was 0.82±0.09 mM (n=56). Stimulation of magfura-2-AM-loaded single pancreatic acinar cells with CCK-8 (10 pM) resulted in an initial sharp transient increase in $[Mg^{2+}]_i$, followed by a decrease to a new-steady-state of 0.5-0.6 mM. This level is maintained in the continuous presence of the agonist (Figure 1A). Removal of CCK-8 resulted in an increase in $[Mg^{2+}]_i$ reaching the

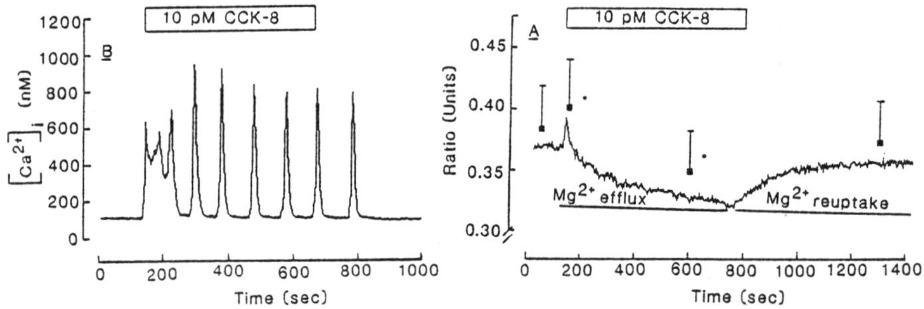

Figure 1A: Time course changes in magfura-2 ratio in the absence and presence of 10 pM CCK-8 in single mouse pancreatic acinar cells. solid squares represent the mean and vertical bars S.E.M. values (n=56) at four points: basal value, undershoot value and post stimulatory value of magfura-2 ratio. * indicates significant difference at a level of P>0.05 from basal value. (B) Time course change in $[Ca^{2+}]_i$ in single pancreatic acinar cells loaded with fura-2-AM. This trace is typical of 7 such experiments.

original basal value. In this study the decrease in $[Mg^{2+}]_i$ following CCK-8 stimulation is referred to as Mg^{2+} efflux where as the subsequent increase following the removal of CCK-8 is referred to as influx (reuptake) of Mg^{2+}. The effect of 10 pM CCK-8 on $[Ca^{2+}]_i$ in fura-2-AM loaded acinar cells is shown in Figure 1B for comparison. CCK-8 at physiological concentration evoked the classical oscillations of cellular Ca^{2+}.Previous studies by Hurley *et al*[14], showed a possible interference of magfura-2-AM with free Ca^{2+}. Thus, cells were co-loaded with a Ca^{2+} chelator 1,2-bis (2-aminophenoxy) ethan-N,N,N',N'-tetraacetic acid (BAPTA). In BAPTA (10 μM) co-loaded acinar cells CCK-8 evoked a small rise of about 5-10 nM in $[Ca^{2+}]_i$ (Figure 2A). Under these conditions in the magfura-2-AM loaded cells the initial increase in $[Mg^{2+}]_i$ evoked by CCK-8 was no longer detectable but however, the sustained decrease was faster (Figure 2B). This observation suggests that the initial peak of the response seems to be due to hormonal-induced Ca^{2+} response which binds to the magfura-2-AM. Whereas, the secondary decrease indicates a real change in $[Mg^{2+}]_i$ in response to the secretagogue. Moreover, the results also suggest that CCK-8 induced Mg^{2+} decrease may be dependent upon cellular Ca^{2+} mobilisation. In order to test this observation we employ thapsigargin, a potent and irreversible inhibitor of the ATP-dependent Ca^{2+} pump of the sarcoplasmic reticulum[15]. Application of thapsigargin (0.5 μM) to magfura-2-AM loaded cells resulted in a sustained decrease in $[Mg^{2+}]_I$ (Figure 3A). On removal of thapsigargin and superfusion of the acinar cells with normal physiological solution for over 30 min $[Mg^{2+}]_i$ failed to return to the basal levels. Subsequent stimulation had no detectable effect, suggesting that the release of intracellular Ca^{2+} from stores is necessary for the observed Mg^{2+} changes under CCK-8

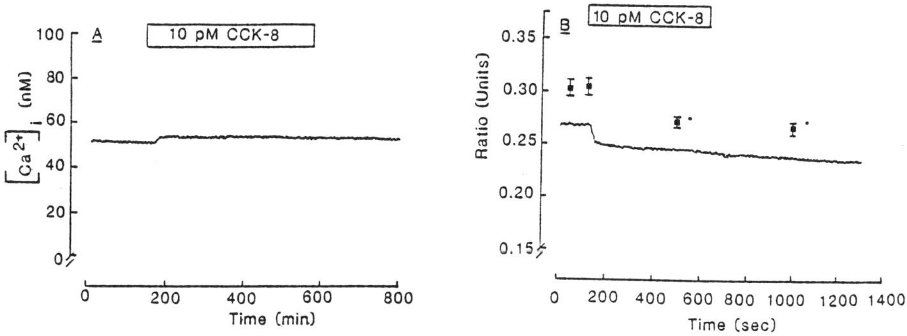

Figure 2. (A)Time course of CCK-8-evoked $[Ca^{2+}]_i$ response of BAPTA-AM co-loaded cell; the trace is typical of 5 such experiments. (B): CCK-8 (10 pM)-evoked time course change in magfura-2 ratio in single mouse pancreatic acinar cells coloaded with BAPTA-AM (10 μM). Solid squares represent the mean and vertical bars S.E.M. values (n=5) at four points: basal value, peak value, undershoot value and post stimulatory value of magfura-2 ratio. * indicates significant difference at a level of P>0.05 from basal.

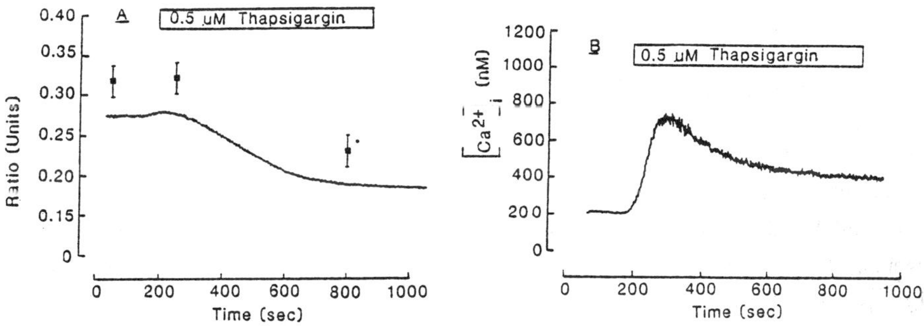

Figure 3 (A): Time course change in $[Mg^{2+}]_i$ following perfusion of magfura-2-loaded cells with thapsigargin (0.5 μM). Solid squares represent the mean and vertical bars S.E.M. values (n=8) at four points: basal value, peak value, undershoot value and post stimulatory value of magfura-2 ratio. * indicates significant difference at a level of P>0.05 from basal. (B) effect of 0.5 μM thapsigargin on $[Ca^{2+}]_i$ from fura-2-loaded acinar cells. The trace is typical of 6 such experiments.

stimulation. The effect of thapsigargin (0.5 μM) on $[Ca^{2+}]_i$ is shown in Figure 3B for comparison. Cells were next perfused with Ca^{2+} free solutions containing 1 mM EGTA to avoid any Ca^{2+} contamination from the extracellular medium. Application of ionomycin (5 uM) to magfura-2-AM loaded cells resulting in a sustained decrease in $[Mg^{2+}]_i$ (Figure 4A) which returned to near pre-stimulatory values after changing to normal buffer. These results confirm our earlier observation that a change in cytosolic Ca^{2+} is associated with agonist-evoked Mg^{2+} efflux. Ionomycin also evoked a transient elevation in $[Ca^{2+}]_i$. (Figure 4B).

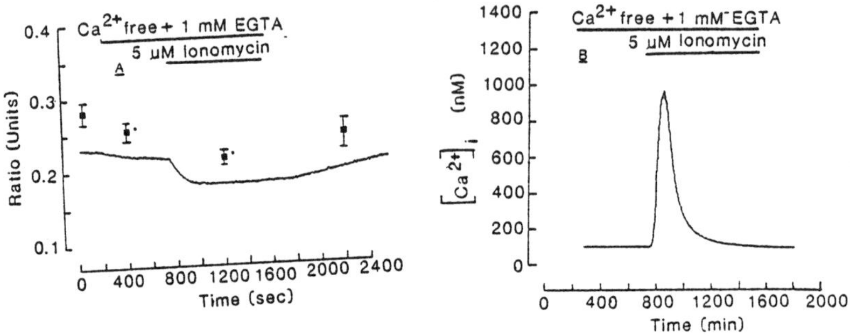

Figure 4 (A): Time course changes in $[Mg^{2+}]_i$ following perfusion of magfura-loaded pancreatic acinar cells with ionomycin (5 µM). Ionomycin was added to Ca^{2+} free solutions in the presence of 1 mM EGTA. Solid squares represent the mean and vertical bars S.E.M. values (n=8) at four points: basal value, peak value, undershoot value and post stimulatory value of magfura-2 ratio. * indicates significant difference at a level of P>0.05 from basal. (B) Effect of 5 µM ionomycin on $[Ca^{2+}]_i$ in fura-loaded pancreatic acinar cells. The trace is typical of 6 such experiments.

Intracellular signalling pathways associated with $[Mg^{2+}]_i$

We have also investigated the intracellular mediators which may be associated with Mg^{2+} transport in magfura-2-AM loaded acinar cells employing a wide range inhibitors and activators including genestein (10 µM), TPA (1 µM), staurosporine (1 µM), 8 Br. cyclic GMP (100 µM), N-LA (2 mM) and forskolin (10 µM) in the absence and presence of CCK-8 (10 pM). Figure 5 shows the mean (±SEM) percentage decrease in $[Mg^{2+}]_i$ evoked by CCK-8 in the presence and absence of the various inhibitors and activators. All the substances tested here had no significant effect on the magfura ratio when employed individually. However, combining CCK-8 with either staurosporine, an inhibitor of protein kinase C , 8 Br. cyclic GMP N-LA, an inhibitor of nitric oxide synthetase or forskolin, an activator of adenylate cyclase resulted in a significant (P<0.05) inhibition in the CCK-8-evoked Mg^{2+} efflux compared to the response obtained with CCK-8 alone. In contrast, genestein, a tyrosine kinase inhibitor and TPA, a protein kinase activator enhanced the CCK-8 induced decrease in $[Mg^{2+}]_i$. These results clearly demonstrate that Mg^{2+} transport in pancreatic acinar cells is modulated by intracellular messengers.

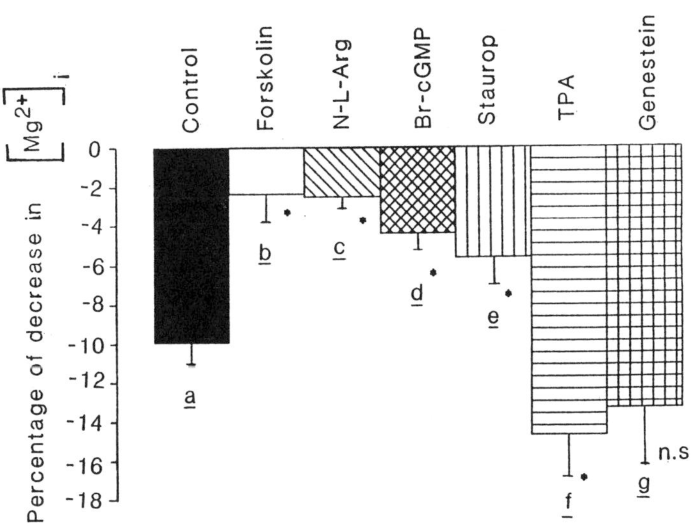

Figure 5: Histograms showing the effect of forskolin (10 μM), N^W-nitro-L-arginine (N-LA) (2 mM), 8-bromo-guanosine monophosphate (Br-cGMP), (100 μM), staurosporine (1 μM), 12-0-tetradecanoyl phorbol 13-α-acetate (TPA) (1 μM) and genestein (10 μM) on the mean (± SEM) percentage of the CCK-8-evoked decrease in Mg^{2+}. Basal $[Mg^{2+}]_i$ was taken as 100%.* indicates significant difference at a level of $P<.0.05$ from control. n=57 for all experiments.

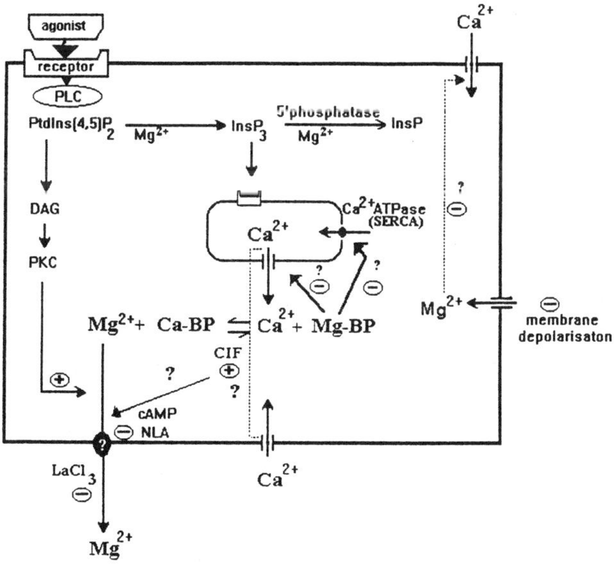

Figure 6 summarises the main conclusions of the study. The results show marked relationship between Mg^{2+} and Ca^{2+} mobilisation in pancreatic acinar cells. Mg^{2+-} is proposed to regulate Ca^{2+} influx from extracellular medium and its release from the ER. Mg^{2+} is also a co-factor for a number of enzymes which are associated with Ca^{2+} homeostasis. Furthermore, secretagogue-evoked Mg^{2+} efflux seems to be associated with several cellular mediators while Mg^{2+} uptake into pancreatic acinar cells is associated with membrane depolarisation.

7. DISCUSSION

The recent development of the fluorescent probe magfura-2-AM has now enabled investigators to study Mg^{2+} regulation[16,17]. Current evidence indicates that $[Mg^{2+}]_i$ is altered by receptor-mediated mechanisms that may have significant physiological effects in various cell types[18]. However, there is still a lack of knowledge in Mg^{2+} homeostasis during cellular regulation. In this study, direct measurement was made of $[Mg^{2+}]_i$ in mouse pancreatic acini. The resting $[Mg^{2+}]_i$ was found to be 0.82 ±0.09 mM, which is within the range of cytosolic free Mg^{2+} found in most cells[12,13]. Stimulation of acini with CCK-8, resulted in a biphasic response consisting of an initial increase in $[Mg^{2+}]_i$ followed by a sustained decrease reaching a new steady-state of about 0.6-0.5 mM after 4-5 min below the basal values. Removal of CCK-8 resulted in Mg^{2+} influx (reuptake) towards the original basal $[Mg^{2+}]_i$ level prior to agonist stimulation (Figure 1). Hurley *et al* [14] have reported that changes in $[Ca^{2+}]_i$ can interfere with measurements of $[Mg^{2+}]_i$ using magfura 2 AM. Thus, $[Ca^{2+}]_i$ was buffered using BAPTA. Figure 2A demonstrates that 10 μM BAPTA reduced the CCK-8-evoked Ca^{2+} mobilisation and inhibited the initial peak seen with the magfura-2-AM loaded cells (Figure 2B). These results suggest that an increase in $[Ca^{2+}]_i$ is required for a CCK-8-evoked decrease in $[Mg^{2+}]_i$. This decrease in $[Mg^{2+}]_i$ can be due to either Mg^{2+} been taken up into stores or due to Mg^{2+} efflux. The latter seems to be the most likely explanation since previous experiments have demonstrated that stimulation of pancreatic segments with either ACh or CCK-8 can result in a significant increase in Mg^{2+} efflux measured by atomic absorbance spectroscopy[12]. However, it cannot be completely excluded that a small percentage can be taken up into stores. If an increase in $[Ca^{2+}]_i$ is associated with this decrease in $[Mg^{2+}]_i$ evoked by CCK-8, then it is very important to understand how $[Ca^{2+}]_i$ is mediating the response.

Mg^{2+} is the second most abundant divalent cation inside the cells, but only approximately 5-10 % is free within the cytosol[20,21]. The remaining 90-95% is present in various cell compartments or is bound to negatively charged ligands such as ATP, citrate, ADP, proteins, RNA and DNA. Depending on the cell type, free and bound Mg^{2+} are completely or partly exchangeable[20,21]. On the other hand $[Ca^{2+}]_i$ is regulated on a very low level (approximately 150 nM) by active Ca^{2+} extrusion mechanisms. Ca^{2+} is transported either extracellularly or into intracellular compartments. Around 25% of total cell Ca^{2+} is mobilizable upon hormonal stimulation resulting in a measured increase of $[Ca^{2+}]_i$ up to 1-1.5 μM. However, the calculated increase should be much higher at around 1 mM . This demonstrates the enormous Ca^{2+} buffer capacity of the cytosol by a number of different calcium binding proteins[11]. Since both Mg^{2+} and Ca^{2+} are divalent ions of comparable size then there should be a competition for certain binding proteins and a displacement out of binding in case of increasing concentration of either ion. Therefore, it is suggested that an indirect action of CCK-8 on the observed decrease in $[Mg^{2+}]_i$ in this study occurs. The hormonal induced increase in $[Ca^{2+}]_i$ leads to a displacement of Mg^{2+} from certain binding sites (e.g. proteins, ADP, ADP etc.) to intracellular chelating agents.

Once Mg^{2+} is released from its binding sites, it is extruded to the outside (see Figure 6). The depletion of intracellular stores by thapsigargin prevented a further hormonal induced Ca^{2+} increase. In magfura loaded acinar cells the same manoeuvre prevented an effect on $[Mg^{2+}]_i$ too (Figure 3A). Therefore, an increase of $[Ca^{2+}]_i$ seemed to be a prerequisite for further Mg^{2+} movements. This together with the BAPTA experiments suggests a strong coupling between an increase in $[Ca^{2+}]_i$ and a decrease in $[Mg^{2+}]_i$. Another mechanism could be a calcium-activated Mg^{2+} channel. But since so far a Mg^{2+} channel has not been demonstrated unequivocally this view seemed to be more speculative.

The relationship between Mg^{2+} and Ca^{2+} homeostasis in pancreatic acinar cell is illustrated in Figure 6. Mg^{2+} is known to regulate over 300 enzymes by a variety of mechanism[20,21]. The activation of these Mg^{2+}-dependent enzymes follows a bell shaped curve consisting of both activation and inhibition. Mg^{2+} is also believed to be closely linked with the mobilisation of Ca^{2+} both from cellular stores and from the extracellular medium[22]. The metabolism of IP_3 which stimulates the release of Ca^{2+} from rough endoplasmic reticulum depends upon the Mg^{2+}-dependent enzyme 5' phosphatase[5]. In conditions where little Mg^{2+} was present this enzyme would be inhibited in its hydrolysis of IP_3. As a result increasing IP_3 levels would stimulate Ca^{2+} mobilisation. This would promote secretion and provide increased $[Ca^{2+}]_i$ for extrusion. In high Mg^{2+} conditions reduced IP_3 and therefore $[Ca^{2+}]_i$ would be expected. Indeed elevated extracellular $[Mg^{2+}]_o$ concentration inhibits $^{45}Ca^{2+}$ uptake, changes in $[Ca^{2+}]_i$ and the amplitude and frequency of Ca^{2+} oscillations whereas low $[Mg^{2+}]_o$ facilitates $^{45}Ca^{2+}$ uptake, increase in $[Ca^{2+}]_i$ and Ca^{2+} oscillations[23,13]. The major Ca^{2+} release in pancreatic acinar cells is now believed to occur through the caffeine-sensitive Ca^{2+} channel[24]. In sarcoplasmic reticulum of skeletal muscle this channel is highly sensitive to $[Mg^{2+}]_i$ [25]. Recent work on the pancreas also suggests that the channel is Mg^{2+}-dependent and is inhibited by Mg^{2+} in the millimolar range [26]. Conversely, in low Mg^{2+} conditions the open state of this channel is promoted [24,25]. Taken together, these observations clearly indicate that Mg^{2+} may play an important regulatory role in the mobilisation of Ca^{2+} during the exocytotic process in the exocrine pancreas. Thus, during agonist stimulation a decrease $[Mg^{2+}]_i$ is required to obtain a full Ca^{2+} response.

The results of this study have also demonstrated that Mg^{2+} transport in pancreatic acinar cells may be associated with the metabolism of a number of intracellular mediators including protein kinase C, cyclic AMP, cyclic GMP, tyrosine kinase and nitric oxide[6-10]. Substances which are known to regulate the metabolism of these cellular mediators had marked effects on the CCK-8 evoked Mg^{2+} efflux. TPA, an activator of protein kinase C [27] and, genestein, a specific tyrosine kinase inhibitor which regulate capacitative Ca^{2+} entry into pancreatic acinar cells[28] enhance the CCK-8-evoked Mg^{2+} efflux. In contrast, either 8-Br cyclic GMP, forskolin, staurosporine or NLA attenuated the CCK-8-induced Mg^{2+} efflux. The precise mechanisms whereby the intracellular mediators act to regulate secretagogue-evoked changes in $[Mg^{2+}]_i$ is still unclear. However, since the results of our study have demonstrated a close interaction between Mg^{2+} efflux and Ca^{2+} signalling then it is tempted to suggest that the intracellular mediators may regulate the mobilisation of cellular Ca^{2+} which in turn regulates Mg^{2+} efflux (see Figure 6). Experiments are now in progress to elucidiate the precise signalling mechanisms involved in Mg^{2+} transport in pancreatic acinar cells.

Acknowledgement. This work was supplied by the British Council-German ARC and the Wellcome Trust.

References

1. Zhang, G.H. and Melvin, J.E. (1994) Intracellular Mg^{2+} movement during muscarinic stimulation. *J.Biol Chem* **269**, 10352-10356.
2. Petersen, O. H. (1992). Stimulus-secretion coupling: cytoplasmic calcium signals and control of ion channels in exocrine acinar cells. *J. Physiol.* (Lond). **448**, 1-51.
3. Flatman, P.W. (1993). The role of magnesium in regulating ion transport: In: *Magnesium and Cell* (Ed. M.J. Birch), Academic Press, London, p.p. 197-216.
4. Gunther, T. Vormann, J. and Hollriegl, V. (1990). Characterisation of a sodium-dependent Mg^{2+} efflux from Magnesium loaded erythocytes. *Biochem. Biophys. Acta.* **1055**, 82-86.
5. Streb, H. and Schulz, I. (1983) Regulation of cytosolic free Ca^{2+} concentration in acinar cells of rat pancreas. *Am J Physiol* **245**, G6347-G6357.
6. Nishizuka, Y.(1988) The molecular heterogeneity of protein kinase C and its implications for cellular regulation, *Nature*, **308**, 661-665.
7. Pandol, S.J., Gukovskaya, A., Bahnson, T.D., and Dionne, V.E. (1994). Cellular mechanism mediating agonist-stimulated calcium influx in the pancreatic acinar cell. *Ann-N-Y-Acad-Sci.* **713**, 41-48.
8. Gukovskaya, A.S. and Pandol S.J. (1995). Dual regulation of cGMP formation by calcium in pancreatic acinar cells. *Am. J. Physiol* **268**, G900-G907.
9. Khare, S., Wilson, D.M., Tien, X-Y and Brasitus, T.A. (1993). M_3 muscarinic receptors on rat colonocytes are coupled to particulate guanylate cyclase activation. *Biochem.Bioph..Acta.* **1179,** 234-237.
10. Duan, R., Wagner, A.C.C., Yule, D.I. and Williams, J.A. (1994). Multiple inhibitory effects of genestein on stimulus secretion coupling in rat pancreatic acini. *Am. J. Pysiol.* **266**, G303-310.
11. Clapham, D.E (1995) Calcium signalling. *Cell.* **80**, 259-268.
12. Singh, J. and Wisdom, D.M. (1995). Second messenger role of magnesium in pancreatic acinar cells of the rat. *M. Cell. Biochem.* **149/150**, 175-182
13. Mooren, F. Ch., Wisdom, D.M., Domschke, W., Geada, M.R., and Singh, J. (1997). Magnesium homeostasis in the exocrine pancreas and its possible role in the treatment of pancreatitis. *Adaptation Biol. Med.* **1**, 284-298.
14. Hurley, T.W, Ryan, M.P., and Brinck, R.W. (1992). Changes of cytosolic Ca^{2+} interfere with measurements of cytosolic Mg^{2+} using magfura-2. *Am.J.Physiol.* **263**, C300-C307.
15. Bian J., Ghoshk, T., Wang, J.C. and Gill, D.L. (1991). Identification of intracellular Ca^{2+} pools. *J. Biol.Chem.* **266, (14)** 8801-8806.
16. Raju, B., Murphy, E., Levy, L.A., Hall, R.D., and London, R.E. (1989). A fluorescent indicator for measuring cytosolic free magnesium. *Am.J. Physiol.* **256**, C540-C548.
17. Lennard, R. and Singh, J.: Secretagogue-evoked changes in intracellular free magnesium concentrations in rat pancreatic acinar cells *J Physiol (Lond.)* 435: 483-492, 1992.
18. Birch, N.J. (1993). *Magnesium and the Cell.* Academic Press, London.
19. Flatman, P.W. (1991). Mechanism of magnesium transport. *Ann. Rev. Physiol.* **431**, 11-25
20. Günther, T. (1990). Functional compartmentation of intracellular magnesium. In H.Sigel And A.Sigel eds.: *Metal Ions in Biological systems*, Vol.26: Compedium on Magnesium and its Role in Biology, Nutrition and Physiology, Marcel Dekker,New York.
21. Wacker, W.E.C. (1968). The Biochemistry and Physiology of Magnesium. *Ann.New York Acad. Sci.* **162**,717-726
22. Case, R.M. (1984). The role of Ca^{2+} stores in secretion. *Cell Calcium* **5** 89-110.

23. Francis, L.P. Lennard, R. and Singh, J (1990). Cellular mechanism of action of magnesium on acetylcholine-evoked secretory responses in the rat pancreas. *Exp. Physiol.* **75**, 669-680.
24. Meissner,G., Darling, E., and Eveleth, J. (1986). kinetics of rapid Ca^{2+} release by sarcoplasmic reticulum. Effects of Ca^{2+}, Mg^{2+} and adenine nucleotides. *Biochemistry.* **25**, 236-244.
25. Osipchuk, Y.V., Wakui, M., Yule, D.I., Gallacher, D.V. and Petersen, O.H. (1990). Cytoplasmic Ca^{2+} oscillations evoked by receptor stimulation, G protein activation, internal application of inositol trisphosphate or Ca^{2+} simultaneous microfluorimeter and Ca^{2+} dependent Cl^- current recording in single pancreatic acinar cells. *EMBO* **9**, 697-704.
26. Capurro, D.E., and Dormer, R.L. (1989). Ca^{2+}-induced Ca^{2+} release from endoplasmic reticular membranes of rat pancreas. *Digestion.* **43**, 37.
27. Boneh, A. and Tenenhouse, H.S. (1990). Calcium-activated, phospholipid-dependent protein kinase (protein kinase C): General aspects and experimental considerations. *Isr. J.Med. Sci.* **26**, 293-300.
28. Yule, D.I., Kim, E.T. and Williams, J.A. (1994). Tyrosine kinase inhibitors attenuate "Capacitative' Ca^{2+} influx in rat pancreatic acinar cells. *Biochem.Biophys.Res.Commun.* **202 (3)** 1697-1704.

EXTRACELLULAR MG^{2+} EFFECTS INTRACELLULAR SIGNALLING PROCESSES AND ACID SECRETION IN RAT PARIETAL CELLS

FRANK CH. MOOREN[1], MARIA MORENO GEADA[2], JAIPAUL SINGH[2], WOLFGANG BEIL[3] AND WOLFRAM DOMSCHKE[1]
*[1]Department of Medicine B, Westfälische Wilhelms-Universität, Münster, Germany, [2]Department of Applied Biology, University of Central Lancashire, UK and * Department of General Pharmacology, Hannover Medical School, Hannover, Germany*

1. INTRODUCTION

Ulcerogenesis is a multifactoriral process despite the undoubtedly important role of Helicobacter pylori. Previously it has been demonstrated that magnesium deficiency in combination with stress evokes a significant increase in ulcer manifestations in rats [2]. There is growing evidence in the literature that extracellular magnesium deprivation enhances stimulus-secretion coupling processes in various cell types [3, 5] suggesting a role of magnesium as modulator in cell metabolism and function [1]. Therefore in the present study the effects of extracellular magnesium concentration ($[Mg^{2+}]_{ex}$) on intracellular signalling processes and on acid secretion were investigated in rat parietal cells in vitro.

2. METHODS:

Parietal cells were isolated using an enzymatic digestion method. Intracellular free concentrations of Mg^{2+} ($[Mg^{2+}]_{in}$)and Ca^{2+} ($[Ca^{2+}]_{in}$) in single cells were determined using the ion sensitive fluorescent dye magfura-2 and fura-2, respectively. Cells were loaded at room temperature with the membrane permeant ester fura-2-AM and magfura-2-AM at a concentration of 5µmol/l for 30 min. Intracellular ion concentration were determined according to the following equation: $[Ca^{2+}]_{in}$ or $[Mg^{2+}]_{in} = (R-R_{min})/(R_{max}-R) * K_d * F$ [4].
The calibration parameters were determined in separate experiments under calcium/magnesium saturated und free conditions. Fluorescence measurements were performed on single cells using a PTI Deltascan (PTI GmbH, Holstein/Wedel, Germany) attached to a Nikon TMD diaphot microscope. Excitation wavelengths were adjusted to 340 nm and 380 nm and the emitted light was collected at 509 nm.

T. Theophanides and J. Anastassopoulou (eds.), Magnesium: Current Status and New Developments, 29-31.

Quantification of acid secretion was performed by measuring [^{14}C]-aminopyrin uptake, as previously described. For each experiment 10^6 parietal cells and an [^{14}C]-aminopyrin concentration of 8.3 μmol/l were used. Adenosine 3`,5` cyclic monophosphate (cyclicAMP) was measured using an ELISA kit.

3. RESULTS

Incubation of parietal cells in low (0.2 mM), control (1.2 mM) and high (10 mM) Mg^{2+} resulted in consecutive changes of the intracellular free magnesium concentration. Carbachol at a concentration of 100 μM induced the well known biphasic calcium response. Initially calcium increased to a peak value of around 432 $\pm$ 30 nmol/l followed by a sustained plateau phase at a mean value of around 323 $\pm$ 29 nmol/l. Both peak and plateau phases were increased under low $[Mg^{2+}]_{ex}$, whereas 10 mM Mg^{2+} outside had the opposite effect. The same effect was observed after application of thapsigargin which is known to increase intracellular calcium without effects on intracellular inositol 1,4,5- trisphosphate levels. Additionally extracellular magnesium reduction resulted in an improved filling state of intracellular calcium stores. Next we investigated the effect of histamine on changes in intracellular cyclicAMP levels during perturbations of $[Mg^{2+}]_{ex}$ at three different times, after 15, 30 and 45 min. There was no significant effect of low $[Mg^{2+}]_{ex}$ observed at any time compared to normal $[Mg^{2+}]_{ex}$. However, high $[Mg^{2+}]_{ex}$ inhibited the production of cyclicAMP significantly at 30 and 45 min after the onset of histamine stimulation compared to normal $[Mg^{2+}]_{ex}$.
Finally we studied the influence of $[Mg^{2+}]_{ex}$ on acid secretion properties. Incubation of single parietal cells in solutions of different magnesium concentrations effected spontanaeous acid secretion with high $[Mg^{2+}]_{ex}$ inhibiting spontanaeous acid secretion and vice versa. In the same way carbachol induced acid secretion was effected, whereas histamine stimulated acid secretion was not affected.

4. CONCLUSIONS

These findings indicate a modulatory role of $[Mg^{2+}]_{ex}$ on the intracellular signalling processes and acid secretory properties in rat parietal cells. These effects seemed to be mediated probably by regulating (1) calcium loading capacity of intracellular stores (2) the permeability of the calcium influx pathway and (3) the formation of cyclicAMP.
Finally it should be mentioned that these results might be of clinical relevance, as indicated by the study of Classen et al. [2]. One example might be the stress ulcer in intensive care medicine. Although a few parameters like reduced local perfusion rates have been addressed to be involved in ulcerogenesis the above results give evidence for a potential role of extracellular magnesium concentration. This is supported by the fact, that hypomagnesiemia is associated in a high percentage with other serum electrolyte abnormalities, often detected in critically ill patients. Naturally, further studies need to be performed to confirm this hypothesis.

References

1. Beyenbach, K.W. (1990) Transport of magnesium across biological membranes. *Magnesium Trace Elem.* **9**, 233-254.
2. Classen, H.G., G. Fischer, L. Dobler, C. Guigas, B. Herold, H. Hirneth, R. Jacob & C.T. Rieg. (1985) Modification of acute stress reactions by the actual magnesium status and synthetic calcium antagonists.In Halpern, Durlach Eds, *Magnesium Deficiency*,., 24-29,
3. Francis, L.P., R. Lennard & J. Singh. (1990) Mechanism of action of magnesium on acetylcholine-evoked secretory responses in isolated rat pancreas. *Exp.Physiol.* **75**, 669-680.
4. Grynkiewicz, G. & R. Tsien (1985) *J.Biol.Chem.* **260,** 3440-3450.
5. Zhiming, Zhu, M. Tepel, C. Spieker & W. Zidek (1995) Effect of extracellular Mg^{2+} concentration on agonist-induced cytosolic free Ca^{2+} transients. *Biochim.Biophys.Acta* **1265**, 89-92.

CHARACTERIZATION OF MAGNESIUM TRANSPORT FROM RAT, PIG, COW AND HUMAN ERYTHROCYTES

S. TAYLOR AND J. SINGH
Department of Applied Biology, University of Central Lancashire, Preston, PR1 2HE, Lancashire, England.

1. SUMMARY

This study investigates the mechanism of magnesium (Mg^{2+}) transport in erythrocytes of rat, cow, pig and human by the method of atomic absorption spectroscopy. Erythrocytes of rat, pig, and cow, which were loaded with Mg^{2+} employing 6μM A23187 and 12 mM $MgCl_2$, released Mg^{2+} in a time-dependent manner with maximal efflux occurring within 40-50 min. Pretreatment of preloaded erythrocytes with either 10^{-3} amiloride, or replacement of extracellular sodium with *N*-methyl-D-glucamine (NMDG), resulted in a significant inhibition of Mg^{2+} release compared to erythrocytes incubated in normal buffer. Conversely, in human erythrocytes both amiloride and NMDG had no effect on Mg^{2+} efflux compared to erythrocytes incubated in normal buffer. The results indicate that Mg^{2+} efflux from erythrocytes of rat, pig, and cow are Na^+ dependent in contrast to efflux from human erythrocytes which appears to be non-Na^+ dependent.

2. INTRODUCTION

The abundant divalent cation Mg^{2+} is involved in several biochemical and physiological processes including the activation and inhibition of several transport systems depending on the transporter (1,2). For example, in erythrocytes Mg^{2+} can regulate Na^+:K^+ATPase and Ca^{2+}ATPase activity, the Na^+:K^+:Cl^- and K^+:Cl^- co-transporters, and the Na^+/H^+ and Cl^-/HCO_3^- exchangers (1,2). Since Mg^{2+} is involved with several membrane transport systems, with defects in Mg^{2+} metabolism or transporter activity leading to diseased states (3), then it is important to understand how erythrocytes can regulate the homeostasis of Mg^{2+}. This study was designed to characterize Mg^{2+} efflux from Mg^{2+} loaded erythrocytes taken from rat, pig, cow and human in order to ascertain any difference or similarity in the characteristics of the Mg^{2+} transport in each species.

T. Theophanides and J. Anastassopoulou (eds.), Magnesium: Current Status and New Developments, 33-35.

3. METHODS

Samples of cow and pig blood were obtained from the local abattoir, whilst rat blood was obtained from the anaesthetised animal by heart puncture. Human blood was authorised and obtained from the vein of volunteers. Both male and females were used in this study. Following centrifugation (3000*g* for 10 min) the erythrocytes were separated from the plasma and buffy coat and loaded with Mg^{2+} using the established method of incubating in a KCl medium containing 6μM A23187 and 12 mM $MgCl_2$ for 30 min at 37°C (4,5). Following loading, the suspension was centrifuged and the erythrocytes separated from the supernatant. Mg^{2+} efflux was measured by reincubating a 10% suspension of the cells in a Mg^{2+}-free medium at 37°C. At specific time intervals, the Mg^{2+} content of the effluent was determined by atomic absorption spectroscopy (5). The erythrocyte protein content was assayed and the results expressed as μmol Mg^{2+}(100mg cell protein)$^{-1}$. Values are presented as mean (± standard error of the mean) and compared with a Student's t test for significance.

4. RESULTS AND DISCUSSION

The mean plasma Mg^{2+} concentration (*n*=4) was significantly ($P<0.05$) lower in human (1.225 ± 0.075 mM) and cow (1.692 ± 0.026 mM) compared with pig (2.015 ± 0.071 mM) and rat (2.104 ± 0.207 mM). Similarly, the mean Mg^{2+} content of pig (*n*=5) and rat erythrocytes (*n*=3) (0.972 ± 0.061 and 1.004 ± 0.27 μmol (100mg protein)$^{-1}$, respectively) was higher than that of cow (*n*=5) and human erythrocytes (*n*=4) (0.457±

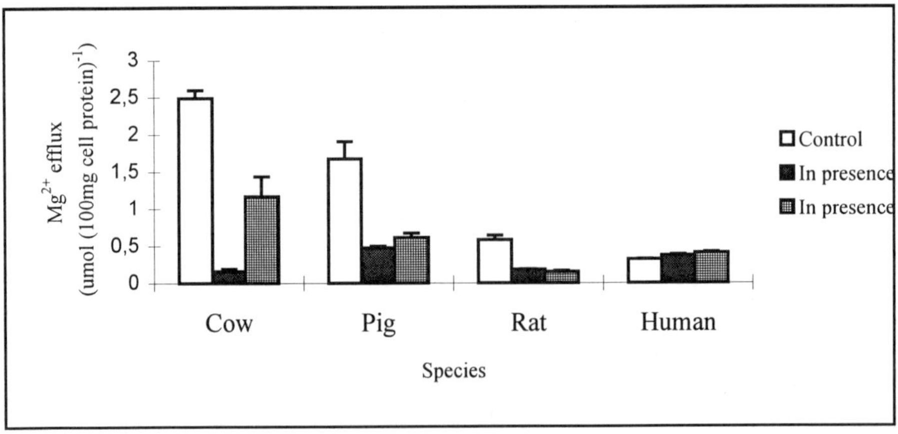

Figure. 1. Histograms showing mean Mg^{2+} efflux (with standard error of mean) from preloaded rat (n=3), cow (n=5), pig (n=5) and human (n=4) erythrocytes over a 50 min period in the absence and presence of either amiloride (10^{-3}) or NMDG.

0.034 and 0.482 ± 0.151 μmol (100mg protein)$^{-1}$, respectively). Incubation of the cells in a KCl medium containing A23187 resulted in a significant ($P<0.05$) elevation in Mg^{2+} content compared to that of the unloaded erythrocytes from each species, the greatest being in rat erythrocytes (4.724 ± 0.391μmol (100mg protein)$^{-1}$, n=3) and the lowest being in cow erythrocytes (0.714± 0.085μmol (100mg protein)$^{-1}$, n=5).
Incubation of the Mg^{2+}-loaded erythrocytes from each species in a Mg^{2+}-free physiological salt solution resulted in a large time-dependent release of Mg^{2+} with maximal efflux occurring within 40-50 min of incubation (Fig. 1). Pretreatment of the preloaded erythrocytes from rat, cow, and pig with either amiloride (10^{-3}M) or replacement of extracellular sodium with NMDG resulted in a significant ($P<0.001$) inhibition of Mg^{2+} release compared to erythrocytes incubated in normal buffer. In contrast, both NMDG and amiloride (10^{-3}M) enhanced Mg^{2+} efflux in human erythrocytes (Fig. 1). The results indicate that Mg^{2+} transport (efflux) from rat, pig and cow erythrocytes is associated with a Na^{+}-sensitive Mg^{2+} transport system although there appears to have been some Na^{+}-independent Mg^{2+} efflux as it was not totally abolished by either amiloride or NMDG. In contrast, the Mg^{2+} efflux from human erythrocytes did not appear to be sensitive to either amiloride or NMDG suggesting differences in the characteristics of Mg^{2+} transport from one species to another.

(Supported by the Nuffield Foundation).

References

1. Flatmann, P.W. (1988) The Control of Red Cell Magnesium. *Magnesium Res.*, **1**, 5-11.
2. Flatmann P.W. (1991) Mechanisms of Magnesium Transport. *Ann. Rev. of Physiol.*, **53**, 259-271.
3. Birch N.J. (1993) *Magnesium and the Cell.* Academic Press (London).
4. Gunther T., Vormann J., Hollriegl, V. (1990) Characterization of Na^{+}-dependent Mg^{2+} efflux from Mg^{2+}-loaded rat erythrocytes. *Biochim. Biophysic Acta,* **1023**, 455-461.
5. Wisdom, D.M., Geada, M.M., Singh, J. (1996). Characterisation of a sodium dependent magnesium efflux from magnesium loaded-rat pancreatic acinar cells. *Exp. Physiol.* **81**, 367-374.

CHEMICAL ASPECTS OF HETEROGENEOUS Mg-CONTAINING APATITES

M. OKAZAKI, M. TAIRA AND J. TAKAHASHI
Osaka University Faculty of Dentistry, Osaka 565, JAPAN

Magnesium, which contributes to various important reactions in human body as a catalyst and medical care of cardiovascular diseases as a useful agent1), is a typical trace element in biological apatites2). Mg-containing hydroxyapatites have been investigated and it has been reported that magnesium ions can substitute into apatite crystals to a limited extent (Fig.1,3), although magnesium also exists on the surface of apatite crystallites, in a nonapatitic mineral phase and with organic components. It is well known that magnesium retards the crystal growth of hydroxyapatites and accelerates the dissolution.

Recent focus of apatite formation in biological systems is heterogeneous precipitation. To investigate such a heterogeneous precipitation, in our previous study we adopted two simple fluoride supply system. During the initial or final half of the experimental period, fluoride containing solution was supplied. From the results, we speculated that two different types of heterogeneous fluoridated hydroxyapatite were formed, one is hydroxyapatite (HA) covered with fluorapatite (FA), another is FA covered with HA (Fig.2)4). In this study, effect of magnesium as a trace element on the physicochemical properties of heterogeneous hydroxyapatites was examined.

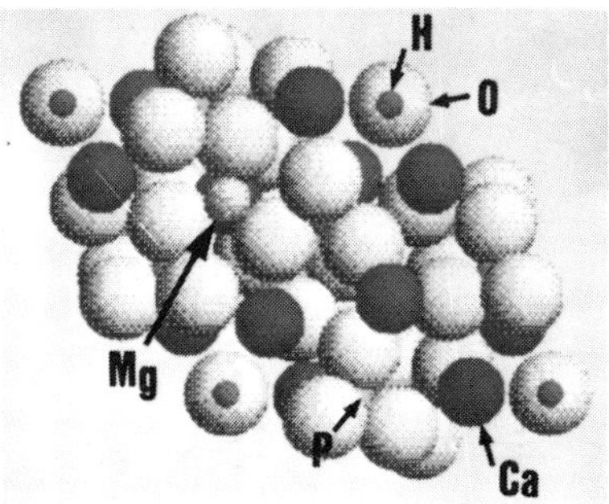

Figure1. Computer graphic of heterogeneous fluoridated apatites

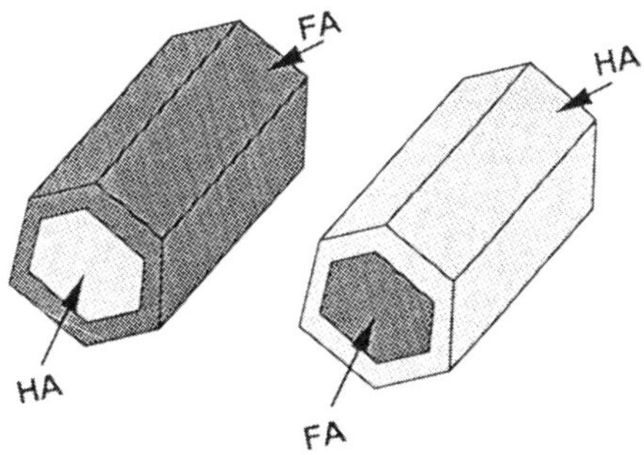

Figure.2. Two schematic models of Mg-containing hydroxyapatite

1. MATERIALS AND METHODS

Magnesium containing apatites were synthesized heterogeneously at 80°C and pH 7.4 in the presence or absence of magnesium. H-Mg type apatite was made by supplying 0.25

T. Theophanides and J. Anastassopoulou (eds.), Magnesium: Current Status and New Developments, 37-40.

L of 100 mmol/L $Ca(CH_3COO)_2H_2O$ solution during the initial half of the experimental period and then 0.25 L of 100 mmol/L $Ca(CH_3COO)_2H_2O$ solution containing 10 mmol/L $Mg(CH_3COO)_2H_2O$ during the remainder of the experiment, and 0.5 L of 60 mmol/L $NH_4H_2PO_4$ solution during the whole experiment period. The two solutions were supplied into 1 L of mechanically stirred 1.3 mol/L CH_3COONH_4 solution. The other Mg-H type apatite was synthesized under the reverse condition.

X-ray diffraction (Rigaku Ultima+) was employed to identify precipitates and estimate crystallinity. Scanning electron micrographs of crystals were obtained with a Nihon Denshi JSM-5400, and the magnesium content of each crystal was analyzed quantitatively by wavelength dispersive spectroscopy (WDS) coupled with SEM. Calcium and magnesium concentrations were determined by atomic absorption spectrophotometry. Total phosphate concentrations were determined by the UV spectrophotometric method. To estimate solubility, samples were left to stand in 50 ml of 0.5 mol/L acetate buffer solution (pH 4.0) at 37 °C for 1 month, then the calcium and phosphate concentrations in the solution were determined.

2. RESULTS AND DISCUSSION

The total magnesium contents of both heterogeneous Mg-containing apatites were

A B

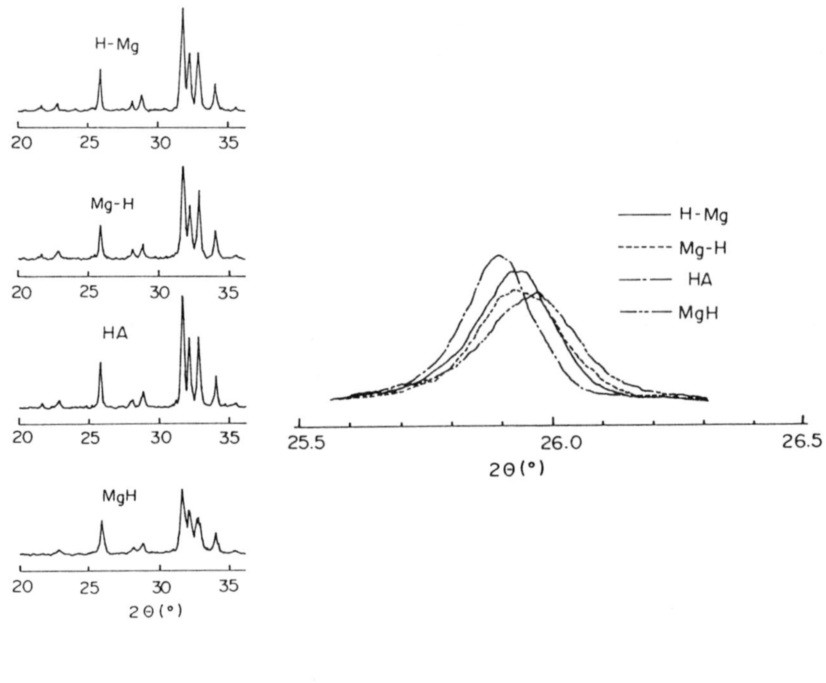

Fig. 3. X-ray diffraction patterns (A) and the (002) reflections (B) of Mg-containing hydroxyapatites, together with homogeneous apatites(HA and MgH).

The (002) peak positions of both H-Mg and Mg-H shifted to higher angles and approached that of homogeneous MgH (Fig.3B). This peak shift suggests that Mg2+ ions may be partially substituted into the apatite crystals.

H-Mg **Mg-H**

Fig.4. Scanning electron micrographs of heterogeneous Mg-containing hydroxyapatites.

Scanning electron microscopy showed different tendency of crystal shapes: plate-like crystals of H-Mg and needle-like crystals of Mg-H (Fig.4). Both crystals were inhibited in their growth in comparison with Mg-free hydroxyapatite, HA. Wavelength dispersive spectroscopy coupled with SEM estimated that the magnesium intensity of H-Mg was higher than that of Mg-H (Fig.5).

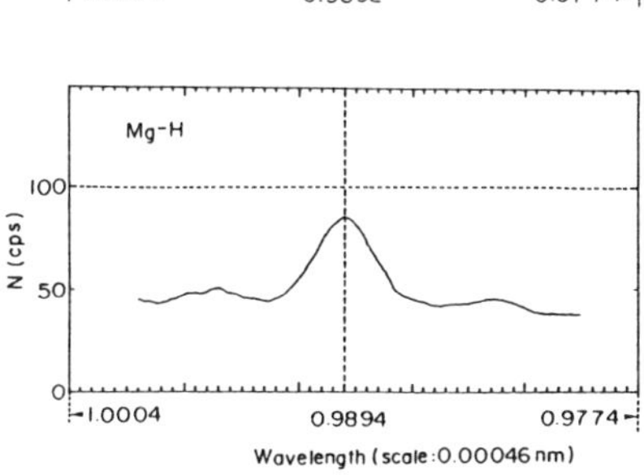

Fig. 5. Wavelength dispersive spectroscopy of heterogeneous Mg-containing hydroxyapatites

In general, magnesium accelerates the dissolution of apatite. Contrary to expectations, however, the apparent solubility of H-Mg, which may be Mg-rich at the surface, was lower than that of Mg-H in 0.5 mol/L acetate buffer solution (pH4.0) at 37°C. Since Mg-H crystal growth was inhibited, small needle-like crystals, which had the relatively larger crystal surface area, might be more soluble than the Mg-rich outer shell of H-Mg type apatite.

These results suggest that two different types of heterogeneous Mg-containing apatites, outer Mg-rich and inner Mg-rich apatites, may be formed.

In addition, Mg-containing fluoridated CO_3apatite was synthesized heterogeneously by supply magnesium and carbonate during the whole experimental period and fluoride during the initial half of the experimental period. Its crystal shape was round hexagon covered with small crystals (Fig.6). Thus, it could be suggested that heterogeneous apatites change their features dramatically in the presence of other trace elements.

Fig. 6. Scanning electron micrograph of heterogeneous Mg-containing fluoridated CO_3apatites.

References

1) Itokawa Y. and Durlach J: *Magnesium in Health and Disease*, John Libbey, 1989.
2) Miles AEW: *Structural and Chemical Organization of Teeth*, Vol.II, Academic Press, New York, 1967.
3) Okazaki M, Takahashi J and Kimura H: Caries Res 20: 324-331, 1986.
4) Okazaki M: *Biomaterials* 13: 749-754, 1992.

FT-IR STUDIES OF PROPOFOL - MAGNESIUM INTERACTIONS

J. ANASTASSOPOULOU, V. KORBAKI, H. NASTOU, M. PALTA AND V. SARROU
National Technical University of Athens, Chemical Engineering Department, Radiation Chemistry and Biospectroscopy, Zografou Campus, Zografou 15780, Athens, Greece

1. Introduction

Propofol is a total intravenous anaesthetic drug, chemically unrelated to other anaesthetics and is now widely used[1]. Propofol is a phenolic ring with 2 iso-propyls in the 2 and 6 carbon of the ring, which make this anasthetic agent chemically unrelated to others. In previous studies we found fluctuations in serum Mg^{++} and Ca^{++} levels after the administration of propofol. It was interesting that the serum magnesium levels appeared to decrease during total anaesthesia with propofol, an event which was eliminated 24h after the administration of the drug[2-3]. In order to investigate the role of propofol to magnesium levels we mixed, *in vitro*, the liquid propofol with $MgCl_2.6H_2O$ by using Fourier Transform Infrared Spectroscopy.

2. Materials and Methods

Hexahydrate magnesium chloride was purchased from Merck and propofol (2,6-diisopropylphenol) from Aldrich. Hexahydrate magnesium chloride was heated to dryness at 110°C for 1 ½h. The FT-IR spectra were obtained with a BOMEM MB-Series FT-IR spectrophotometer, with 4 cm^{-1} resolution and 25 scans. All the spectra were obtained in liquid on KRS5 crystals windows.

3. Results and Discussion

The most important changes, occured in the region between 1800 cm^{-1} and 600 cm^{-1} (Fig. 1). The band at 1638 cm^{-1} of $MgCl_2.6H_2O$, which corresponds to the bending vibration of the H_2O molecules[4], that are coordinated with Mg, seem to change after mixing the compounds, creating a broad band at 1638 cm^{-1}. This change might immply a possible binding of propofol to the magnesium hydrated ion, through the water molecules. There were also observed intensity changes at 1309, 1200 and 1148 cm^{-1},

T. Theophanides and J. Anastassopoulou (eds.), Magnesium: Current Status and New Developments, 41-42.

which correspond to the absorptions of -C-O-H bending, C-O stretching and C-OH bending vibrations[5], respectively. These changes are due to the binding of magnesium ions to propofol molecule, to C-O-H group site. Intensity changes were also observed in the region 1461-1441cm^{-1}, where the vibrations of the ring, as well as of the group -CH_3 appear. The FT-IR spectra, in the region 1800-600 cm^{-1}, of propofol (a), of the complex propofol-$MgCl_2.4H_2O$ (b), and $MgCl_2.6H_2O$(c) are shown in Figure 1.

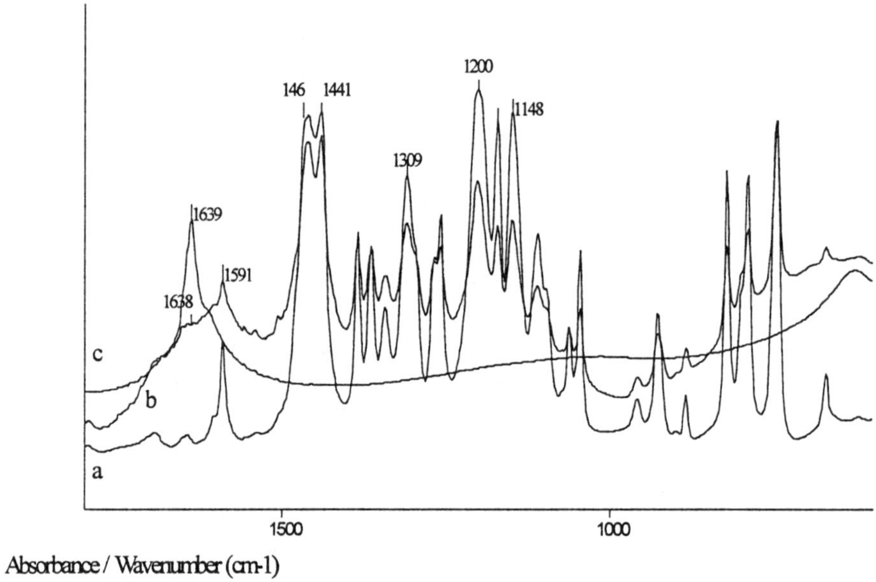

Figure 1. FT-IR spectra of propofol (a), propofol-$MgCl_2$ (b) and $MgCl_2.6H_2O$(c)in the region 1800-600cm^{-1}.

There were also found a few changes in the region 4000-1800cm^{-1}, which show that the hydrogen bond is stronger after the mixing of the compounds.
All these results can go further by studying more the physicochemical properties of the compounds.

References

1. Langley M.S., Heel C.: *Drug Evaluation*, ADIS Press Ltd., England, 1988.
2. Nastou H., Sarros G., Palta M., Nastos A., Sarrou V. and Anastassopoulou J. (1996) Fluctuations in serum magnesium levels during total intravenous anaesthesia with propofol, *Magnes. Res.*, **9**, 47-53.
3. Nastou H., Sarros G., Palta M., Nastos A., Drakopoulou A., Anastassopoulou J.: (1996) The effect of general anaesthesia in serum magnesium and calcium levels, in Ph. Collery, J. Corbela, J. Domingo & J.M. Llobet (eds.), Paris, *Metal ion in Biology and Medicine*, John Libbey, pp. 550-552.
4. Bellamy L.J.,: *The Infra-red spectra of complex molecules*, Methuen & Co LTD, London, 1957.
1. Silverstein R.M., Basler G.C., Morrill T.C.,: *Spectrometric identification of organic compounds*, John Wiley and Sons, INC., Canada, 1991.

MAGNESIUM IONS - GUANOSINE-5'-MONOPHOSPHATE COMPLEXES: A SPECTROSCOPIC STUDY

J. ANASTASSOPOULOU, C. BARBAROSSOU, V. KORBAKI, H. NASTOU, N. Th. RAKINTZIS AND T. THEOPHANIDES
National Technical University of Athens, Chemical Engineering Department, Radiation Chemistry and Biospectroscopy, Zografou Campus, Zografou 15780, Athens Greece.

1. Introduction

Magnesium is the second most abundant mineral in the organism, and cells depend directly on its concentration. It also participates as a catalyst or activator in countless physical activities. Magnesium cations are extremely important nucleic acid stabilizers[1]. The aim of this work is to study the reaction of Mg^{2+} cations to guanosine-5'-monophosphate disodium salt (5'-$GMPNa_2$), in aqueous solutions, using Fourier Transform Infrared Spectroscopy and to see also the role of SO_4^{2-} anions, upon complexation.

2. Material and Methods

Magnesium sulfate heptahydrated ($MgSO_4.7H_2O$) and trihydrated guanosine-5'-monophosphate disodium salt (5'-$GMPNa_2.3H_2O$ =L, ligand) were purchased from Carlo Erba and Aldrich Chemical, respectively. They both were used without further purification. The Fourier Transform Infrared (FT-IR) spectra were obtained with a BOMEM MB-Series spectrophotometer, with 4 cm^{-1} resolution and 25 scans, in KCl pellets. All the spectra were recorded in the region 4000-400 cm^{-1}. The complexes were prepared by mixing aqueous solutions of the compounds in molar ratios $[Mg^{2+}]:[L]=1:1$, 2:1, 4:1 and 8:1.

3. Results and Discussion

The FT-IR spectra of 5'-$GMPNa_2$ and the $[Mg^{2+}]:[L]$ in 1:1 and 8:1 complexes are shown in Figure 1a,b and c, respectively, in the region 1800-600 cm^{-1}, where the most important changes are observed after complexation. The band at 1628 cm^{-1}, which is assigned to bending vibration of the nucleotide's crystal water [2,3] is disappeared after complexetion. On the other hand the appearance of a shoulder band at about 1638 cm^{-1}

T. Theophanides and J. Anastassopoulou (eds.), Magnesium: Current Status and New Developments, 43-45.

is due to the magnesium coordinated water. A lower shift of 8 cm^{-1} of the band at 1492 cm^{-1}, which is assigned to δC8-H and νC8-N7 vibrations confirms the binding of the Mg^{2+} ions to N7 site of the guanine molecule, indicating the delocalization of the

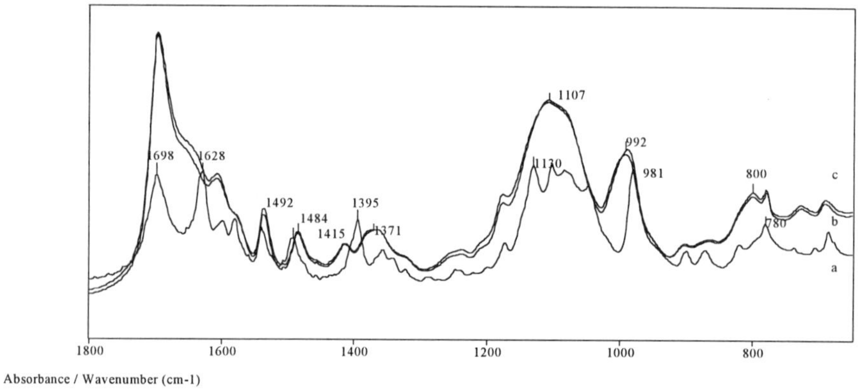

*Figure 1.*FT-IR Spectra of: (a) 5'-GMP, (b) Mg-5'-GMP, M:L=1:1 and (c) Mg-5'-GMP, M:L=8:1 in the region 1800-600cm^{-1}.

electronic cloud in the ring [3]. Considerable changes are also observed at the region 800-1200 cm^{-1}, where the sugar and phosphate groups absorb. These findings are attributed with the fact that Mg^{2+} ions interact with phosphate group. The shift from 820 to 800 cm^{-1} corresponds to the conformational change from *C2'-endo,anti* to *C3'-endo,anti*, which is the characteristic "marker band" of the sugar and indicates a model change from B- DNA to A-DNA [3,4]. Finally, it was found that the spectra remains the same as the ratio [Mg^{2+}]:[L] increases from 1 to 8.

In Table 1 are given the most characteristic bands of the nucleotide, their assignments, as well as the changes after complexation with magnesium ions for ratio [Mg^{2+}]:[L] 1 and 8.

TABLE 1. FT-IR spectra of magnesium-5'-GMP complexes and band assignments,

5'-GMP cm^{-1}	M:L 1:1	M:L 8:1	Δν cm^{-1}	possible assignment
1698	1695	1695	3	νC6=O
1628	shoulder	shoulder		coordinated H_2O, δNH_2
1492	1484	1484	8	δC8-H>νC8-N7
1395	1415	1415	20	purine ring vib.
1130				νC-O,of sugar ring
981	992	992	11	νsym PO_3^{-2}
820	800	800	20	C3'-endo
780	780	780		νPO

References

1. Missailidis, S. Anastassopoulou, J,. Theophanides, T. (1996), FT-IR spectra of oligonucleotide-Mg++ complexes, in Ph. Collery, J.L. Domingo, J.-C.Etienne and J.M. Llobet, *Metal Ions in Biology and Medicine* John Libbey, Eurotext, Paris.**Vo 4**, pp. 39-41.
2. Theophanides, T. Polissiou, M.(1986) Magnesium-Nucleic Acid Conformational Changes and Cancer, *Magnesium* **5**, 221-233.
3. Theophanides, T. (1984) FT-IR spectra of nucleic acids and the effect of metal ions, in T. Theophanides (ed.), *Fourier Transform Infrared Spectroscopy,* Reidel Publishing Co., Dordrecht, pp. 105-124.
4. Anastassopoulou, J., Polissiou, M., Manfait, M and Theophanides, T (1992), Magnesium-nucleic acid interactions in water, in J. Anastassopoulou, Ph. Collery, J.-C.Etienne and T. Theophanides (eds.*), Metal ions in Biology and Medicine,* John Libbey, Eurotext, Paris, **Vo 2**, pp. 26-31.

SUPRAMOLECULAR MG(II) COMPLEXES OF CYTOSINE AND 1-METHYLCYTOSINE

D. ARMENTANO[1], G.DE MUNNO[1], M. REGINA[1],
J. ANASTASSOPOULOU[2] AND T. THEOPHANIDES[2]
[1]*Dipartimento di Chimica, Università della Calabria, Arcavacata di Rende, Consenza, Italy.*
[2]*National Technical University of Athens, Chemical Engineering Department, Radiation Chemistry and Biospectroscopy, Zografou Campus, Zografou 15780, Athens, Greece.*

1. INTRODUCTION

The interaction of metal ions with nucleic acids has been a subject of study for many years[1,2]. The binding of magnesium to biological molecules, in cells, in the body, in plants and its role in many living processes is of great interest. In this paper we present some results on crystal structures of Mg^{2+} with nucleic constituents. Cytosine (Cyt) and 1-methylcytosine (1-Mecyt) were used as nucleic bases to interact with magnesium salts. Some of these complexes formed could be regarded as supermolecules.

2. RESULTS AND DISCUSSION

The magnesium complexes are illustrated in Figures 1 and 2. They were crystallized from $Mg(ClO_4)_2$ and 1-Mecyt from water solutions.
In these two supermolecules the methylcytosine molecule behaves differently, since it binds indirectly to the magnesium ion through hydrogen bonds (Figure 1) and directly (Figure 2).

Recently [3] we have obtained some more crystal structures of Mg(II) complexes with cytosine and 1-Mecyt using MCl_2 and $MgBr_2$ salts. The chemical formulae of these complexes are: $[Mg(H_2O)_6(1\text{-Mecyt})_2]X_2$ **1**, $[Mg(H_2O)_4(cyt)_2]X_2$ **2**, and $[Mg(H_2O)_2(cyt)_4]X_2$ **3**, where X=Cl, Br. These are octahedrally coordinated in which some water molecules have been substituted by the nucleobase. In complex **1** the nucleobase is not directly coordinated to the metal ion, but through hydrogen bonds in which the water molecules are

T. Theophanides and J. Anastassopoulou (eds.), Magnesium: Current Status and New Developments, 47-48.

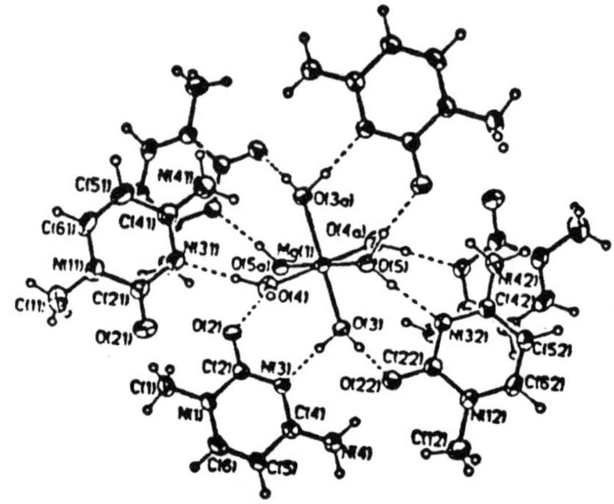

Figure 1. The supramolecular complex $[Mg(H_2O)_6\ (1\text{-}Mecyt)_2]X_2$, where X= Cl, Br.

involved. In complex **2** two cyt molecules are directly coordinated to the metal ion. The structure of **3** is obtainted from complex **2** due to further coordination of another two base molecules.

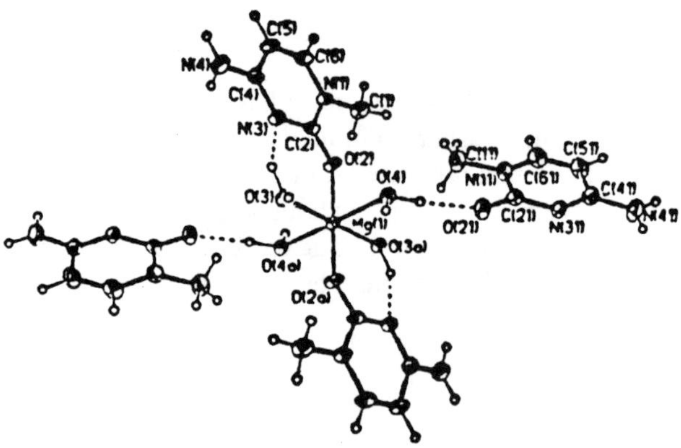

Figure 2. The supramolecular complex $[Mg(H_2O)_2\ (1\text{-}Mecyt)_4\,]X_2$, where X= Cl, Br.

References

1. Eichhorn, G.L., (1973), *Inorganic chemistry*, Elsevier, Amsterdam, Vo 1, 2, Chapts, 33, 34, pp1207.
2. Theophanides, T., (1981), Interactions des acides nuclèiques avec les mètaux, *Can. J. Spectroscopy*, **26**, 165-179
3. Gaday, M.A., De Munno, G., Mrdaglia, M., Anastassopoulou, J., Theophanides, T. (1997) First Isolation of Supramolecular Assemblies containing Nucleic Bases and Magnesium (II) Hexahydrate Ions, *Angewante Chemie, Int. Ed. Engl.*, **36**, 511.

SYNTHESIS AND NMR SPECTROSCOPIC STUDIES OF A NEW MAGNESIUM ANTIBIOTIC COMPOUND

MARGARITA PETROLIAGI,[a] OLGA MARKOPOULOU[a,*] and JOHN MARKOPOULOS[b]
[a] *Organic Chemistry Laboratory, National Technical University of Athens, Zografou Campus, GR-15773*
[b] *Inorganic Chemistry Laboratory, University of Athens, Greece*

The combinatorial approach to drug discovery has led to focus on heterocyclic nitrogen frameworks enjoying a splendid therapeutic record. A special branch of such compounds is the **3-alkanoyl-5-arylidene tetramic acids, pyrrolidine-2,4-diones**. These natural products possessing an enolic β,β'-tricarbonyl moiety, a hydrophobic 5-substituent and a lipophilic 3-alkanoyl-substituent, are interest complexation hosts for various metal ions such as Mg(II), Pt(II) [1]. Their biological activity is related to their complexation properties and the mechanisms responsible for it is affected by the polarity of these complexes to transport through cell boundary.

For instance, magnesidin **I**, a compound of the covalent magnesium chelates of the 3-alkanoyl tetramic acid derivatives, inhibits various Gram-positive bacteria (MIC 2 - 7 $mgmL^{-1}$) and prevents decay of foodstuffs caused by spore generating compounds [2]. The contribution of the magnesium atom to the antibiotic potency is also marginal as indicated by structure - activity relationships.

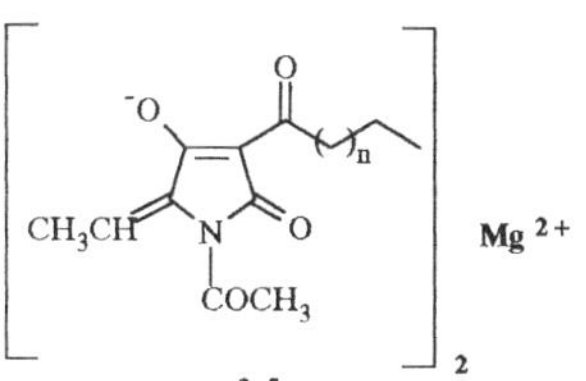

n = 3,5

I

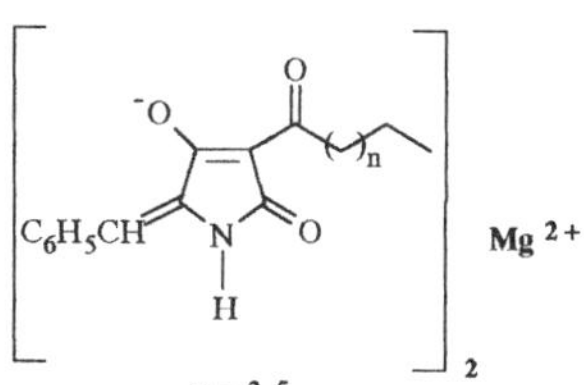

n = 3,5

II

In the course of our research programme on the design of novel complexes of 3-substituted tetramic acids which could increase liposolubility and permeability

T. Theophanides and J. Anastassopoulou (eds.), Magnesium: Current Status and New Developments, 49-50.

through cell walls, we have developed a new synthetic route for their construction [3].

In this paper we present the synthesis of Magnesidin related compounds **(II)** via a new synthetic sequence (Scheme 1).

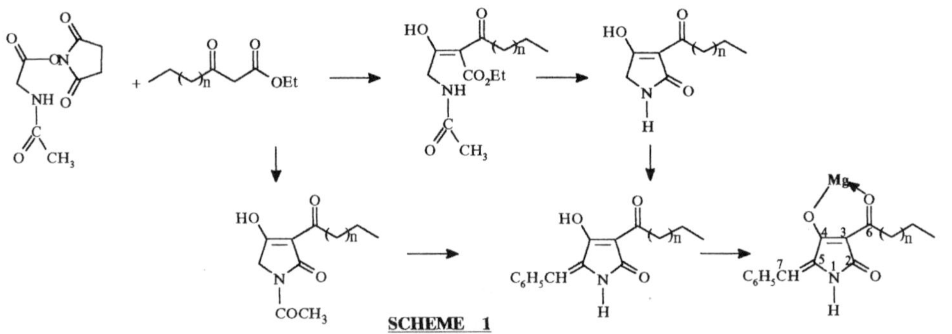

SCHEME 1

The structure of the novel 5-benzylidene-3-alkanoyl tetramic acids **(II)** was confirmed by ^{1}H, ^{13}C NMR and IR spectroscopy. According to the ^{13}C NMR and IR spectroscopy there was observed a significant transposition of the signals of the carbonyls C-2, C-6. The ^{1}H spectroscopy revealed the existence of two isomers according to the double signal of the proton of C-7.

With the aid of this method, the application of our strategy to the synthesis of magnesium complexes of 5-arylidene-3-alkanoyl tetramic acids, as well as the investigation of their structure - activity relationships is currently being pursued.

References

1. Matsunaga, K., Hasegawa, S., and Muto, M., Jpn.Pat. 01/313,488 [*CA* **113**, 51507 (1990)]; Hasegawa, S., Matsunaga, K., and Muto, M., Jpn.Pat. 02/48,591 [*CA* **114**, 16607 (1991)].
2. Imamura, N., Adachi, K., and Sano, H. (1994) *J.Antibiotics*, **47**(2), 257-261; Bhat, S.V., Kohl, H., Ganguli, B.N., and DE Souza, N.J. (1977) *Eur. J. Med. Chem. - Chimica Therapeutica*, **12**(1), 53-57; Hoechst A.G., Brit. 1,478,643 [*CA* **88**, 168480 (1978)].
3. Markopoulou, O., Markopoulos, J., and Nicholls, D. (1990) *J.Inorg.Biochem.*, **39**, 307-316; Heaton, B.T., Jacob, C., Markopoulos, J., Markopolou, O., Nahring, J., Skylaris, C.-K., and Smith, A.K. (1996) *J. Chem. Soc., Dalton Trans*, 1701-1706.

SPECTROPHOTOMETRIC DETERMINATION OF MAGNESIUM IN SERUM USING XYLIDYL BLUE REAGENT IN MICELLAR MEDIUM

D. STANKOV[1], T. JOVANOVIĆ[2], AND M. JELIKIĆ-STANKOV[2]
[1]*Hospital "Dr Dragiša Mišović", Biochemistry Department,* [2]*Faculty of Pharmacy*
[1]*Milana Tepića 1, 11000 Belgrade,* [2]*Institute of Analytical Chemistry, Vojvode Stepe 450, 11000, Belgrade, Yugoslavia*

Methods for the determination of magnesium have been based on atomic absorption spectrometry (AAS), molecular spectrophotometry using xylidyl blue reagent [1-azo-2-hydroxy-3-(2,4-dimethylcarboxanilido)-naphthalene-1'-(2-hydroxybenzene-4-sul phonate)], titanium yellow, 2,2':6',2" terpyridine and calmagite [1-4].

The purpose of our work was the optimization of the most frequently applied spectrophotometric methods for determination of magnesium usimg Mg-xylidyl blue complex in serum and by adding non-ionic surface-active compound TRITON X-100. The optimization of the method consisted of an increased linearity of the method up to 3.00 mmol/L of Mg^{2+} ions. Due to the intensive colouring of the reagent, experimentally was estimated the optimal concentration of the reagent xylidyl blue for the given method to be $1.46 \cdot 10^{-4}$ mol/L. To avoid the interfering effect of Ca^{2+} ion, usually masked reagent EGTA [Ethylene Glycol-bis(3-aminoethyl Ether)N,N,N'N'-Tetracetic Acid] was used as it unselectively complexes Ca^{2+} ions. In the proposed method, using carbonate buffer pH 11.30 in micellar medium, the influence of Ca^{2+} ion was minimized so that the determination of magnesium was done without the addition of EGTA. It was established that with the addition of TRITON X-100 in a concentration range of 0.1-0.3% an optimal increase of linearity was achieved along with analytically unimportant influence of Ca^{2+} ion.

The addition of tensids caused hypsochromically shift of the maximum absorption of Mg-xylidyl blue complex to 520 nm, decreasing in that way the influence of the reagent blank. Because of the high sensitivity of the method , only small volumes of samples were used (1:100) in relation to the reagent, thus obviating the utilization of the sample blank.

To determine Mg^{2+} ion in the serum, a calibration curve was prepared using standard

T. Theophanides and J. Anastassopoulou (eds.), Magnesium: Current Status and New Developments, 51-52.

solution of Mg^{2+} ions of linear dilution up to 3.00 m mol/L. The obtained correlation coefficient was r=0.999, intercept a=0.004 and slope b=0.287.
The results of the determination of magnesium by the proposed method in serum were compared with the results obtained by AAS, the correlation coefficient being r=0.998, slope b=1.05 and intercept a=0.006. The influence of Ca^{2+} ion on the determination of magnesium was tested and the obtained results are presented in Table 1.

TABLE 1. Effect of calcium on the determination of magnesium in serum

Metal added in serum concentration (mmol/L)		Recovery (%)
Mg	Ca	
1.00	-	98
2.00	-	97
3.00	-	100
2.00	1.50	101
2.00	2.50	100
2.00	3.50	103

Using the proposed method, magnesium may be determined in the serum, heparinated plasma, urine and cerebrospinal liquid. Owing to its simplicity, the method may be utilized in manual work as well as on automatic biochemical analysers.The method have been satisfactory applied to serum with recoveries between 97-103 %.
As it is a one-component reagent, different from most previously used reagents, when the working reagent had been prepared by mixing buffer with the solution of chromogen, calculations of the unknown concentration of magnesium may be carried out not only by the standard of magnesium, but also by the calibration curve.

Acknowledgement. This research was financially supported by the Ministry for Science and Technology of Serbia.

References

1. Glindler et al. *Clin. Chem.* 17(1971) 662
2. Tietz N.W. *Clinical Guide to laboratory Tests*. W.B. Saunders Co (1983)
3. Wimmer, M.C. Artiss, J.D. i Zak, *B. Clin. Chem.* 32(1986) 632
4. Peerzada N. and Kozlik E. *Anal. Letters*, 23(6)(1990) 1087-1093

INFLUENCE OF FOUR NON-STEROID ANTI-INFLAMMATORY DRUGS (NAPROXEN, TOLMETIN, TOLFENAMIC ACID AND VOLTAREN) AND ONE ANTIBIOTIC (MONENSIN) ON MAGNESIUM CONCENTRATIONS IN GUINEA PIG OR RAT TISSUES

V.-P. KOTSAKI-KOVATSI.
Laboratory of Pharmacology, Faculty of Veterinary Medicine, Aristotle University, Thessaloniki 540 06, Macedonia Greece.

1. Summary

In the present study, the effect of the administration of naproxen, tolmetin, tolfenamic acid and voltaren (non-steroid anti-inflammatory drugs) as well as monensin (antibiotic) on the magnesium concentration in guinea pig or rat tissues, was studied. Naproxen significantly decreased the magnesium concentration in the spleen and heart of the treated guinea pigs. Although tolmetin increased the magnesium concentration in some tissues of guinea pigs (brain, heart, lung, kidney, adrenals and testis) and decreased it in some others (spleen and thigh bone), the changes were not statistically significant. Spleen, ovaries and bone magnesium concentrations were increased in guinea pigs fed tolfenamice acid. Voltaren significantly decreased the magnesium concentration in the heart, liver, kidney and testis and increased it in the spleen of guinea pigs. Monensin significantly decreased the magnesium concentration in the brain, heart, lung, spleen and testis of rats and increased it in the thigh bone.

2. Introduction

Several studies [1,7] have shown that drugs which are potentially capable, from the chemical point of view, of forming complex compounds with metal ions, and specially those which are chronically administered, can cause translocation and/or elimination of these ions from the living organism.

We chose to study the translocation and or elimination from the organism of magnesium because the importance of it's presence for the normal heart function is very well known. It is also known that tolmetin and naproxen can form complexes with copper and zinc ions in vitro [3,17] and there is a strong possibility, from the chemical point of view, that similar complexes with magnesium ions may be formed. It is also known that the antibiotic monensin binds numerous mono-and divalent cations,

T. Theophanides and J. Anastassopoulou (eds.), Magnesium: Current Status and New Developments, 53-58.

known that the antibiotic monensin binds numerous mono-and divalent cations, primarily in dimeric complexes which facilitate the passage of metal ions through hydrophobic lipid membranes [14,15,16]

Even though there are several reports about the interactions of monensin with mono- and divalent cations in the organism [4,5,19] there is surprisingly little information about the translocation and/or elimination that it may cause to important for the growth and development of the organism cations.

The rest two of the examined N-SAIDS (voltaren and tolfenamic acid) were chosen because the study of their chemical structure showed that there is a strong possibility to form complexes with metal ions.

The aim of this study was to determine whether the chronic administration of the above mentioned drugs may cause drug-induced magnesium redistribution and/or elimination from guinea pig or rat tissues.

3. Material and methods

The fact that the administration of the studied substances results in translocation and or elimination of magnesium from the organism is of great importance because free, unbound and therefore biologically available Mg^{2+} plays a major role in numerous intracellular processes. Magnesium activates approximately 300 enzyme systems [6] in the human organism, including most of the enzymes responsible for the tranport, storage and use of energy [20-24].

For the study of naproxen twenty eight (twenty test and eight control) female Hartley-Albino guinea pigs six months old, 450-550 g each in weight, were used. Naproxen was administered i.p. 100 mg/kg b.w. /week [9,10].

For the study of tolmetin twelve (8 test and 4 control) adult male Hartley-Albino guinea pigs six months old and 450-550 g each in weight were used. Tolmetin was administered i.p. 105 mg/kg b.w./week [12]

Tolfenamic acid was studied on twelve 7 test and 5 control adult female Hartley-Albino guinea pigs eight months old with average weight 500 g each. Tolfenamic acid was incorporated to the diet. The offered amount of the diet was calculated so that each animal received daily 8,5 mg/kg b.w./day [13].

For the study of voltaren fourteen 9 test and 5 control adult male Hartley-Albino guinea pigs six months old and 450-550 g each in weight were used. Voltaren was administered i.p. 7 mg/kg b.w./week [11].

Monensin was studied on sixteen (12 test and 4 control) adult male Wistar rats five months old with average weight 350 g each. Monensin was incorporated to the diet. The offered amount of the diet was calculated so that each animal received 4 mg/kg b.w./day [8].

4. Results and discussion

In table 1, the results of naproxen, tolmetin, tolfenamic acid, voltaren and monensin administration on magnesium concentrations in the examined tissues, are shown. A statistically significant decrease of the magnesium concentration in the spleen and heart of the naproxen treated animals was observed. Although the magnesium concentration was increased in some tissues (brain, heart, lung, kidney, adrenal and testis) and decreased in some others (spleen and thigh bone), the changes were not statistically significant after tolmetin administration. Spleen, ovaries and thigh bone magnesium concentrations were increased in guinea pigs fed tolfenamic acid. The result of voltaren administration to guinea pigs, was a statistically significant decrease of magnesium concentration in the heart, liver, kidney and testis and a statistically significant increase of magnesium concentration in the spleen. The study of the results of monensin administration to rats, showed a statistically significant decrease of the magnesium concentration in the brain, heart, lung, spleen and testis and a statistically significant increase of it in the thigh bone.

Our results indicate that chronic administration of drugs which are potentially capable of forming complexes with metal ions, can cause translocation and/or elimination of these ions in the organism. The selected dose of all the substances administered was equivalent to the human highest therapeutic one. Selection of magnesium for analysis was made because of the importance of it's presence for the normal heart function and it's ability (from the chemical point of view) to form complexes with the administered substances.

There are several reports of drugs binding competitively with essential elements. These drugs form complexes with divalent cations in the organism and in this way, biologically important metal ions can be displaced or excreted from the organism [8,10]. Many side effects observed after drug treatment and specially with drugs used for long term treatment, could be explained by the induced depletion of essential ions (e.g. Mg^{2+})

That means that the observed alterations in the concentration of magnesium in the several tissues and in the blood could cause a significant change in the biochemical profile of the organism and further more induce the appearance of lesions or clinical signs.

TABLE 1: Influence of naproxen, tolmetin, tolfenamic acid voltaren and monensin on magnesium concentrations in guinea pig or rat tissues (μg/g wet weight tissue).
+: p< 0,05, ++: p< 0,01, +++: p< 0,001 N.S.: non significant
*: mg/g wet weight tissue

Organs		Naproxen		Tolmetin		Tolfenamic acid		Voltaren		Monensin (rat)	
		Control	test	Control	test	Control	test	Control	test	Control	test
	x	229,8	193,92	301,36	347,2	295,54	330,25	884,29	395,49	777,1	287,4
Brain	SD	17,86	60,14	126,02	129,28	76,34	108,39	739,64	181,7	208,3	146,8
	P	N.S.		N.S.		N.S.		N.S.		+++	
	x	222,26	186,42	280,62	311,19	388,09	428,23	1483,6	397,85	1045	337,9
Heart	SD	20,15	28,52	24,91	169,71	124,12	91,51	714,24	158,64	239,93	105,2
	P		+	N.S.		N.S.		+++		+++	
	x	214,51	216,6	197,03	414,56	414,8	301,55	815,99	654,03	825,21	221,1
Lung	SD	5,41	86,97	7,59	208,82	114,52	98,15	167,7	226,76	265,29	111,5
	P	N.S.		N.S.		N.S.		N.S.		+++	
	x	325,21	299,53	256,65	252,26	334,21	383,9	580,63	384,39	430,13	536,6
Liver	SD	22,01	71,85	61,8	212	99,79	152,07	107,09	109,16	118,8	242,8
	P	N.S.		N.S.		N.S.		++		N.S.	
	x	273,67	245,91	234,96	252,32	353,26	370,88	1740,2	278,05	586,27	373
Kidney	SD	40,95	95,41	40,51	141,28	131,93	144,61	704,03	41,16	191,69	142,3
	P	N.S.		N.S.		N.S.		+++		N.S.	
	x	199,78	226,95	153,8	295,92	250,04	480,88	2882,2	5966,3		
Adrenals	SD	20,12	60,58	40,08	183,94	99,11	219,54	3423,7	3526,9		
	P	N.S.		N.S.		N.S.		N.S.			
	x	373,88	232,11	408,13	330,38	474,71	1187,1	345,91	1052,1	913,59	373
Spleen	SD	39,72	41,76	146,85	177,1	137,25	377,14	141,13	304,86	148,25	142,3
	P	+++		N.S.		+		+++		+++	
	x					453,76	772,87				
Stomach	SD					321,44	245,29				
	P					N.S.					
	x					483,01	477,66				
Nistis	SD					291,27	181,7				
	P					N.S.					
	x	187,67	236,34			221,48	359,66				
Uterus	SD	10,08	57,27			89,49	128,52				
	P	N.S.				N.S.					
	x					196,1	2376,8				
Ovaries	SD					97,52	437,54				
	P					+++					
	x			184,4	241,08			695,62	264,53	887,21	409,8
Testis	SD			37	138,97			117,5	108,8	461,96	259,6
	P			N.S.				+++		+	
	x	**3,56**	**3,56**	4471	3217,6	3044,2	4090,1	3989,2	4055	1941,6	4061
Bone	SD	**0,41**	**0,47**	761,14	1607,3	358,74	278,09	484,29	429,29	634,99	979,6
	P	N.S.		N.S.		+++		N.S.		+	
	x					82,51	59,17				
blood	SD					9,38	23,39				
	P					N.S.					

5. References

1. Cantilena L.R. and E.D. Klaasen (1982). The effect of chelating agents on the excretion of endogenous metals. *Toxicol. Appl. Pharmacol.*, **63**, 344-350.

2. Chebowski J.F. and J.E. Coleman (1976). Zinc and its role in enzymes. *In chart of metal ions in biological systems.* H. Sigel, Marcel Dekker Inc., N.Y. **6**, 118.

3. Dendrinou-Samara E., D. Kessisoglou, G. Manoussakis, D. Mentzafos and A. Terzis (1990). Copper (II) complexes with anti-inflammatory drugs as ligands. Molecular and crystal structure of bis (dimethyl sulphoxide) tetrakis (6-methoxy-a-methyl-2-naphthaleneacetate dicopper (II) and bis (dimethyl sulphoxide) tetrakis [1-met yl-5-(p-toluoyl)- H-pyrrole- 2aeitato) dicopper (II). *Journal of the Chemical Society.* Dalton Transactions, **1**, 959-965.

4. Green L.W., G.T. Schelling and F.M. Byers (1986). Effects of dietary monensin and potassium on apparent absorption of magnesium and other macroelements in sheep. *J. Anim. Sci.*, **63**, 1960.

5. Greene L.W., B.J May, G.T. Schelling and F.M. Byers (1988). Site and extent of apparent magnesium and calcium absorption in steers fed monensin. *J. Anim. Sci.*, **66**, 2987.

6. Gunter T. (1981). Biochemistry and pathobiochemistry of magnesium. *Magn. Bull.*, **3-1**, 91-101.

7. Hamilton Smith C., and W.R. Bidlack (1980). Interrelationship of dietary ascorbic acid and iron on the tissue distribution of ascorbic acid, iron and copper in female guinea pigs. *J. Nutr.* **110**, 1398-1408.

8. Kotsaki - Kovatsi V-P, A. Kovatsis, G. Rozos, A. Vafiadou and G. Batzias (1992). Effects of monensin on zinc, copper, magnesium and calcium tissue concentrations in rats. *Proceedings of the 30th TIAFT international meeting* in Fukuoka, Japan.

9. Kotsaki-Kovatsi, V -P., G. Batzias, G. Rozos, and A. Kovatsis (1993a). Influence of naproxen on the concentration of iron, potassium and sodium in guinea pig tissues. *Bulletin of the Hellenic Veterinary Medical Society.* **44**, 19-22.

10. Kotsaki-Kovatsi, V.-P., E. Alexaki -Tzivanidou, G. Batzias, G. Rozos, and A. Kovatsis (1993b). Influence of naproxen on the zinc, copper magnesium and calcium content of guinea pig tissues. *J. Vet. Med.* **A 40**, 555-58.

11. Kotsaki-Kovatsi, V.-P., A. Kovatsis, A. Dega, S. Tzika and G. Rozos (1993 c). Influence of voltaren on metal ion distribution in guinea pig tissues. *Proceedings of 8th Medical Congress of Northern Greece*, Thessaloniki, Greece.

12. Kotsaki-Kovatsi, V.-P., G. Rozos, G. Batzias, A. Vafiadou and A. Kovatsis (1994a). Influence of tolmetin on the zinc, copper and magnesium content of guinea pig tissues. *J. Vet. Pharmacol. Therap.* **17**, 396-398.

13. Kotsaki-Kovatsi, V P., A. Kovatsis, G. Rozos, and G. Samouilidou-Keller (1994 b). Influence of tolfenamic acid on metal ion distribution in guinea pig tissues. *Presented at Tiaft-Soft Congress* in Tampa, Florida, USA.

14. Langer, P. (1972). Carier mediated ion transport. *Science,* **179**, 24

15. Ovchinnikov Y.A. (1979). Physico-chemical basis of ion transport through biological membranes. Ionophores and ion channels. *European J. Biochem.*, **94**, 321.

16. Pressman B.C.(1973). Properties of ionophores with a broad range of cation selectivity. *Fed. Proc.* **32**,1696.

17. Rao C.M., K.V. Ramesh, K.L.Bairy and D.R. Kullkarni (1990). Comparative evaluation of tolmetin and tolmetin-zinc on wound- repair and inflammation. *Indian Journal of Medical Research.* **92**, 205-208.

18. Rossowska M.J., C. Dinh, S.B. Gottschalk, M. Yazddani, F.S. Sutton and T. Nakamoto (1990). Interaction between caffeine intake and heart zinc concentrations in the rats. *British Journal of Nutrition.* **64**, 561-567.

19. Southern, L.L. and D. H Baker (1982). Eimeria acervulina infection in chicks fed excess dietary methionine. *J. Anim. Sci.* **54**, 989.

20. Touyz, R. M., Milne, F. J. and Reinach, S. G. [1992] : Platelet and erythrocyte Mg^{2+}, Ca^{2+}, Na^{+}, K^{+} and cell membrane adenosine triphosphate activity in essential hypertension in blacks. *J. Hypertens.* **10**, 571-578.

21. Risco, F. and Traba, M. C. [1994] : Possible involvement of a magnesium dependent mitochondrial alkaline phosphatase in the regulation ot the 25-hydroxyvitamin D3-1a- and 25-hydroxyvitamin D3-24R-hydroxylases in LLC-PK1 cells. *Magnes. Res.* 7, 169-178.

22. Durlach, J. [1988] : In : *Magnesium in clinical practice*, pp. 9-10. London : John Libbey.

23. Dhalla, N. S. and Zhao, D. [1989] : Possible role of sarcolemmal Ca^{2+}/ Mg^{2+} ATPase in heart function. *Magnes. Res.* **2**, 161-172.

24. Wacker, W. E. C. [1969] : The biochemistry of magnesium. *Ann. N. Y. Acad. Sci.* **162**, 717.

INFLUENCE OF INOSITOLHEXAPHOSPHORIC ACID (PHYTIC ACID) ON THE DISTRIBUTION AND THE EXCRETION OF MAGNESIUM IN RAT TISSUES

V.P.KOTSAKI-KOVATSI*, L.KOVATSI, TH. KIRIAKIDIS, AND G. KOEHLER-SAMOUILIDOU.
*Lab. of Biochemistry and Toxicology and *Lab. of Pharmacology, Faculty of Veterinary Medicine, Aristotle University, Thessaloniki 54006, Macedonia, Greece.*

1. Summary

In the present investigation, the effect of the administration of phytic acid on magnesium (Mg) concentrations in rat tissues was studied. The animals were divided into three groups: Group A (received a diet supplemented with 2% phytic acid), group B (received a diet supplemented with 10% phytic acid) and group C (control). At the end of the experiment, the animals were sacrificed and the concentration of magnesium was determined in several tissues.
Phytic acid significantly increased Mg concentration in the duodenum and the ovary of the animals of both groups. Magnesium concentration was also increased in the heart, liver, jejunum and blood of the animals of group B as well as in the lung, adrenals and spleen of the animals of group A.
On the other hand, the lung and stomach magnesium concentrations of the animals of group B as well as the colon and hair magnesium concentrations of the animals of group A were decreased. Magnesium excretion through feces was significantly decreased in the animals of both groups, while the excretion through urine was only decreased in the animals of group B. No significant changes were observed in the brain, ocular tissue, kidney, uterus and thigh bone concentrations of magnesium.
In conclusion, the administration of phytic acid can produce translocation and/ or elimination of magnesium in several tissues of rats.

Keywords: Magnesium, rats, phytic acid.

2. Introduction

Phytic acid is present in many plants, constituting 1 to 5% of the weight of many cereals and legumes. It's molecular stucture makes possible the formation of complexes with several elements in vitro and we believe that the same can happen in

T. Theophanides and J. Anastassopoulou (eds.), Magnesium: Current Status and New Developments, 59-63.

vivo. Similar hypotheses have been made and experimentally confirmed with other drugs [1,2,3]. Concern about the presence of phytic acid in the diet, arises from evidence that it decreases the bioavailability of many essential minerals by interacting with multivalent cations to form complexes. These complexes appear to be responsible for the decreased bioavailability of the complexed minerals [4-6].
The aim of the present study was to investigate the effect of phytic acid on the concentration of Mg in several rat tissues, based on the facts that phytic acid is a strong calcium binding agent [7,8] and that it's molecular structure makes possible the formation of complexes with other elements both in vitro and in vivo.

3. Material and methods

Twenty four female Wistar rats, six months old, weighing 200-250 g were used. The animals were housed in polypropylene cages, maintained at 22-24^0 C, with lighting conditions of 12h of light and 12h of darkness.
We divided our animals into three groups A, B and C (8 animals in each group). The animals of group C (control) received nothing. They were provided daily with water and a balanced, pelleted diet *ad libitum*. The metal content of the diet was as follows: 12 g Ca/kg, 50 mg Zn/kg, 7 mg Cu/kg, 50 mg Mg/kg.
The animals of group A received a diet supplemented with 2% phytic acid for 4 weeks.
The animals of group B at first received a diet supplemented with 2% phytic acid for 8 weeks and then a diet supplemented with 10% phytic acid for another 4 weeks.
Phytic acid was purchased from Aldrich, as a syrupy, aqueous solution 40% per wt.
At the end of the experiment, the animals were kept in metabolic cages for 24h in order to collect urine and feces. Urine and feces were collected in glass tubes and stored at -20^0 C until analysis. The next day the animals were sacrificed.
Immediately after death, blood was collected directly from the heart, through cardiac puncture, using heparinized syringes. Brain, ocular tissue, heart, lung, liver, spleen, kidney, adrenals, ovaries, uterus, a portion of the stomach, duodenum, jejunum, colon, thigh bone and about 2 g of hair from the neck were removed. The tissues were blotted dry, weighed and kept frozen (-20^0 C) until further analysis. For the determination of the Mg concentrations in the tissues, blood, urine and feces, flame atomic absorption spectroscopy was used. Student's t-test was used for statistical analysis with $p<0.05$ being considered significant.

4. Results and Discussion

In tables 1 and 2, we present the results of phytic acid administration on the concentration of Mg in the examined tissues, hair, blood, urine and feces. Mg concentrations in the duodenum and in the ovary were significantly increased in the animals of both groups ($p<0.05$). Magnesium concentration was also significantly

increased in the the heart, liver, jejunum and blood of the animals of group B as well as in the lung, adrenals and spleen of the animals of group A ($p<0.05$).
On the other hand, the lung and stomach magnesium concentrations of the animals of group B as well as the colon and hair magnesium concentrations of the animals of group A were significantly decreased ($p<0.05$).
Magnesium excretion through feces was significantly decreased in the animals of both groups, while excretion through urine was only decreased in the animals of group B. No significant changes were observed in the ocular tissue, brain, kidney, uterus and thigh bone concentrations of magnesium.
In conclusion, the administration of phytic acid to rats, can produce translocation and/or elimination of magnesium ions in several tissues.
The above results support our opinion [1,2,3,9,10,11] that substances which are potentially capable, from the chemical point of view, of forming complex compounds with metal ions, and specially those which are chronically administered, can cause translocation and/or elimination of these ions from the living organism.

TABLE 1. Magnesium concentrations in tissues (μg/g wet weight)
Data represent mean value ± SD
+: values significantly different ($P<0.05$)
NS: values not significantly different ($P>0.05$)

Organs		Control	Test A	Control	Test B	Test A	Test B
BRAIN	X	223.66	265.68	223.66	203.04	265.68	203.04
	SD	39.21	16.67	39.21	12.82	16.67	12.82
	P	NS		NS		+	
OCULAR TISSUE	X	523.56	633.36	523.56	667.2	633.36	667.2
	SD	93.69	122.24	93.69	144.16	122.24	144.16
	P	NS		NS		NS	
HEART	X	151.39	168.41	151.39	227.86	168.41	227.86
	SD	4.21	22.62	4.21	21.74	22.62	21.74
	P	NS		+		+	
LUNG	X	183.14	235.61	183.14	139.04	235.61	139.04
	SD	7.52	24.45	7.52	11.66	24.45	11.66
	P	+		+		+	
LIVER	X	661.68	709.72	661.68	1098.83	709.72	1098.83
	SD	44.85	61.22	44.85	205.61	61.22	205.61
	P	NS		+		+	
SPLEEN	X	118.44	143.66	118.44	120.54	143.66	120.54
	SD	4.82	14.88	4.82	9.41	14.88	9.41
	P	+		NS		+	
KIDNEY	X	106.78	122.48	106.78	101.43	122.48	101.43
	SD	6.93	19.56	6.93	6.3	19.56	6.3
	P	NS		NS		NS	
ADRENALS	X	66.59	96.4	66.59	71.85	96.4	71.85
	SD	8.2	5.74	8.2	10.47	5.74	10.47
	P	+		NS		+	

TABLE 2. Magnesium concentrations in tissues (μg/g wet weight), blood (μg/ml), urine (μg/24h) and feces (mg/g).
Data represent mean value ± SD
+: values significantly different ($P<0.05$)
NS: values not significantly different ($P>0.05$)

Organs		Control	Test A	Control	Test B	Test A	Test B
OVARIES	X	259.53	312.86	259.53	388.79	312.86	388.79
	SD	5.14	29.52	5.14	13.83	29.52	13.83
	P	+		+		+	
UTERUS	X	168.97	175.3	168.97	177.17	175.3	177.17
	SD	21.93	21.47	21.93	14.91	21.47	14.91
	P	NS		NS		NS	
STOMACH	X	171.59	130.54	171.59	95.73	130.54	95.73
	SD	28.21	32.17	28.21	12.79	32.17	12.79
	P	NS		+		NS	
DUODENUM	X	122.22	171.8	122.22	281.46	171.8	281.46
	SD	27.43	33.82	27.43	36.34	33.82	36.34
	P	+		+		NS	
JEJUNUM	X	136.73	144.89	136.73	191.84	144.89	191.84
	SD	9.63	28.76	9.63	34.44	28.76	34.44
	P	NS		+		+	
COLON	X	264.25	166.26	264.25	336.65	166.26	336.65
	SD	33.28	13.23	33.28	98.08	13.23	98.08
	P	+		NS		+	
THIGH BONE	X	4319.99	7421.87	4319.99	5760.49	7421.87	5760.49
	SD	1411.33	2675.14	1411.33	2466.25	2675.14	2466.25
	P	NS		NS		NS	
HAIR	X	513.78	256.75	513.78	523.84	256.75	523.84
	SD	176.34	56.92	176.34	120.37	56.92	120.37
	P	+		NS		+	
BLOOD	X	430.66	445.52	430.66	526.26	445.52	526.26
	SD	45.01	93.6	45.01	16.5	93.6	16.5
	P	NS		+		NS	
URINE	X	4035	3680	4035	2800	3680	2800
	SD	297.7	481.66	297.7	734.85	481.66	734.85
	P	NS		+		NS	
FECES	X	39.85	20.15	39.85	22.49	20.15	22.49
	SD	8.35	2.15	8.35	3.64	2.15	3.64
	P	+		+		NS	

5. References

1. Kotsaki-Kovatsi, V. P., Vafiadou, A., Rozos, G., Batzias, G. and Kovatsis, A. [1992] : Influence of Tolmetin on the zinc, copper and magnesium content of guinea pig tissues. *Toxicology letters suppl.* p 245.

2. Kotsaki-Kovatsi, V. P., Alexaki-Tzivanidou, E., Batzias, G., Rozos, G. and Kovatsis, A. [1993] : Influence of naproxen on zinc, copper, magnesium and calcium content of guinea pig tissues. *J. of Med. Vet.* **A 40**, 555-558.

3. Kotsaki-Kovatsi, V. P., Koehler-Samouilidis, G., Kovatsis, A. and Rozos, G. [1997] : Fluctuation of Zinc, Copper, Magnesium and Calcium concentrations in Guinea Pig Tissues after administration of Captopril [SQ 14225]. *J. Trace Elements Med. Biol.* **11**, 32-36.

4. Kies, C. and Umoren, J. [1989] : Inhibitors of copper bioutilization: fiber, lead, phytate and tannins. *Adv Exp Med Biol.* **258**, 81-93.

5. Rose, H. E. and Quarterman, J. [1984] : Effects of dietary phytic acid on lead and cadmium uptake and depletion in rats. *Environ Res.* **35**, 482-489.

6. Sandstraom, B., Cederblad, A., Stenquist, B. and Andersson, H. [1990] : Effect of inositol hexaphosphate on retention of zinc and calcium from the human colon. *Eur J Clin Nutr.* **44**, 705-708.

7. Martin, C. J. and Evans, W. J. [1986] : Phytic acid-metal ion interactions. II. The effect of pH on Ca [II] binding. *J. Inorg. Biochem.* **27**, 17-30.

8. Gersonde, K. and Weiner, M. [1981-82] : The influence of infusion rate on the acute intravenous toxicity of phytic acid, a calcium-binding agent. *Toxicology.* **22**, 279-86.

9. Kotsaki-Kovatsi, V. P., Kovatsis, A., Rozos, G. and Keller-Samouilidou, G. [1994] : Effect of DMSA on the distribution and excretion of metal ions in rat tissues intoxicated by lead ro not. *Proceedings of the 9th Medical Congress of Northern Greece, Thessaloniki.*

10. Kotsaki-Kovatsi, V. P., Kovatsis, A., Rozos, G., Vafiadou, A. and Batzias, G. [1992] : Effects of monensin on zinc, copper, magnesium and calcium tissue concentrations in rats. *Proceedings of the 30th International Meeting of TIAFT, Japan, Fukuoka.*

11. Kotsaki-Kovatsi, V. P., Kovatsis, A., Rozos, G. and Keller-Samouilidou, G. [1994] : Effect of DMSA on the distribution and excretion of metal ions in rat tissues intoxicated by lead or not. *Proceedings of the 9th Medical Congress of Northern Greece, Thessaloniki.*

CHANGES OF MAGNESIUM CONCENTRATIONS IN RAT TISSUES AFTER THE ADMINISTRATION OF DRUGS USED IN HYPERTENSION THERAPY

V.P. KOTSAKI-KOVATSI*, E. BACOYANNI, G. KOEHLER-SAMOUILIDOU, A. VAFIADOU AND AN. KOVATSIS.
*Lab. of Biochemistry and Toxicology and *Lab. of Pharmacology, Faculty of Veterinary Medicine, Aristotle University, Thessaloniki 54006, Macedonia , Greece.*

1.Introduction

The importance of the presence of the trace elements and their physiological distribution in the organism as well as the appearance of symptoms related to their redistribution and/or elimination are well known. Some drugs have the ability to form complexes with metal ions in vitro and we believe that the same happens in vivo. We also believe that through that mechanism they mobilize the essential for the normal function of the organism trace elements, and the result is their redistribution and elimination from the organism.[1,2,3]. It is therefore important to know how the administration of drugs and specially those which are administered chronically affects the concentrations of essential trace metals in biological systems.
Hydrochlorothiazide (HCTZ) belongs to the thiazide group of compounds which are used as diuretics, alone or in combination formulations, for the treatment of essential hypertension, management of edema associated with congestive heart failure and edema associated with cirrhosis of the liver [4]. An adverse effect of HCTZ administration is an increase in magnesium excretion [4]. It is well known that amiloride provides additive antidiuretic effects, when used in combination with thiazides and also reduces the thiazide-induced hyperkaliuria and hypokalemia [5]. So the aim of this work was to study the influence of simultaneous administration of HCTZ and amiloride hydrochloride on the magnesium concentrations in several tissues, urine and feces of rats.

2.Material and Methods

For the study of HCTZ and amiloride hydrochloride eleven [six test (T) and five control (C)] female Wistar rats, four months old and 200-250g in weight were used. HCTZ and amiloride hydrochloride were incorporated to the diet. The offered amount of the diet was calculated so that each animal received a daily dose of 4 mg HCTZ / kg b.w. and 0.4 mg amiloride-HCl / kg b.w. for seven weeks. Samples of urine and feces were collected one day before the beginning of the experiment and at the end of it. Blood samples were taken from the animals through cardiac puncture under light ether anesthesia, just before their sacrifice. Brain, ocular tissues, heart, lung, liver, spleen, kidney, ovaries, uterus, stomach, small intestine, jejunum, colon, , thigh bone and hair were removed immediately after death. For the determination of the Mg concentrations in the tissues, urine and feces, flame atomic absorption spectrophotometry was used. Student's t-test was used for statistical analysis with $p<0.05$ being considered significant.

3.Results and discussion.

T. Theophanides and J. Anastassopoulou (eds.), Magnesium: Current Status and New Developments, 65-66.

	brain	ocular tissues	lung	heart	stomach	liver	kidney	spleen	duodenum
C	94.8 ±15	207.2 ±34	88.3 ±8.3	125.8 ±6.7	235.1 ±85.7	438.1 ±71.8	197.3 ±67.4	162.2 ±5.6	171.2 ±55.6
T	387.7 ±109 *	417.6 ±95 *	213.5 ±84 *	198.6 85.5	523.1 ±90.5 *	393 ±41.1	384.9 ±155 *	297.8 ±141 *	244.1 ±136

	jeju-num	colon	uterus	ovaries	bone	blood	hair	urine	feces
C	190.6 ±77.1	167.5 ±92.6	168.1 ±69.5	435.5 ±122	5518 ±1256	28.4 ±3.2	128.4 ±31.1	2875 1607	7.43ɑ 1.42
T	228.7 ±89.5	387.3 ±139 *	144.1 ±27.7	303.3 ±41.1	4251 ±656	35.5 ±6.1	301.8 ±22.8	2911 974	17.3ɑ 5.1 *

table 1. Magnesium concentrations in tissues (μg/g wet weight), blood (μg/ml), urine(μg/24h) and feces.. Data represent mean value ± SD. ɑ : mg/gr
* : values significantly different $p<0.05$.

The results of HCTZ and amiloride-HCl administration on magnesium concentrations in the examined tissues are shown in table 1. A statistically significant increase of the magnesium concentration was found in the brain, ocular tissues, lung, spleen, kidney, stomach, colon and hair. A significant increase was observed of the magnesium concentration in the feces at the end of the experiment.

Our finding of the unchanged Mg concentration in the blood is in agreement with the findings of Kisters et al. who used thiazides combined with potassium-sparing substances during diuretic therapy and found out that the combination of the above substances is a useful means to avoid Mg^{++} loss in blood [6]. The observed increase of Mg concentration in several tissues may be attributed to increased absorption of Mg through gastrointestinal tract.

We also noticed an increased elimination of Mg through feces which may be an effort of the organism to eliminate the increased concentrations of Mg in the tissues. The normal rate of the excretion through urine is due to the fact of simultaneous administration of HCTZ and amiloride-HCl.

The above mentioned results support our opinion that some diuretic drugs [7] or/and drugs which are capable to form complexes with metals [1,2,3] and specially those which are chronically administered may cause redistribution and possibly elimination of metal ions from the organism.

4.References.

1.Kotsaki-Kovatsi V.-P., A.Vafiadou, G.Rozos, G Batzias &. A.Kovatsis (1994). Influence of tolmentin on the zinc, copper, magnesium and calcium concentrations in guinea pig tissues. *J. of Veterinary Pharmacology and Therapiutics*, 17, 396-398.

2.Kotsaki-Kovatsi V.-P.,G.Koehler-Samouilidis, A.Kovatsis &G.Rozos. (1997). Fluctuations of zinc, copper, magnesium and calcium concentrations in guinea pig tissues after administration of captopril. *J.Trace Elements Med.Biol.*, 11, 32-36.

3.Rang H.P &.M.M.Dale (1987). *Pharmacology* by Churchill Livingstone, U.K., 323-325.

4.Alon U., D.Wellons & J.C.M.Chan, (1985).Hydroclorothiazide, amiloride and tolmentin in the treatment of diabetes insipidus of brattleboro rats. *Medical Biology* 63, 117-122.

5.Kisters K., K.H.Rahn, W.Zidek. (1992). Intracellular Mg++ concentrations under diuretic treatment in mild essential hypertensive patients. In *Metal ions in biology and medicine*, Vol 2, by John Libbey Eurotext, Paris, 224-225.

6.Mountokalakis T.D. 1992. Zinc deficiency: an ignored side effect of antihypertensive treatment. In *Metal ions in biology and medicine*, Vol 2, by John Libbey Eurotext. Paris, 209-214.

MAGNESIUM IN RABBITS SUBMITTED TO PROLONGED LEAD OR CADMIUM INTOXICATION

V. MATOVIC, Z. STOJANOVIC, D. VUJANOVIC & D. SOLDATOVIC
Dept. of Toxicology, Faculty of Pharmacy, University of Belgrade
11221 Belgrade, Vojvode Stepe 450, Yugoslavia

1. Introduction

Besides being industrial toxicants, lead and cadmium are environmental pollutants of considerable importance. These metals, with cumulative properties, may cause wide variety of negative effects in the organism, from long-ago described kidney damages to now-a-days established data of their cocarcinogenic and carcinogenic activities. Complex actions of toxic metals in the organism can not be explained without profound knowledge of their effects on metabolism of bioelements, Mg being one of them.

Our own earlier studies explored the influence of chronic lead intoxication on blood and tissue magnesium in rabbits [1,2], and in workers occupationally exposed to lead [3]. Later experiments were conducted in order to determine the effect of lead on Mg elimination via urine [4].

The purpose of this work was to study the effect of prolonged Cd intoxication on magnesium content in rabbits and its elimination via urine, and to compare these results with data obtained for magnesium in chronic lead intoxication.

2. Materials and Methods

Experiments were conducted with rabbits *Oryctolagus cuniculus-Belgian hare,* weighing 3-3.5 kg and kept under standard laboratory conditions. After the adaptation to our conditions, rabbits were divided into groups, of 10 rabbits each.

Pb intoxication: one group of animals served as control, and the other one was given orally, by orogastric tube, every day for 4 weeks, 20 mg Pb/kg b.w. as aqueous solution of $Pb(NO_3)_2$.

Cd intoxication: with the exception of the controls, rabbits were treated orally, for 4 weeks, with aqueous solution of $CdCl_2$ to give cadmium dose level of 10 mg Cd/kg b.w. per day.

Prior and during the intoxication, blood samples were taken from the ear artery using a cannula and 24 h urine was collected in metabolic cages. Investigated organs (brain, heart, lungs, liver, kidneys, sceletal muscle and bone) were excised from sacrificed

T. Theophanides and J. Anastassopoulou (eds.), Magnesium: Current Status and New Developments, 67-68.

animals, before and at the end of the experiment.
Concentration of magnesium was determined in mineralized samples by AAS, and statistical comparisons were performed using Student's t-test.

3. Results and Discussion

Similar to lead, prolonged cadmium intoxication induced significant decrease of blood Mg. Concentration of blood Mg in non-intoxicated rabbits was 2.2 mmol/l, while a level of 1.5 mmol Mg/l was determined at the end of Cd intoxication.
No significant changes of magnesium, during lead intoxication, were observed in heart, brain and sceletal muscles, although a significant Mg increase was obtained in lungs, liver, kidneys and bone. On the other hand, prolonged cadmium intoxication elevated magnesium level in liver and reduced Mg content in sceletal muscles.
Both lead and cadmium caused significant elevation of urinary Mg elimination: Pb after one week of intoxication till the end of the experiment, and Cd during the last week of poisoning. Compared with the controls, cca 2-fold increase of Mg content in urine was observed at the end of intoxication with lead or cadmium.
We can conclude that prolonged lead and cadmium intoxications have similar effects on blood and urinary Mg, as well as on Mg content in liver.

The obtained results are in agreement with literature data indicating antagonism of Mg and Pb or Cd [1-6]. Our own investigation of Pb-Mg antagonism encouriges us to believe that these findings could be used in clinical practice [7,8], i.e. that Mg treatment could replace chelating agents in therapy of chronic intoxications by lead, and maybe cadmium.

4. References

1. Soldatovic, D., Bresjanac, M., and Petrovic, C.: Changement du taux du calcium et du magnésium dans le serum des animaux soumis aux conditions d'intoxication chronique par le plomb, *Arh. Farm.* **1** (1965), 1-6.
2. Soldatovic, D., Nedeljkovic, M., and Farah, G. (1980): Changes in the quantity of Mg in organs of rabbits submitted to chronic intoxication by some mineral poisons (saturnism, fluorosis and manganism), in M. Cantin and M.S. Seelig (eds.), *Magnesium in Health and Disease,* Spectrum Press, New York, pp.961-965.
3. Soldatovic, D., and Milic, S. (1973): L'influence de l'intoxication chronique par le plomb sur la teneur en Mg dans le serum chez les personnes professionnellement exposées au plomb, in J. Durlach (ed), *1er Symp. Intern. Defic. Magn. Path. Hum.,* SGEMV, Vittel, pp.569-572.
4. Matovic, V., Vujanovic, D., and Soldatovic, D.: Elimination of bioelements copper, zinc and magnesium via urine in rabbits exposed to chronic lead intoxication, *Arh. Farm.***44,** 1-2 (1994), 434-435.
5. Speich, M., Auget, J.-L., and Arnaud, P.: Correlation between magnesium and heavy metals in blood and sixteen tissues of rabbits, *Magnesium Res.* **2,3** (1989), 179-182.
6. Guiet-Bara, A., Bara, M., and Durlach, J.: Magnesium: a competitive inhibitor of lead and cadmium. Ultrastructural studies of the human amniotic epithelial cell, *Magnesium Res.* **3,** 1 (1990), 31-36.
7. Soldatovic, D., Matovic, V., and Vujanovic, D.: Prophylactic effect of high magnesium intake in rabbits exposed to prolonged lead intoxication, *Magnesium Res.* **6**, 2 (1993), 145-148.
8. Matovic, V., Soldatovic, D., and Vujanovic, D. (1995): Magnesium against lead: New approach to prophylaxis and therapy of chronic lead intoxication, in V.P. Kotsaki-Kovatsi and A.J. Vafiadou (eds), *Aspects on Environmental Toxicology*, Tecnika Studio, Thessaloniki, pp.78-82.

INFLUENCE OF NORFLOXACIN ON MAGNESIUM CONCENTRATIONS IN RAT TISSUES AND URINE

A.J. VAFIADOU, V.P. KOTSAKI-KOVATSI*, G. KOEHLER-SAMOUILIDOU AND AN. KOVATSIS,
*Lab. of Biochemistry-Toxicology and *Lab. of Pharmacology, Faculty of Veterinary Medicine, Aristotle University, Thessaloniki 54006, Macedonia , Greece.*

1. Introduction

There are several reports about drugs, which bind competitively with the organism's essential elements and they form complexes with them [1]. Consequently biologically important metal ions can be redistributed or excreted from the organism, under the influence of the above mentioned category of drugs. Thus, many of the observed adverse effects of these drugs could be explained.
Norfloxacin is a synthetic antibacterial agent that is undergoing extensive investigation for both human and veterinary use for the treatment of a variety of bacterial infections. Norfloxacin has a complex forming ability with magnesium ions through carboxyl group at the position 3 and carbonyl group at the position 4 of the quinolone ring [2].Magnesium also participates to the process of binding of norfloxacin with DNA, forming a ternary complex [3].
According to our opinion drugs capable to form complexes with metal ions in vitro can cause redistribution /elimination of these ions from the organism because they may also form complexes in vivo [4,5].

2. Material and Methods.

For the study of norfloxacin sixteen male [eight test (T-M) and eight control (C-M)] and sixteen female [eight test (T-F) and eight control (C-F)] Wistar rats four months old (weighing 170-200 g and 260-290 g respectively) were used. Norfloxacin was incorporated to the diet. The offered amount of the diet was calculated so that each animal received a daily dose of 11 mg/kg b.w./day for 21 days. Twenty four hours urine samples were collected once a week. Twelve hours after the last dose blood samples were taken from the animals through cardiac puncture under light ether anesthesia, just before their sacrifice. Brain, ocular tissues (posterior third of eyeball, including the optic nerve coloboma), heart, lung, liver, spleen, kidney, adrenals, testis, epididymis, uterus, ovaries, stomach, small intestine, jejunum, colon, thigh bone and hair were removed immediately after death. For the determination of magnesium concentration in the tissues, urine and whole blood, flame atomic absorption spectrophotometry was used. The data were analyzed by one way analysis of variance and in some cases the data used were transformed in to logs. The level of significance was set at $p<0.1$.

3. Results.

T. Theophanides and J. Anastassopoulou (eds.), Magnesium: Current Status and New Developments, 69-70.

	brain	ocular tissues	lung	heart	stomach	liver	kidney	spleen	duo-denum
C-m	180.9 ±8.5	106.9 ±12.9	115.7 ±5.0	229.0 ±40.4	243.9 ±18.4	219.4 ±6.7	269.6 ±5.97	196.0 ±7.5	129.2 ±5.7
T-m	165.2 ±15.4 *	102.9 ±14.6	130.5 ±10.5*	259.1 ±12.5*	254.1 ±49.0	207.6 ±8.8	236.6 ±25.6	206.0 ±7.7	179.9 ±23.3
C-f	165,1 ±12,5	103,9 ±9,4	121,9 ±7,4	337,3 ±44.6	259,1 ±19,6	228,8 ±5,7	290,5 ±23,8	214,5 ±9,3	133,1 ±9,4
T-f	155,9 ±11,2	115,6 ±10,6	148,8 ±13,7*	301,0 ±15,8*	269,4 ±34,7	228,7 ±22,5	279,9 ±31,0	268,0 ±9,0*	201,1 ±22,1

	jeju-num	colon	uterus/ testis	ova-ries	epidi-dymis	adre-nal	bone	blood	hair
C-m	141.2 ±20.5	127.2 ±10.1	144.6 ±8.1		133.6 ±3.6	89,9 11,4	1958 ±112	25.3 ±2.8	143.8 ±13.7
T-m	173.2 ±10.7*	167.8 ±17.5*	145.6 ±6.5		145.5 ±5.8*	77,6 4,2 *	2341 ±74*	29.3 ±3.8*	197.5 ±9.6
C-f	127,1 ±15,7	141,0 ±13,2	147,8 ±11,58	216,2 ±6,3		110,6 ±8,2	2533 ±91	23,1 ±1,9	170,3 ±11,8
T-f	187,5 ±20,8*	190,9 ±11,3*	139,5 ±23,6	168,3 ±14,9*		93,4 ±11,2*	2778 ±366	31,4 ±3,5*	230,8 ±8.7*

table 1. Magnesium concentrations in tissues (μg/g wet weight) blood (μg/ml). Data represent mean values ± SD. * : values significantly different p<0.1.

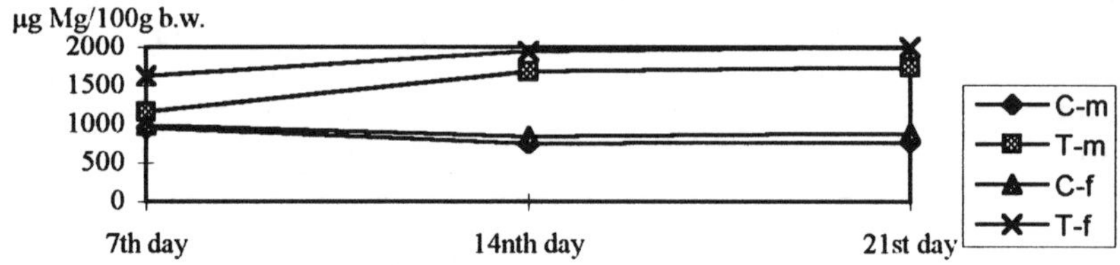

Figure1. Magnesium concentration in urine.

The results of norfloxacin administration on magnesium concentration of the examined tissues are shown in table 1. A statistically significant increase of magnesium concentration was found in the lung, jejunum, colon and blood of male and female rats, in the spleen and hair of female rats and heart, epididymis and bone of male rats. Also a statistically significant decrease of magnesium concentration was found in the heart, ovaries and the adrenals of the female and in the brain and adrenals of the male rats. The urinary excretion of magnesium which is significantly higher every week, in both male and female rats, after the administration of norfloxacin, is shown in figure 1.

4. References.

1. Hamilton S.C., W.R. Bildlack. Interrelationship of dietary ascorbic acid, iron and copper in female guinea pigs. *J.Nutr.* **110** (1980), 1398-1408.
2. Lecomte S., M.H.Baron, M.T. Chenon, C.C.Oupry & N.J. Moreau. *Antimicrobial Agents and Chemotherapy*, Vol **38,** No12 (1994), 2810-2816.
3. Palu C., S.Valisena, G.Ciarrocchi, B.Gatto and M.Palumbo. Quinolone binding to DNA is mediated by magnesium ions. *Proc. Natl. Acad. Sci.* USA.Vol **89** (1992), 9671-9675.
4. Kotsaki-Kovatsi V.-P., A.Kovatsis, A.Vafiadou,.G.Rozos, & G Batzias. (1993). Effects of monensin on zinc, copper, magnesium and calcium concentrations in rats. *Proceedings of 30th International Meeting of TIAFT, Fukuoka Japan.*, 158-163.
5. Kotsaki-Kovatsi V.-P., A.Vafiadou,.G.Rozos, G Batzias &. A.Kovatsis (1994). Influence of tolmentin on the zinc, copper, magnesium and calcium concentrations in guinea pig tissues. *J. of Veterinary Pharmacology and Therapeutics*, **17**, 396-398.

MAGNESIUM LIGANDS IN ILEAL DIGESTA OF PIGLETS FED SKIM MILK AND SOYBEAN FLOUR

T. MATSUI and H. YANO

Division of Applied Biosciences, Graduate School of Agriculture, Kyoto University
Sakyo-ku, Kyoto, 606-01, Japan.

1. Introduction

Magnesium bioavailability is higher from diets consisting of dairy products than from diets consisting of soybean products. Soybean products are known to contain substantial amounts of phytate which decreases magnesium solubility in vitro (Champagne and Phillippy, 1989). Ros (1962) suggested that magnesium absorption depended on the concentration of soluble magnesium. Thus, phytate in soy products has been considered to suppress magnesium absorption through lowering magnesium solubility in the digestive tract. In fact, the previous report suggested that phytate in defatted soybean flour decreased the solubility of magnesium in the lower small intestine of rats and suppressed magnesium absorption because the removal of phytate by fermentation raised magnesium solubility in the digesta of the small intestine and increased magnesium availability (Hirabayashi et al., 1995).

Although the solubility is an important factor affecting magnesium absorption, there may be several soluble forms of magnesium, i.e., ionized magnesium and magnesium complexes in digesta and these forms of magnesium may not possess similar absorbability. Lindberg et al. (1990) reported that citric acid made soluble complex with magnesium and this complex was easily absorbed. Magnesium might make soluble complex with fructose in the ileal digesta, which stimulated magnesium absorption (Heijden, et al., 1994). Triptic hydrolysates of casein, casein phosphopeptides, were reported to stimulate magnesium absorption (Goto, et al., 1995), which suggested that these peptides possibly make soluble complexes with magnesium and these complexes might be easily absorbed. However, few investigations have tried to study the soluble forms of magnesium in digesta.

This experiment was conducted to study magnesium forms in the digesta of ileum, which is the major site of magnesium absorption, in piglets fed skim milk (SM) or defatted soybean flour (SF).

T. Theophanides and J. Anastassopoulou (eds.), Magnesium: Current Status and New Developments, 71-76.

2. Materials and Methods

Eight weaned piglets were fed SM-based diet containing 45 % of SM or SF-based diet containing 43 % of SF and 20 % of lactose. The piglets were fed 400 g of diet twice a day for 30 days and were given free access to deionized water. Feces were collected during the last 5 days of the feeding trial. At the end of the feeding trial, the animals were killed by exsanguination from the jugular vein and carotid arteries. The ileum (between 1 m and 2 m anterior to the ileocecal valve) was removed from the abdominal cavity. Digesta was collected from the ileum within 30 min after slaughter. The digesta was centrifuged at 8000 × g for 30 min and pH were immediately determined in the supernatant.

Magnesium concentrations were measured in diets, feces, whole digesta and the supernatants of digesta using an atomic absorption spectrophotometer after digestion with concentrated nitric acid and 6 mol/L perchloric acid. Soluble magnesium and magnesium solubility were defined as magnesium in supernatant and the ratio of soluble magnesium to total magnesium, respectively.

Fifty μL of pooled supernatant from each dietary group was submitted to a size-exclusion high-performance chromatography separation on a Asahipak GS-220 column (250 × 7.6 mm, Showa Denko Co., Tokyo) and eluted with 5mM Tris-HCL buffer (pH 7) at a flow rate of 1.5 mL/min. The concentration of MgCL2 in elute was changed between 0 and 120 μM. The eluted solution was monitored for UV absorbance at 280 nm by spectrophotometer (SPD-10A, Shimazu, Kyoto) and for magnesium concentration by atomic absorption spectrophotometer (AA-6600, Shimazu, Kyoto). Bile, 0.1% of lactose, fructose and citric acid, and 2 mM MgCL2 were also submitted to this column. To estimate MW of magnesium complexes, polyethylene glycol (MW 1000 and MW 3000), lactose and fructose were eluted by the solution with 40 μM magnesium and monitored by refractive index detector (RID-6A, Shimazu, Kyoto).

Data were expressed as means±SD. Student's t-test was used to compare the two dietary groups. The data expressed as percentages were transformed by arcsin calculation before the analyses to normalize their distributions. Differences were considered significant at $P<0.05$.

3. Results and Discussion

The apparent magnesium absorption was significantly higher in the SM group than in the SF group (TABLE 1). These results were in agreement with some reports showing that magnesium bioavailability was higher from a diet consisting of dairy products than from a diet consisting of soybean products. Brink et al. (1992) indicated that lactose in cow's milk increased magnesium absorption in rats. However, factors other than lactose caused the differences between the two groups because lactose was added to the SF diet to adjust dietary lactose.

Total magnesium concentration in the ileal digesta did not differ between the dietary groups. The pH value in liquid phase of digesta of each group was 7.1. Soluble magnesium concentration and magnesium solubility was higher (concentration, P=0.06; solubility, P<0.05) in the SM group than in the SF group. We previously reported that apparent absorption of magnesium and magnesium solubility in the lower small intestine were less in the rats fed SF than in the rats fed the fermented SF which contained negligible amount of phytate (Hirabayashi, 1995). Phytate in SF probably decreased magnesium solubility in the ileum of pigs, which was considered to, at least partly, decrease magnesium absorption in the SF group because magnesium absorption was reported to depend on the concentration of soluble magnesium (Ros, 1962).

TABLE 1. Apparent magnesium absorption and magnesium solubility in ileal digesta of pigs

Diets	Skim milk	Soy flour
Mg intake and absorption (mg/d)		
Intake	672	736
Fecal excretion	329±51	466±63
Apparent absorption	343±51	270±63
Absorption (% of intake)	51.0±7.6	36.7±8.6*
Magnesium concentration in digesta (mmol/L)		
Total Mg	17.56±0.97	16.72±4.67
Soluble Mg	8.17±1,25	4.74±1.56
Solubility (%)	46.84±8.71	28.92±7.81*

Means ± SD, n=4
*; Significantly (P<.05) different from the skim milk group.

When the digesta of the SM group was eluted by the solution containing 10 to 40 μM magnesium, the chromatogram showed two minor peaks (Peak 1 and Peak 2) and two major peaks (Peak 3 and Peak 4) of Mg in the SM group (*Figure 1a*). Molecular weights of peaks 2 and 3 were approximately estimated as 690 and 300, respectively. Peak 1 was eluted as fast as the void volume shown by the elution of polyethylene glycol 3000. Peaks 2, 3 and 4 were also found in the SF group but Peak 1 was not clearly shown (*Figure 1b*). Peaks 2 and 3 in both groups were increased with increasing magnesium concentration in elute (*Figure 2*). However, peak 4 in each group became unclear at 80 μM of magnesium in elute and was disappeared at 120 μM of magnesium. Because the chromatogram of bile showed one peak of magnesium which was eluted at the same time as peak 3 (Data were not shown), magnesium ligand in peak 3 possibly originated from excreted bile into the digestive tract.

Inorganic magnesium (MgCL2) was eluted at the same time as peak 4 in digesta. Furthermore the chromatogram using elusion without magnesium showed that peak 4 was found but the other peaks were trace (Data were not shown). Thus, peak 4 was considered to be ionized magnesium. The recovery of magnesium was calculated to be 86% and 91% from the area of peak 4 by elusion without magnesium in the SF group

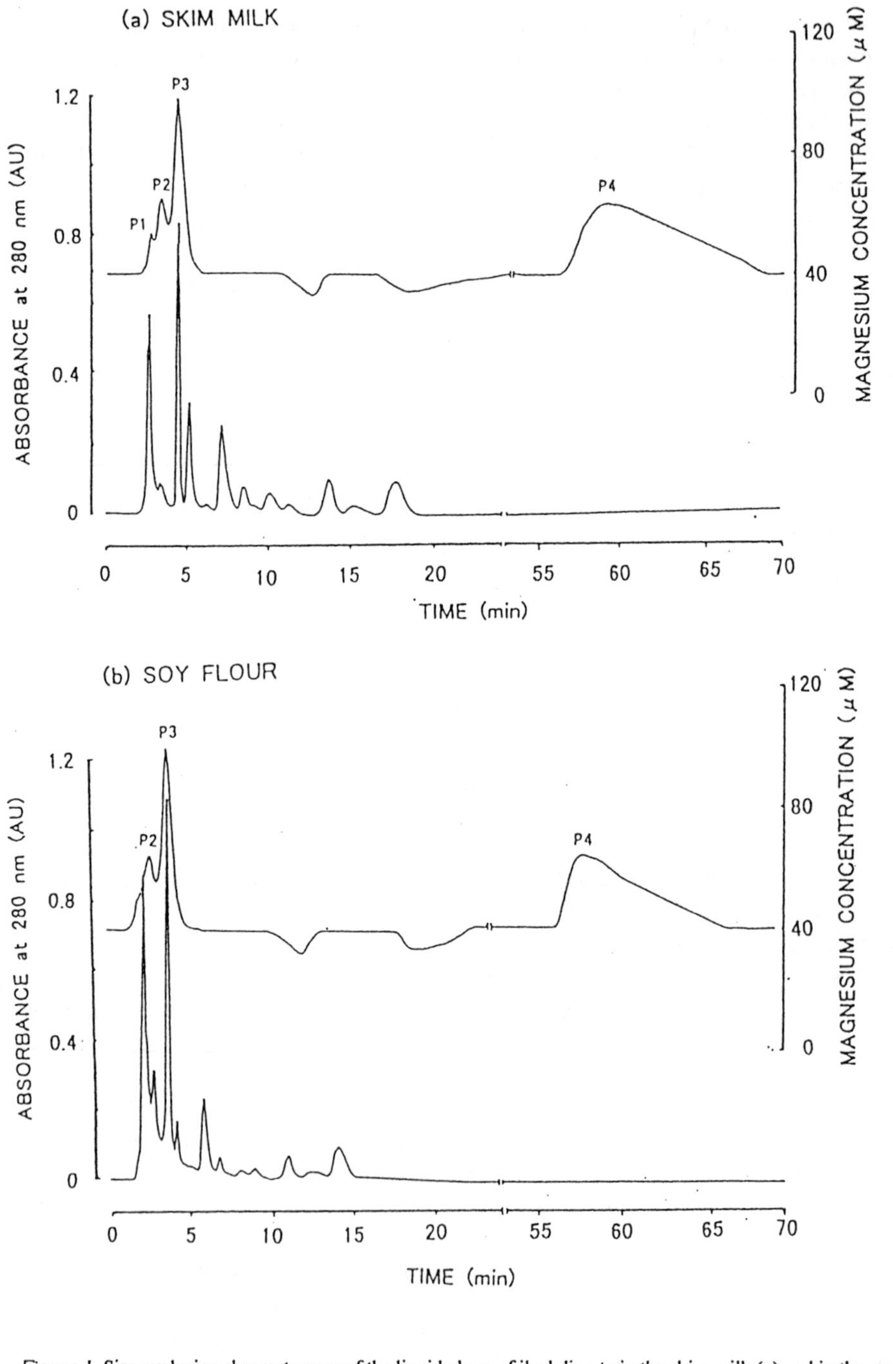

Figure 1. Size-exclusion chromatogram of the liquid phase of ileal digesta in the skim milk (a) and in the soy flour group (b). The elute consisted of 5 mM Tris-HCl buffer (pH 7) with 40 μM of MgCL2.

and SM group, respectively.

When lactose and fructose were eluted by the solution with 40 μM magnesium, no peak of magnesium was found. On the other hand, there was a peak of magnesium when citric acid was submitted to the column. These results suggest that these sugars do not make soluble complex with magnesium and that citric acid possibly bind magnesium in digesta. However, magnesium complex with citric acid was not found in the chromatogram of soluble digesta in this experiment.

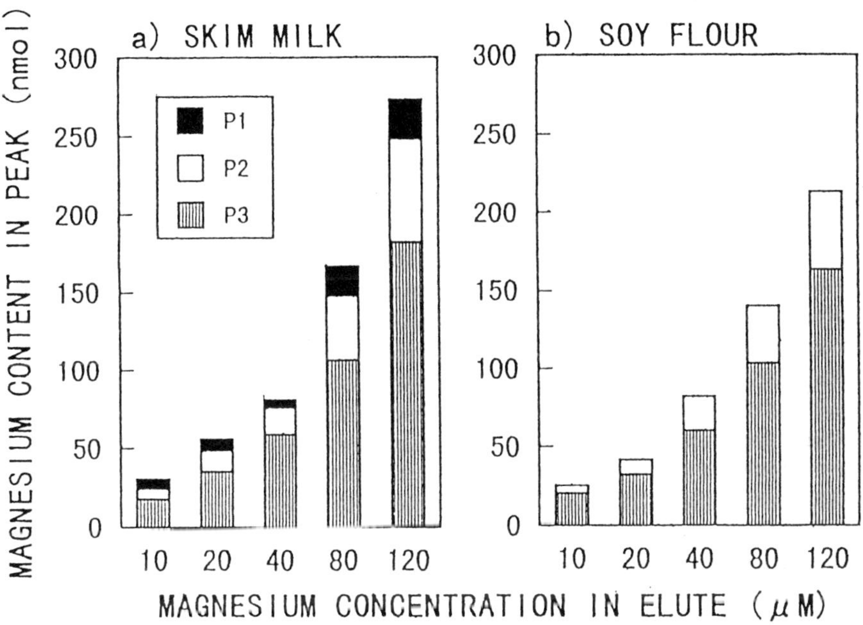

Figure 2. Magnesium bound to ligands in different concentrations of magnesium of elutes. The amount of magnesium in a peak was calculated from the area of magnesium and the base line, and expressed as magnesium content in 50 μL.

Peak 1 was found in the SM group but not in the SF group. The previous report suggested that casein phosphopeptides were produced by dietary SF in piglets and calcium complexes with casein phosphopeptides existed in the fraction of ileal digsta which was not ultrafiltrable by a 3000 MW cutoff membrane (Matsui et al., 1997). Casein phosphopeptides possibly bind magnesium in digesta because phosphoserine residue, which exists in casein phosphopeptides, is involved in casein binding to magnesium (Van-Dijk, 1990). Thus, peak 1 may be magnesium complexes with casein phosphopeptides. Casein phosphopeptides are reported to stimulate magnesium absorption (Goto, et al., 1995) although the action of casein phosphopeptides acid is still controversial. On the other hand, magnesium.content in peak 1 was largely less than that in other peaks. These results suggest that casein phosphopeptides are not responsible for the higher absorption of magnesium in the SM group than in the SF group.

The ligands for magnesium in the liquid phase of digesta was not largely affected by the dietary treatment. Furthermore, a large part of magnesium probably made complex with a low-molecular-weight ligand which might be endogenous, irrespective of the diets. These results proposed that magnesium solubility but not its ligands in the liquid phase of digesta was the most important factor affecting magnesium absorption in the ileum.

4. References

Brink, E.J., Dekker, P.R., Van-Beresteijn, E.C.H., and Beynen, A.C. (1992) Bioavailability of magnesium and calcium from cow's milk and soya-bean beverage in rats. British Journal of Nutrition 68, 271-282.

Champagne, E.T., and Phillippy, B.Q. (1989) Effects of pH on calcium, zinc, and phytate solubilities and complexes following in vitro digestions of soy protein isolate. Journal of Food Science 54,587-592.

Goto, T., Yonehara, Y., Tsuchita, H., and Kuwata, T. (1995) Availability of calcium and phosphorus from casein phosphopeptides in growing rats. Journal of Japanese Society of Nutrition and Food Science 48,195-202.

Hirabayashi, M., Matsui, T., and Yano, H. (1995) Fermentation of soybean by Aspergillus Usamii increases magnesium availability in rats. Japanese Journal of Magnesium Research 14,45-53.

Lindberg. J.S., Zobitz, M.M., Poindexter, J.R., and Pak, C.Y.C. (1990) Magnesium bioavailability from magnesium citrate and magnesium oxide. Journal of American College Nutrition 9,48-55.

Matsui, T., Kawakita, Y., and Yano, H. (1997) Dietary skim milk and defatted soybean flour with calcium phosphate show different calcium distribution in the small intestine of piglets. Journal of Nutrition in press.

Ros, D.B. (1962) In vitro studies on the transport of magnesium across the intestinal walls of rats. Journal of Physiology (London) 160,417-428.

Van der Heijden, A., Van den Berg, G.J., Lemmens, A.G., and Beynen, A.C. (1994) Dietary fructose v. glucose in rats raises urinary excretion, true absorption and ileal solubility of magnesium but decreases magnesium retention. British Journal of Nutrition 72,567-577.

Van Dijk, H.J.M. (1990) The properties of casein micelles: 2. Formation and degradation of the micellar calcium phosphate. Netherlands Milk Dairy Journal 44,111-124.

MAGNESIUM IONS HINDER THE GROWTH OF COLEOPTYL IN DEUTERIUM DEPLETED WATER

SANDOR A. KISS[1], ZOLTAN GALBACS[2], GABOR GALBACS[2]
[1]Hungarian Magnesium Society, 6726 Szeged, Fυ fasor 73A/2, Hungary
[2]Jσzsef A. Univ., Dept. of Inorg. and Anal. Chem., 6720 Szeged, Dσm tιr 7., Hungary

1. ABSTRACT

The length and mass growth of coleoptyl and root of germinating seeds is a function of the deuterium concentration of water. If the deuterium concentration is either higher or lower than that of natural waters (ca. 150 ppm D), the growth of coleoptyl and root is hindered. If we add 0.01 mole/dm^3 magnesium ions to the deuterium depleted (20 ppm D) water, the hindering effect is enhanced. According to our findings, the shorter coleoptyls and roots contain less sodium and potassium than lengthier ones of the same age. The magnesium concentration changes in the opposite way to the sodium and potassium concentration, which results in a smaller Mg/Na and Mg/K ratios for lengthier coleoptyls and roots. A probable explanation for these findings can be, that an appropriate pH is needed for mitosis, which is supported by the Na/H and/or K/H antiport. An increase in the magnesium concentration hinders the Na/H and K/H antiport and so the settling of the necessary pH.

Keywords: coleoptyl, magnesium, sodium, potassium, antiport

2. INTRODUCTION

Deuterium is an isotope of hydrogen with a mass number of two. As a consequence of the 100% mass difference existing between the two hydrogen isotopes (^{1}H and ^{2}D), which is more than the difference between the isotopes of any other element, significant differences can be observed in the chemical and physical properties of their compounds. These deviations can be also found in biochemical relations, like mitosis, germination, etc. [1,2] That is why Galbαcs et al. [3] consider deuterium as a "new" microelement.
According to the literature [4-6], until now the investigations on the biological effects of deuterium were performed on D-enriched (30-90% D_2O containing) media. In these works it was found, that the increasing D_2O content of water hinders the plant growth. As a contrast to this, we investigated the impact of D-depleted water (20 ppm D) on the growth of the coleoptyls of germinating seeds [2]. Our intention was to demonstrate, that waters possessing D deficiency with respect to the natural concentration (ca. 150 ppm

T. Theophanides and J. Anastassopoulou (eds.), Magnesium: Current Status and New Developments, 77-80.

D) also inhibit the plant growth. The basis of our assumption is the results of a study of Somlyai et al. [7], who showed the inhibiting effect on animal cell mitosis.

In his former works Kiss [8,9] pointed out, that magnesium treatments promote the germination and growth of plants. This finding led us to investigate the possibility to balance the inhibitory effect of D deficiency with magnesium dosing. The current study reports on the results of these investigations.

3. MATERIALS AND METHODS

Furio type corn (*Zea mays*) seeds were used in the germination experiments. During the experiments, 50-50 seeds were germinated in 6-6 repetitions. The control series was handled with distilled water of the natural, 150 ppm D content. The other series were treated with D-depleted (20 ppm D) distilled water, D-enriched (280 ppm D) distilled water and with a 0.01 M solution of $MgSO_4$ prepared in 20 ppm D distilled water, respectively. The germination was carried out at 25°C in dark, in capped plastic cups on a cotton-wool base covered with filterpaper. At the end of the 10-14 days germinating period, the roots and coleoptyls of germinating seeds were cut off, and 50 pieces were measured for their mass (roots) and length (coleoptyls). Coleoptyls were also measured for their wet mass on an analytical balance. Experimental results were calculated by averaging.

The plant parts were also taken under composition analysis. For this purpose, the parts were first dried at 80°C in a ventilation hot box and then were digested by a HNO_3-H_2O_2 acid microwave method. The solutions were analysed by flame atomic absorption spectroscopy and flame photometric techniques for their Mg, Na and K content.

4. RESULTS

The results for coleoptyl mass and length measurements are shown in Table 1. The data clearly demonstrate the effect of D concentration on the coleoptyl growth.

Table 1. The effect of distilled waters of different D-content on the germination of Furio type corn seeds (average of 6 series, 50 pieces each)

Treatment	Coleoptyls			
	Length mm	% relative to the length in 150 ppm water	Mass mg	% relative to the mass in 150 ppm water
20 ppm D water	41.4	71	97.8	76
150 ppm D water	58.8	100	129.3	100
280 ppm D water	46.6	80	107.4	83

Table 2. shows the impact of Mg on corn plants in D-depleted aqueous medium. It was found, that Mg reduced the mass of both coleoptyls and roots. The Mg content of plant parts increased and the Na content decreased. The K content decreased in the coleoptyls, while increased in the roots. The Mg/Na and Mg/K ratios changed accordingly. No change was observed in dry matter content.

Table 2. The impact of magnesium addition on the hindering effect of D-depleted waters on the growth and composition of germinating seeds (average of 3 series, 50 pieces each)

Treatment	Mass mg	Mg	Na	K	Mg/Na	Mg/K
		mmol/g				
Coleoptyls						
20 ppm D water	89.6	0.0709	0.0135	0.61	5.27	0.116
20 ppm D water + Mg	60.7	0.0741	0.0123	0.51	6.01	0.126
Change, %	67	105	92	97	114	109
Roots						
20 ppm D water	69.5	0.0409	0.0508	0.55	0.70	0.074
20 ppm D water + Mg	60.0	0.0772	0.0453	0.63	1.70	0.123
Change, %	87	189	72	115	242	166

5. CONCLUSION

Deviation of the D concentration of medium in either directions from the natural value (150 ppm) hinders the growth of corn coleoptyl and root growth, therefore the biochemical and biological processes. From this it can be concluded, that deuterium is an essential, but depending on the concentration, toxic microelement.

Magnesium completion was found to further enhance the inhibition of growth. We believe, that the reason for this effect lies in a change in the functioning of the Na/H antiport system. Molecular biological studies revealed, that the Na/H antiport in the cell membrane is activated before mitosis. As a result of this, H^+ is released and Na^+ is taken up, and the pH of the cell shifts to the basic region[10]. In the lack of Na/H activity, mitosis can not occur. If the D-content of medium deviates from the natural value, then the Na/H antiport activity is decreased, as it was found in investigations made with *Elodea canadiensis*. The extracellular pH of leaves of this weed was found to be lower (more acidic) with a value of 0.9 in 20 ppm D water, than in natural water [11]. The addition of Mg is known to inhibit the Na/H antiport activity both directly[13] and indirectly, via the stimulation of the Na efflux [12], and so also the increase of intracellular pH.

The change of K content of the intracellular space imposes a smaller effect on the pH change, because the K influx increases the H efflux to a smaller extent than Na [14].

Results of Cameron and Smith [15] showed, that the Na concentration of the quickly dividing cells is nearly five times higher than that of cells with low division rate. At the same time, the ratio in K concentration was found to be 1.4 only. As a result of these two processes, the change of intracellular K concentration is a less important factor to influence the pH change than the Na concentration.

References

1. Kiss, A.S., Galbαcs, Z., Galbαcs, G. (1996), *Proc. of the 2nd Symp. on Anal. and Environ. Problems*, Szeged, pp. 171-181.
2. Kiss, A.S., Galbαcs, Z., Kamasz-Sαri, V., Galbαcs, G. (1996), *Proc. of the 6th Hung. Magnesium Symp.*, Balatonszιplak, p.16.
3. Galbαcs, Z., Kiss, A.S., Galbαcs, G. (1996), in *Proc. of the 7th Int. Trace Element Symp.*, Budapest, pp. 243-246.
4. Blake, M.I., Crane, M.I., Uphaus, R.A., Katz., J.J. (1975) *Lloydia* **27**, 254.
5. Uphaus, R.A., Blake, M.I., Katz, J.J. (1975) *Can. J. Bot.* **53**, 2118.
6. Sacchi, G.A. (1992) *Plant. Physiol.* **100**, 1962.
7. Somlyai, G., Jancsσ, G., Jαkli, Gy., Vass, K., Barna, B., Lakics, V., Gaαl, T. (1993) *FEBS* **317**, 1.
8. Kiss, A.S. (1981) *Magnesium Bull.* **1-a**, 6-12.
9. Kiss, A.S. (1983*) Magnιziumtrαgyαzαs, magnιzium a biolσgiαban, Mezυgazdasαgi Kiads*, Budapest, pp. 110-115.
10. Poyssegur, J. Sardet, C., Franchi, A., L'Allemain, G., Paris, S. (1984) *Proc. Natl. Acad. Sci. USA* **81**, 4833-4837.
11. Laskai G. (1996) private communication.
12. Beaugι, L., Rojas, H., Polo, B. (1983) *Bichim. Biophys. Acta* **728**, 463-466.
13. Gónzel, D., Móller, A., Hintz., K., Schlue, W.R. (1996) Magnesium Res. 9, 67-68.
14. Glass, A.D.M., Siddigi, M.Y., Giles, K.I. (1981) *Plant Physiol.* **68**, 457-459.
15. Cameron, I.L., Smith N.K.R. (1989) *Magnesium* **8,** 31-44.

THE EFFECT OF DEUTERIUM DEPLETED MEDIUM ON PLANT TUMORS

SÁNDOR A. KISS[1], IMRE LÁSZLÓ[2], ÉVA SZÕKE[2], ZOLTÁN GALBÁCS[3] AND GÁBOR GALBÁCS[3]

[1]Hungarian Magnesium Society, H-6726 Szeged, Fõ fasor 73A/2, Hungary[2]Inst. of Pharmacognosy, Semmelweis Univ. of Medicine, H-1085 Budapest, Ullõi út 26., Hungary, [3]József Attila Univ., Department of Inorg. and Anal. Chem., H-6720 Szeged, Dóm tér 7., Hungary

1. ABSTRACT

Deuterium in the natural, ca. 150 ppm concentration is essential for plants. A change in deuterium content of the medium in the direction of either lower or higher concentrations hinders the plant growth through the inhibition of mitosis. Healthy cells can adapt themselves to deuterium deficiency, and within 10-16 days catch up with the control in growth. As a contrast to this, tumors can not tolerate deuterium deficiency. In 20 ppm D containing medium, the tumor growth slows down with respect to the control (150 ppm D). This effect can be explained by the influence of deuterium on the Na/H antiport.

Keywords: deuterium, tumor, magnesium, cell cultures, Na/H antiport

2. INTRODUCTION

Several biological and biochemical studies investigated the hindering effect of increased deuterium concentration on the plant growth[1-3]. Somlyai et al.[4] was the first to describe the hindering effect of deuterium depleted water on the growth of mammalian cells, especially tumors. It has been also shown, that the D-depleted water (20 ppm D) hinders the plant growth[5-7].
The present study investigates the different reaction of healthy cells and tumors to D-depleted medium.

3. MATERIALS AND METHODS

The effect of 20 ppm D water on the growth of healthy plant parts was monitored by measuring the growth of coleoptyls of germinating Furio type corn (*Zea mays*) seeds. The experiments were carried out in six repetitions on 50-50 seeds in each series. Coleoptyl lengths were measured daily.

T. Theophanides and J. Anastassopoulou (eds.), Magnesium: Current Status and New Developments, 81-84.

The experiments were carried out in six repetitions on 50-50 seeds in each series. Coleoptyl lengths were measured daily.

The experiments on tumors were conducted at two hungarian research institutes in parallel: NÖVIKI (Research Institute for Plant Protection, Budapest) and Institute of Pharmacognosy of SOTE (Semmelweis Medical University, Budapest). The plant tumors were induced on *Nicotiana tabacum* (Hevesi 276) organised cultures maintained in S solid medium (0.7% agar) containing 20 g/l sucrose (Reanal), 10 mg/l Ca-panthotenate (Reanal), 1.13 g/l Murashige and Skoog basal medium (Sigma), pH 5.6 and in IVS solid medium containing similar components with 1 mg/l indol-butyric-acid (Reanal). The cultures were incubated at 26°C in light (12 hours of light per day).

The tumors were obtained by micro-injection on leaves and stems with *Agrobacterium tumefaciens* strain A348. The isolated tumor was maintaned in Murashige and Skoog (MS) solid medium (1% agar) containing 20 g/l sucrose (Reanal), 4.703 g/l Murashige and Skoog basal medium (Sigma), pH 5.6 and 500 mg/l Claforan (Roussel). The tumor cultures were grown in dark. Subcultures were made at 6-weeks intervals (100 ml flask containing 40 ml medium) with cultures being maintained under the same conditions for 18 weeks. The Claforan content was reduced to 250 mg/l. After 12 weeks the cultures were subcultivated in Claforan free MS medium. First the cultures seemed more compact later turned to slack.

Pieces from 4 weeks old tissue cultures were placed on MS medium containing distilled water with 20 ppm D content (35 ppm D in the experiments made at NÖVIKI). The control contained 150 ppm D in distilled water. Tumor pieces were reweighed three times a week, fresh weight and dry weight (80°C drying) were measured and the factor fw/dw (%) was calculated. The dried samples were also taken under microwave aided digestion with a HNO_3-H_2O_2 acid mixture and then inductively coupled plasma atomic emission spectrometric (ICP-AES) analysis in order to record their Na, K, Mg and Ca content.

4. RESULTS

The growth of healthy corn coleoptyls in length was found to be different in 20 and 150 ppm D containing medium. The initial big difference (10-14 mm) in length was seen to diminish with the days, and on the 12th day the difference practically disappeared (**Figure 1.**)

Experiments made at NÖVIKI showed, that in the 22 days old culture the tumor growth was smaller with 30-40% in the 35 ppm D containing medium than in the medium containing the natural D concentration (ca. 150 ppm). In the SOTE experiments it was found, that the difference between the tumor growth in the two media increased with the time. At the end of the investigated 22 days period 10-14% growth hindering effect was registered in the 20 ppm D medium (**Figure 2.**). The elemental analysis of tumors showed, that the decrease in D concentration caused the Mg and Na concentration to decrease in the tissues (**Table 1.**).

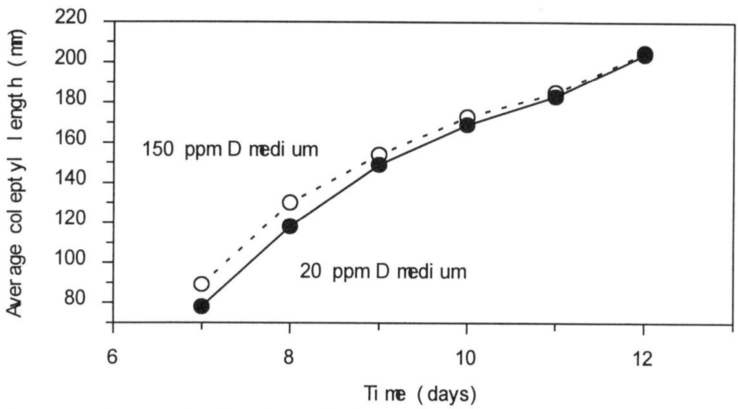

Figure 1. Reaction of growing healthy plant cells (Zea mays coleoptyls) to media of different D content

The decreased concentration of Deuterium, being an essential microelement [7], hinders the mitosis. The reason for this is, that the lower than natural D concentration causes the pH to settle slower (inhibits the settlement) at the critical, basic value needed for mitosis, which is normally the result of the Na/H antiport. This antiport functioning is hindered by the decreased deuterium concentration, as it was demonstrated in the studies of Laskai[8] carried out on *Elodea canadiensis.*

Our former investigations[5] showed, that a D concentration either lower or higher than the natural value (ca. 150 ppm) hinders the plant growth. According to studies by Sacchi[1], the increasing D concentration of medium makes the mitosis slower and decreases the H efflux and K influx. Our measurements (**Table 1.**) revealed, that in D-depleted medium, both the Na and Mg concentration (and similarly the Na/Mg ratio) in tumors is lower than in those tumors cultivated in a medium of natural D concentration. Based on these data it can be assumed, that the growth is inhibited via the hindering of mitosis in media with either higher or lower D concentration than natural.

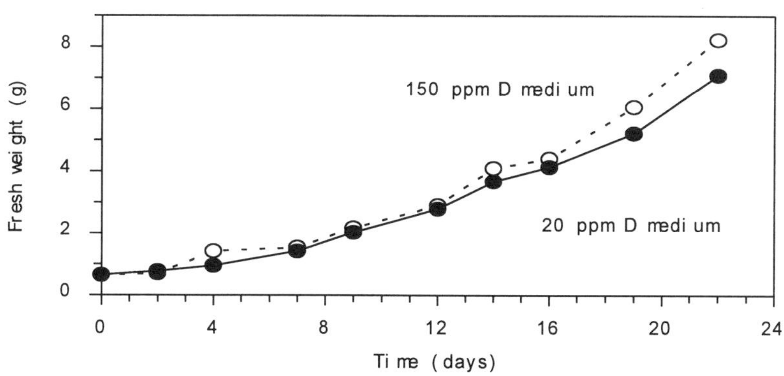

Figure 2. Reaction of plant tumors to media of different D content

TABLE 1. Composition of *Nicotiniana tabacum* tumors on the 20th days of cultivation in 20 ppm and 150 ppm D containing media (average of 5 repetitions)

Treatment	150 ppmD	20 ppmD
Na mmol/kg	286	227
%	100	80
K mmol/kg	1592	1782
%	100	112
Mg mmol/kg	89	78
%	100	88
Ca mmol/kg	179	143
%	100	80
Na/Mg	3.2	2.9

5. CONCLUSION

The change of D concentration of medium in either direction from the natural value hinders the mitosis and the growth of tissue. We see the reason for this effect in the inhibition of the Na/H antiport.

The healthy tissue can adapt themselves in a relatively short period of time to the "disturbance" in conditions caused by the decrease of D concentration of the medium. On the other hand, tumors seem to have lower tolerance than healthy cells, therefore tumor growth slows down, or eventually stops, in D-depleted medium.

***Acknowledgement*:** This work was supported by Reanal Finechemical Co. Carlo Erba Reagenti.

References

1. Sacchi, G.A. (1992) *Plant. Physiol.* **100,** 1962.
2. Blake, M.I., Crane, M.I., Uphaus, R.A., Katz., J.J. (1975) *Lloydia* **27**, 254.
3. Uphaus, R.A., Blake, M.I., Katz, J.J. (1975) *Can. J. Bot.* **53**, 2118.
4. Somlyai, G., Jancsó, G., Jákli, Gy., Vass, K., Barna, B., Lakics, V., Gaál, T. (1993) *FEBS* **317**, 1.
5. Kiss, A.S., Galbács, Z., Galbács, G. (1996), in *Proc. of the 2nd Symp. on Anal. and Environ. Problems*, Szeged, 171-181.
6. Kiss, A.S., Galbács, Z., Kamasz-Sári, V., Galbács, G. (1996), in *Proc. of the 6th Hung. Magnesium Symp.*, Balatonszéplak, 16.
7. Galbács, Z., Kiss, A.S., Galbács, G. (1996),in *Proc. of the 7th Int. Trace Element Symp.*, Budapest, 243-246.
8. Laskai G. (1996) private communication.

CONNECTION BETWEEN THE MAGNESIUM CONTENT OF THE MEDIUM AND THE GROWTH OF THE GENETICALLY TRANSFORMED *LOBELLA INFLATA L.* CULTURES

I. BALVANYOS, E. SZOKE, L. KURINSKI
Semmelweis University of Medicine, Institute of Pharmacognosy
H-1085 Budapest, Ulloi ut 26, Hungary

1. Introduction

The herb of *Lobelia inflata L.* contains more than 20 piperidine alkaloids. Its main alkaloid is L-lobeline (/-/cis-β,10-diphenyl-lobelionol) though lobelanine, norlobelanine, lobelanidine, etc. occur in considerable quantities too. The herb is applied in cases of asthma, collapse, gas- and narcotic poisoning. Lobeline also causes nausea. In large doses its effect resembles the effect of nicotine, therefore it is used in antismoking preperations. Lobelinum hydrochloricum, Lobeliae herba or Tinctura lobeliae are official preparations in many dispensatories [1].

Recently some new results were published about the alkaloid production of the *Lobelia inflata L* hairy root cultures [2, 3]. Polyacetilenes were also shown out in the genetically transformed cultures of *Lobelia inflata L.* [4,5].

Studies on the biomass and alkaloid production of in vitro cultures from *Lobelia inflata L.* proved that these organized cultures are able to synthesize the characteristic alkaloids of the intact plant [1, 6]. In addition to callus and organized cultures, hairy root clones transformed by genetically modified *Agrobacterium rhizogenes* R-1601 branch were also produced.

2. Materials and Methods

The sterilized seeds of *Lobelia inflata L.* were grown in light after germination in a solid culture medium consisting of a half dilution Murashige-Skoog (MS) salts and vitamins [7] with 3% sucrose. Transformed root cultures of *Lobelia inflata L.* were obtained by infection of the sterile, two-months old plants with *Agrobacterium rhizogenes* strain R-1601. The plant cells were infected by microinjection. More than 20 clones were isolated after 14 days. The antibiotic combination of cefotaxime 250mg/l (Sigma) and ampicillin 1000mg/l (Sigma) was added for several subcultures to eliminate bacteria. The hairy roots-exempted from bacteria- were cultivated in solid, hormonefree culture media consisting of Murashige-Skoog (MS) - or Gamborg

T. Theophanides and J. Anastassopoulou (eds.), Magnesium: Current Status and New Developments, 85-87.

B5(B50) [8] salts and vitamins with 2% sucrose. The applied $MgSO_4$ (Reanal) concentrations changed from 0 to 1000 mg/l in the B5 solid medium containing 2% sucrose.

3. Results

The cultivation of plant cells transformed by *Agrobacterium rhizogenes* is a new possibiliy to produce pharmacologically active secondary metabolites in vivo and in vitro. The T-DNS of *Agrobacterium rhizogenes* -which is part of the Ri-plasmid- transforms the plant cells.

After injection of *Lobelia inflata L.* with *Agrobacterium rhizogenes* R-1601 branch we obtained the twenty cell growing hairy root clones. We studied the growth rate of the alkaloid production of the cultures in different media in light and in dark as well. Because of the selection of medium we measured the linear growth rate (Figure 1) and fresh weught (Figure 2) of three hairy root clones (8009/g9, 8009/g3 and 8009/h2).

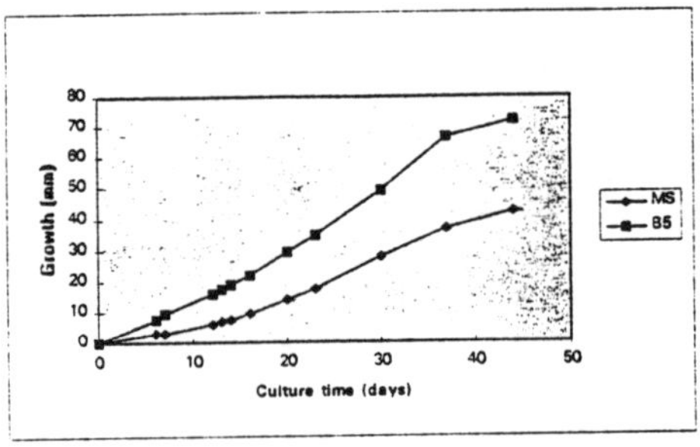

Figure 1 Time course for growth of *Lobelia inflata L.* hairy root (8009/g9 clone) in B5 basal medium in dark.

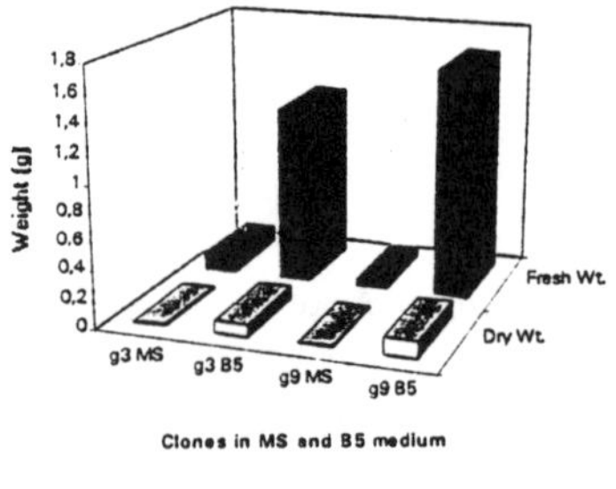

Figure 2 Dry weight and fresh weight of *Lobelia inflata L.* hairy root cultures (8009/g9 and 8009/g3 clones) in MS and B5 basal medium, in dark.

The clones maintained in B5 medium were gron more intensively than in MS medium but the number of ramification was more in B5 medium. Therefore we used B5 solid medium for further examinations. We studied the linear groth rate every three days in B5 solid medium containing different $MgSO_4$ concentrations (0 mg/l, 125 mg/l, 250 mg/l, 500 mg/l, 1000 mg/l). At the end of the cultivation eriod we measures the fresh weight and dry weight too. For the determination of dry weight the cultures were freez dried at 40°C. The results were compared to control (B5(B50)). The presence of the alkaloids was detected by TLC method. The results demonstrate that the hairy roots are able to produce lobeline. The identification of the active compounds using HPLC techniques is in progress.

4. Conclusion

We obtained the twenty well growing hairy root clones and studied the growth in B5 and MS basal medium. The clones grew more intensively in B5 medium than MS medium and the biomass production was also higher in B5 medium.

We studied the linear growth rate in Be solid medium containing different $MgSO_4$ concentrations and measured the fresh-and dry weight. The presence of the alkaloids was detected by TLC - and HPLC method.

5. Acknowledgements

This research work was supported by Renal Finechemical Co. Carlo Erba Reagenti and PFP-0227/1997.

References

1. Szoke, E. (1994) In vitro culture and the production of lobeline and other related secondary metabolites, *Medicinal and Aromatic Plants VII,* Springer, Berlin Heidelberg.
2. Yonemitsu, H. Shimomura, K. Satake, M. Mochida, S. Tanaka, M. Endo, T. Kaji, A. (1990) Lobeline production by hairy root culture of *Lobelia inflata L., Plant Cell Rep* **9**, 307-310.
3. Ishimaru, K. Ikeda, Y. Kuranari, Y. Shimomura, K. (199) Growth and lobaline production of *Lobelia inflata* hairy roots, *Shoykugaku Zasshi* **46**, 265-267.
4. Ishimaru, K. Yonemitsu, H. Shimomura, K. (1991) Lobetyolin and lobetyol from hairy root culture of *Lobelia inflata , Phytochemistry* **30**, 2255-2257.
5. Ishimaru, K. Sadoshima, S. Neera, S. Koyama, K. Takahashi, K. Shimomura, K. (1992) A poly acetylene gentiobioside from hairy roots of *Lobelia inflata, Phytochemistry* **31**, 1577-1579.
6. Szoke, E. Krajewska, A. Neszmelyi, A. Pomazi, A. Meszaros, A. (1992) In vitro multiplication of *Lobelia inflata* plants and their alkaloid production, *Planta Med* **58**, 625-626.

MAGNESIUM IN LIVER CIRRHOSIS

G. CRIPPA and M. GIORGI-PIERFRANCESCHI
Department of Internal Medicine and Hypertension Unit,
Piacenza Civil Hospital, 29100 Piacenza Italy

1. Introduction

Cirrhosis of the liver is frequently associated with total bodily magnesium deficiency [1,2,3,4] which is difficult to disclose and to quantify since neither total nor ionised plasma magnesium concentrations may reflect the real magnesium status of the patient. In fact there is not a sufficient correlation between plasma and intracellular magnesium concentrations and thus the finding of normal plasma magnesium levels may not exclude an even relevant magnesium deficiency [4,5,6].

The kidney controls the magnesium excretion over a wide range and when magnesium intake and/or intestinal absorption are reduced the urinary excretion of this cation decreases consistently. Thus the finding of a low urinary magnesium excretion (for example < 0.5 mmol/24 hours) in patients suffering from liver diseases may be interpreted as a homeostatic mechanism of magnesium turnover regulation and as a sign of bodily magnesium deficiency [2].

Intracellular magnesium concentrations still remain the most sensitive tests for the assessment of bodily magnesium status [6] but, in the clinical practice and for clinical purposes, cannot be routinely performed.

The magnesium loading test, which evaluates the urinary magnesium excretion after intravenous administration of a fixed amount of magnesium sulphate represents a valuable alternative for the assessment of bodily magnesium status in cirrhotic patients since there is a good correlation between intracellular magnesium concentration and urinary magnesium excretion after magnesium load [3,7].

2. Causes of magnesium deficiency in cirrhosis of the liver

Magnesium turnover is altered in cirrhotic patients as result of an imbalance between the input (dietary intake and intestinal absorption) and the output (faecal loss and urinary excretion) of this cation. Moreover, commonly used drugs for the treatment of liver cirrhosis (thiazide and loop diuretics) may further increase magnesium losses.

T. Theophanides and J. Anastassopoulou (eds.), Magnesium: Current Status and New Developments, 89-94.

2.1. REDUCED DIETARY INTAKE

There are many reasons why dietary magnesium intake is generally reduced in cirrhotic patients and particularly in those abusing of alcohol. It is well known that alcoholic patients have the typical habit of buying alcohol instead of food and of reducing the intake of food rich in magnesium (such as green leafy fresh vegetables which are a major dietary source of magnesium). Anorexia is a typical symptom of cirrhosis of the liver due, *inter alia,* to gastric mucosa alterations, changes in biliary secretion etc.. Vomiting, a very frequent symptom in patients affected with hepatic insufficiency, may convey a further loss of the dietary magnesium [8]. Low-sodium and low-protein diets prescribed to patients with advanced liver cirrhosis may further reduce magnesium dietary intake since, in our experience, many of the diets routinely prescribed to cirrhotic patients contain less than 8 mmol of magnesium per day.

2.2. IMPAIRED INTESTINAL ABSORPTION AND INCREASED FAECAL LOSS

Magnesium is absorbed along the entire small bowel and cirrhotic patients may present with many alterations which may reduce magnesium absorption [8]. Particularly the structural and functional alterations of the intestinal mucosa (due to alcohol, portal hypertension and oedema) and the hyperammonogenesis may reduce magnesium absorption. Lactulose is a fundamental therapy for the prevention and treatment of hepatic encephalopathy, but it frequently causes diarrhoea that may significantly increase magnesium loss with the faeces. Moreover, in cirrhotic patients the altered bile secretion and a possible concomitant pancreatic insufficiency may cause steatorrhoea ; an increased amount of fatty acids in the intestinal lumen conveys the formation of insoluble soaps with magnesium which is therefore lost out with the faeces [1,2].

More uncertain is the role of Vitamin D and altered bile-acid synthesis and binding in the impaired magnesium intestinal absorption observed in cirrhotic patients.

2.3. INCREASED URINARY EXCRETION OF MAGNESIUM

The kidney play a fundamental role in magnesium homeostasis since it control magnesium excretion over a wide range (from 1 to 250 mmol per day) and therefore renal magnesium losses are crucial in the pathogenesis of magnesium deficiency.

Hyperaldosteronism, that accompanies the more advanced stages of cirrhosis, may be a relevant cause of increased magnesiuresis since aldosterone has a direct magnesiuretic effect. Moreover severe hypokalaemia may *per se* alter tubular function causing a further renal magnesium wasting [9].

Antidiuretic hormone abates tubular magnesium reabsorption and its reduced hepatic catabolism and increased secretion may be relevant under particular circumstances (delirium tremens and hepatic coma). Furthermore, during such stress situation the increased catecholamine secretion activates lipolysis and increases NEFA which chelate magnesium and increases its urinary excretion [1].

Alcohol intake produce an increased urinary magnesium excretion but the real direct role of alcohol in determining a negative balance of magnesium is questionable. In fact magnesiuresis has been found to be increased only during acute alcoholic abuse while in chronic alcoholism the magnesiuresis is usually reduced [9,10].

3. Diuretics and Magnesium Balance

There are no doubts that diuretics may have deleterious effects on magnesium turnover, especially when used for prolonged period of time and at large doses in patients prone to develop magnesium depletion. Thiazide and thiazide-like diuretics are potent magnesiuretic drugs and the 24-hour magnesiuresis due to a common thiazide formulation such as HCTZ 25 mg is higher than that induced by a common loop diuretic formulation such as furosemide 40 mg. [11]

The potassium-and-proton retaining substances (such as amiloride, triamterene and spironolactone) do not increase urinary magnesium excretion [12]. However, this type of drugs only tend to retain magnesium but are not magnesium retaining substances, or almost to a lesser extent than they retain potassium and when the dose is increased, in order to obtain a certain magnesium retention, they might even cause perilous hyperkalaemia and hyponatraemia [13].

Loop diuretics act at the thick ascending limb of the Henle's loop, where the majority of the filtered magnesium is reabsorbed. However the magnesiuretic effect of loop diuretics is not entirely due to their direct interference with the tubular reabsorption of magnesium. In pathological conditions indirect mechanisms such as hyperaldosteronism and haemodynamic alterations could play an important role in the magnesiuretic effect of diuretics and might explain the differences in the magnitude of the magnesiuresis induced by various loop diuretics [14].

We have evaluated the magnesiuresis induced by a single oral high dose of various loop diuretics in compensated cirrhotic patients to disclose possible differences in their magnesiuretic action. Four homogeneous groups of patients, each consisting of 15 subjects, were considered in an investigative series [14,15]. Each group of patients was treated with one different equipotent diuretic formulation (azosemide 320 mg, muzolimine 90 mg, piretanide 24 mg, torasemide 50 mg) as a single oral dose. The effects of each drug given (in random order 7 days apart from the previous one) were compared in the same patient with those of furosemide 100 mg, taken as reference drug.

Amongst the drugs we evaluated we found significant differences in their diuretic action, particularly with respect to natriuresis and magnesiuresis. (Table 1). Furosemide 100 mg, piretanide 24 mg and azosemide 320 mg showed very similar natriuretic and magnesiuretic effect. The highest magnesiuresis was observed after torasemide 50 mg but it is noteworthy that this formulation also induced the highest urinary sodium excretion. The potency and the pharmacodynamics of torasemide depend on liver function and accordingly we observed that in cirrhotic patients torasemide induced a significantly higher cumulative excretion of water and sodium than that induced in healthy volunteers treated with the same torasemide formulation [14].

TABLE 1. Urinary magnesium (Mg^{2+}) and sodium (Na^{+}) excretion and urinary magnesium/sodium ratio (Mg^{2+}/Na^{+}) in 15 compensated cirrhotic patients. Pre-treatment values between parentheses. Means ± SEM. Contrast versus Furosemide : * = p<0.05, ** = p<0.025

	Furosemide 100 mg	**Azosemide 320 mg**	**Muzolimine 90 mg**	**Piretanide 24 mg**	**Torasemide 50 mg**
Mg^{2+} µmol/m	**3.8 ±0.2 (2.3 ±0.5)**	**4.2 ±0.6 (2.7 ±0.4)**	**2.8 ± 0.3* (2.6 ± 0.4)**	**4.0 ± 0.2 (2.2 ± 0.5)**	**5.6 ± 0.2* (2.7 ± 0.5)**
Na^{+} µmol/m	**157.4 ± 4.5 (64 ±10.4)**	**161.5 ± 17.1 (60 ± 11)**	**168.7 ± 16.9 (61.6 ± 9.9)**	**133 ± 13.5 (60.6 ± 8.9)**	**311.9 ± 19.5** (62.9 ± 11)**
Mg^{2+}/Na^{+}(x1000)	**23.9 ± 3.6**	**25.9 ± 3.5**	**16.8 ± 1.7****	**30.5 ± 2.9**	**18.0 ± 1.9***

Thus, if the urinary magnesium excretion is evaluated in relation to the natriuresis (Na^{+}/Mg^{2+} ratio), the magnesiuretic effect of torasemide 50 mg was shown to be lower than that of azosemide 320 mg, furosemide 100 mg, and piretanide 24 mg.

Unlike the other drugs considered in the study, muzolimine 90 mg did not significantly increased the 24-hour urinary magnesium excretion. These results indicate that the magnesiuretic action of a diuretic is only partly due to the interference with magnesium reabsorption in the main site of action of the drug. The different magnesiuretic effect of drugs acting on the same nephronal segment may be interpreted as the result of a different interference with humoral mechanisms of regulation of magnesium turnover such as aldosterone, PTH, calcium etc. and suggest that the effects of loop diuretics on the cellular and paracellular transfer of magnesium (and calcium) deserve being investigated further.

4. Magnesium status in cirrhotic patients evaluated with magnesium loading test

The magnesium loading test has been used to assess magnesium turnover since is a valuable and repeatable tool which results correlate with the intracellular content of magnesium in normal subjects and in various pathological conditions [6,16,17].

In the last few years the relevance of viral infections in the pathogenesis of liver cirrhosis has been more clearly established. It has been found that the great majority of the so called non-A-non-B hepatitis is in reality due to hepatitis C viruses (HCV) and many liver diseases which were previously interpreted as alcoholic because of undiagnosed viral infection and concomitant history of alcohol use, are in reality due to HCV [18]. Since alcohol is *per se* a cause of magnesium depletion it is very important to differentiate the magnesium status in patient with cirrhosis of alcoholic or viral origin.

In order to disclose possible differences in the magnesium status of patients with alcoholic or HCV related liver cirrhosis we performed the magnesium loading test in 2 homogeneous (as far age, sex distribution, dietary habits, histological and clinical conditions) groups of compensated (Child A or B) cirrhotic patients (16 alcoholic liver cirrhosis and 16 HCV related cirrhosis in teetotal subjects). The magnesium load was carried out as described by Gullestad et al.[16] administering 30 mmol of magnesium sulphate in 1.000 ml isotonic saline as a continuous intravenous infusion during 8 hours (from 8 a.m. to 4 p.m.). A 24-hour urine sample was collected from the start of the load. Magnesium concentration was evaluated in plasma (before and 48 hours after the test) and in urine; the magnesium retention, expressed as percentage of the given dose, was calculated from the administered dose and the 24 hour urinary magnesium excretion. The results were compared with those observed in 12 normal subjects who underwent the same test. Tables 2 summarises the principal outcomes of the study.

Both patients with cirrhosis of viral and alcoholic origin showed highly significant magnesium retention (35.7 % and 53.1 % respectively) in comparison with healthy subjects (14.5 %). The differences between the 3 groups resulted highly significant. Plasma magnesium concentrations in patients with cirrhosis of alcoholic or viral origin resulted significantly lower than that observed in normal subjects. Forty-eight hours after the magnesium load both normal subjects and patients significantly increased their plasma magnesium concentrations and patients affected with alcoholic cirrhosis (who had plasma magnesium mean concentration below the lower reference limit) normalised this value. Our data shows that a magnesium deficiency (as assessed by the magnesium loading test) may be present in cirrhotic patients irrespectively of the alcoholic or HCV origin of the liver disease.

TABLE 2. Magnesium retention and plasma magnesium concentration before and after magnesium loading test in 12 healthy subjects, 16 teetotal HCV+ and 16 alcoholic cirrhotic patients.
Mg^{2+} retention : normal subjects vs. HCV+ cirrhosis and alcoholic cirrhosis $p<0.001$,HCV+ vs. alcoholic cirrhosis $p<0.01$
Plasma Mg^{2+} before loading : normal subjects vs. HCV+ and alcoholic cirrhosis $p<0.001$, HCV+ vs. alcoholic cirrhosis N.S.
Before vs. after loading Plasma Mg^{2+}: normal subjects $p<0.01$, HCV+ and alcoholic cirrhosis $p<0.001$

	Normal Subjects	HCV + cirrhosis	Alcoholic cirrhosis
Mg^{2+}retention (%)	**14.5 ± 3.2**	**35.7 ± 7.5**	**53.1 ± 7.2**
Plasma Mg^{2+} (mmol/L) before loading	**0.86 ⊥ 0.05**	**0.77 ⊥ 0.04**	**0.73 ± .06**
Plasma Mg^{2+} (mmol/L) after loading	**0.89 ± 0.04**	**0.91 ± 0.05**	**0.87 ± 0.08**

The fact that patients with alcoholic cirrhosis had lower mean plasma magnesium concentration and higher mean retention of magnesium after loading test emphasises the role of alcohol abuse as adjunctive factor in the pathogenesis of bodily magnesium deficiency in liver diseases. However, liver cirrhosis, even in teetotal patients, is *per se* a condition associated with magnesium imbalance.

5. Conclusion

-Cirrhosis of the liver, even in early stage, is a clinical condition associated with magnesium deficiency, which becomes more relevant in alcoholic patients. Thus, all the possible causes of depletion and/or reduced intake of this cation should be prevented and treated through: dietary intervention (high-magnesium diet, alcohol withdrawal etc.), oral supplementation (magnesium salts), reduction of faecal magnesium loss (steatorrhoea), abatement of diuretic need (particularly through a dietary sodium curtailment).

-Treatment with common diuretically-active formulations of loop and early distal tubule diuretics definitely worsens magnesium status in patients with cirrhosis of the liver. The final magnesiuretic effect of magnesium wasting diuretics also account for the possible interference with the humoral mechanisms of regulation of magnesium turnover (aldosterone, PTH, calcium). Potassium-sparing diuretics are not magnesium-retaining substances, or almost not to the extent that they are potassium retaining diuretics.

-The intravenous magnesium loading test is a reliable, simple and inexpensive tool in the estimation of magnesium deficiency in patients suffering from cirrhosis of the liver. Furthermore, the magnesium load itself conveys an almost partial correction of the magnesium deficiency.

6. References

1. Durlach, J. (1988): Le magnésium en pratique clinique, pp 71-220. Paris: J.B. Ballière.
2. Shils, M.E. (1988): Magnesium in health and disease. *Ann. Rev. Nutr.* **8**, 429-460.
3. Lim, P. and Jacob, E. (1972): Magnesium deficiency in liver cirrhosis *Q. J. Med.* **41**, 291-300.
4. Flink, E.B. (1981): Magnesium deficiency. Aetiology and clinical spectrum. *Acta Med. Scand.* **647**, 125-137
5. Reinhart, R.A. (1988): Magnesium metabolism. *Arch. Intern. Med.* **148**, 2415-2420.7.
6. Elin, R.J. (1989): Assessment of magnesium status. In: *Magnesium in health and disease*, eds. Y. Itokawa and J. Durlach, pp 137-146. London, Paris : John Libbey.
7. Holm, C.N., Jepsen, J.M., Sjogaard, G., and Hessob, I. (1987): A magnesium load test in the diagnosis of magnesium deficiency. *Human Nutrition: Clin. Nutrition.* **41C**, 301-306.
8. Tansy, M.F. and Kendall, F.M. (1981): Magnesium and the gastrointestinal tract. *Magnesium-Bull.* **1a**, 55-65
9. Dirko, J.H. (1983): The kidney and magnesium regulation. *Kidney Int.* **23**, 771-776.
10. Flink, E.B.(1986): Magnesium deficiency in alcoholism. *Alcoholism* **10**, 590-594.
11. Leary, W.P. and Reyes, A.J. (1984): Diuretic-induced magnesium losses. *Drugs* **28/1**, 182-187. 20.
12. Mountokalakis, T., Dourakis, S., et al. (1989): Magnesium-sparing diuretics. In: *Magnesium in health and disease*, eds. Y. Itokawa and J. Durlach, pp 409-413. London, Paris : John Libbey.
13. Reyes, A.J., and Leary, W.P. (1984) : Diuretics and magnesium. *Magnesium-Bull.* **6**, 87-99
14. Crippa, G., Cassi, A. and Vergani, F. (1992): Renal excretory pharmacodynamics of diuretics in hepatic diseases. Prog. Pharmacol. Clin. Pharmacol. **9**, 597-617.
15. Crippa, G., Giorgi-Pierfranceschi, M., Sverzellati, et al (1997): Urinary magnesium excretion after very high doses of loop diuretics in cirrhotic patients. *Magnesium Research*, In Press.
16. Gullestad, L., Dolva, L.O., Waage, et al. (1992): Magnesium deficiency diagnosed by an intravenous loading test. *Scand. J. Clin. Lab. Invest.* **52**,245-253.
17. Bohmer, T., Mathiesen, B. (1982): Magnesium deficiency in chronic alcoholic patients uncovered by an intravenous loading test. *Scand. J. Clin. Lab. Invest.* **42**, 633-636.
18. Crippa, G., Giorgi Pierfranceschi, M. (1997): Magnesium in cirrhosis of the liver. *In Advances in Manesium Research,* R. Smetana (ed.), pp 294-301. London, Paris, John Libbey.

ATHEROGENIC LIPID FRACTIONS ARE RELATED TO FREE MAGNESIUM ION STATUS

A. HAENNI, M. ÖHRWALL, H. LITHELL
Dept of Geriatrics, Uppsala University
PO Box 609, S-75125 Uppsala, Sweden

1. Introduction

Magnesium deficiency in experimental animals has been found to cause adverse alterations in the blood lipid composition [1, 2] and to affect lipid infiltration into the endothelium [3]. Diets supplemented with magnesium, on the other hand, have been shown to prevent the development of atherosclerosis by inhibiting lipid accumulation in the aortic wall [4]. In spite of this pattern consistent correlations have not been demonstrated between magnesium and lipid concentrations [5]. It has been questioned whether the total magnesium concentration in the blood reflects the biological activity of magnesium, since only about 60 % of the magnesium in the blood is in a free, unbound state, as determined by indirect methods. More than 30 % of the total serum magnesium is bound to proteins, mainly albumin, and the remaining 10 % is bound to phosphate, citrate, and other substances [6]. It has now become possible, by use of electrodes sensitive to ionized magnesium, to assess the concentration of free magnesium ions in the blood [7]. The aim of the present study was to investigate the associations between variables reflecting lipid metabolism, on the one hand, and the free ion status and total magnesium concentration in the blood, on the other hand.

2. Material and methods

The study comprised 29 men, mean age 72.5 (70.2-75.1) years, participating in a health survey of 70-year old men in Uppsala. None of the participants were being treated with antidiabetic drugs. Four patients were receiving pharmacological treatment: b-blockers in two patients and diuretics in the other two. Three of the participants were smokers. The lipoprotein status, the plasma concentration of free magnesium ions, and the total serum magnesium concentration were determined. All metabolic investigations were performed in the morning, after an overnight fast.

Triglyceride and cholesterol concentrations were measured in serum by enzymatic-colorimetric techniques, using the IL Test Triglyceride Trinder´s method 181610-60 and IL Test Cholesterol method 181618-10 for use in Monarch centrifugal analyzer (Instrumentation Laboratory, Lexington, MA, USA). VLDL, LDL and HDL were isolated by a combination of preparative ultracentrifugation and precipitation with sodium phosphotungstate and magnesium chloride. The concentrations of the serum

T. Theophanides and J. Anastassopoulou (eds.), Magnesium: Current Status and New Developments, 95-98.

apolipoproteins (apo) A1 and B were determined by turbidimetry in a Monarch apparatus using an immunochemical commercial assay (Orion Diagnostica, Espoo, Finland).

The free ion status was determined by the KONE microlyte analyzer with Mg++ (KONE Instruments, Espoo, Finland). Blood samples were taken from an antecubal vein without stasis after the patient had been in the supine position for 30 minutes, immediately before the clamp test. The blood samples were heparinized in accordance with the guidelines given in the instructions for the KONE microlyte analyzer [8]. Only silicon-free laboratory equipement was used. The blood samples were centrifuged for 1 min and the concentration of free magnesium ions in plasma was immediately determined in duplicate and the mean value was used in the calculations. The same magnesium-sensitive electrode was used in the analyses of all plasma samples. No changes over time in the assessed free magnesium concentrations, characteristic for aging electrodes [9], were observed. The adequacy of the free ion analysis was checked by using prefabricated solutions with known ion concentrations (KONE). The coefficient of variation for plasma free magnesium measurement was <10% with this procedure.
The serum total magnesium concentration was determined at the Department of Clinical Chemistry at the University Hospital, Uppsala. The blood samples were collected in vacuum tubes for serum analysis (Vacutainer 367788 BN326SST4, Becton Dickinson Vacutainer Systems, Rutherford, NJ, USA) under conditions similar to those mentioned above, and analyzed during the same day. The serum magnesium concentration was measured by atom absorption photometry (Instrumentation Laboratory, Lexington, MA, USA).

3. Statistics

For variables with a skewed distribution (VLDL triglyceride, LDL cholesterol, HDL cholesterol, LDL/HDL cholesterol ratio, apolipoprotein B/A1 ratio), logarithmic transformation was performed. Relationships between variables were analzed by calculations of Pearson´s correlation coefficient. To elucidate the interdependence of different variables, stepwise multiple regression analysis was applied. Values of $p < 0.05$ were considered statistically significant. All statistical analyses were performed with the statistical program package Statistical Analysis System and JMP (SAS Institute Inc, NC, USA)

4. Results

The free magnesium ion concentration (fMg++) in plasma, total serum magnesium concentration (Mg), and lipid status were assessed in all 29 participants.
The total serum magnesium concentration (mean±SD) was 0.82 (±0.06) mmol/l The free magnesium concentration, fMg++, was 0.52 (±0.04) mmol/l, corresponding to 63 % of Mg, and the non-ionic magnesium was thus 0.30 mmol/l. The ionized magnesium concentration correlated to Mg ($r = 0.50$, $p < 0.01$).

Plasma fMg++ was inversely correlated to serum VLDL cholesterol ($r = -0.39$, $p<0.04$), LDL cholesterol ($r = -0.57$, $p<0.002$), and total cholesterol ($r = -0.55$, $p<0.002$), VLDL triglyceride ($r = -0.36$, $p<0.05$), LDL triglyceride ($r = -0.48$, $p<0.01$), and total triglyceride ($r = -0.41$, $p<0.03$). Total magnesium concentration, Mg, was also inversely correlated to these lipid fractions, but the correlations were less pronounced and did not reach statistical significance. The HDL concentrations did not correlate significantly to any of the magnesium variables. However, fMg++, was closely inversely correlated to the LDL/HDL cholesterol ratio ($r = -0.67$, $p<0.0001$). The corresponding correlation between this ratio and Mg was less pronounced ($r = -0.30$, $p = 0.12$). Further, fMg++ was inversely correlated to apolipoprotein B concentration ($r = -0.56$, $p<0.002$), and to the apolipoprotein B/A1 ratio in serum ($r = -0.58$, $p<0.001$). The corresponding figures for Mg were $r = -0.37$, $p = 0.05$, and $r = -0.37$, $p = 0.05$, respectively.

The correlations between the variables reflecting lipid metabolism and fMg++ remained significant when BMI and waist-hip ratio were included in a stepwise multiple analysis model.
The theoretically calculated nonionized magnesium fraction did not correlate to any of the described metabolic variables.

5. Discussion

The mean fMg++ and serum total magnesiumvalues as well as the ratio of ionized to total magnesium are in accordance with with previous findings [5,9].
We found that fMg++ was fairly closely correlated to lipid fractions, and ratios between different lipid fractions, known to be associated with atherogenicity. Another observation was that these atherogenic lipid fractions were more closely correlated with fMg++ than to Mg, the routin magnesium test at present. Thus, measurement of total Mg seems to be inappropriate in studies of the relationship between magnesium and lipid metabolism.

6. Conclusion

The present study indicates that atherogenic lipid fractions are more closely associated with the ionized magnesium concentration in plasma than with the total magnesium concentration in serum. Routine use of serum total magnesium for evaluating the magnesium status is therefore probably inadequate when studying interrelations between serum electrolytes and metabolic variables, notably lipoproteins.

7. References

1. Rayssiguier, Y.: Magnesium, lipids and vascular diseases. Magnesium **5**(1986),182-190.

2. Gueux, E., Cubizolles, C., Bussiere, L., Mazur, A., Rayssiguier, Y.: Oxidative modification of triglyceride-rich lipoproteins in hypertriglyceridemic rats following magnesium deficiency. Lipids **28**(1993):573-5.

3. Yokoyama, S., Gu, J., Kashima, K., Nishida, H.I., Smith, T.L., Kummerov, F.A.:. Combined effects of magnesium deficiency and the atherogenic level of low density lipoprotein on uptake and metabolism of low density lipoprotein by cultured human endothelial cells. Magnes Res **7**(1994):97-105.

4. Ouchi, Y., Tabata, R., Stergiopoulos, K., Sato, F., Hattori, A., Orimo, H.: Effect of dietary magnesium on development of atherosclerosis in cholesterol-fed rabbits. Arteriosclerosis **10**(1990):732-7.

5. Speich, M., Gelot, S., Auget, J.: Pasma and erytrocyte magnesium, calcium, zinc, copper, phosphorus and cholesterols in a population of 1050 healthy adults. Mag-Bull **17**(1995):62-9.

6. Walser, M.: Magnesium metabolism. Rev Physiol Biochem Exp Pharmacol **59**(1967):185-341.

7. Maj-Zurawska, M., Lewenstam, A.: Fully automated potentiometric determination of ionised magnesium in blood serum. Analytica Chimica Acta, **236**(1990):331-5.

8. KONE: Users Manual to KONE Microlyte with Magnesium. Helsinki, Finland, 1993.

9. Ising, H., Bertschat, F., Günther, T., Jeremias, E., Jeremias, A.: Measurement of free magnesium in blood, serum and plasma with an ion-selective electrode. Eur J Clin Chem Clin Biochem **33**(1995):365-71.

MAGNESIUM AND OXIDATIVE STRESS STATUS IN PATIENTS WITH CHRONIC FATIGUE

B. MANUEL Y KEENOY,G. MOORKENS, S. MELUDU, J. VERTOMMEN, M. NOË AND I. DE LEEUW
Laboratory of Endocrinology, Dept Internal Medicine, Universitaire Instelling Antwerpen, Universiteitsplein 1, B-2610 Wilrijk-Antwerpen, Belgium.

1. INTRODUCTION

Magnesium deficiency has been linked to numerous clinical conditions, in some of which, such as atherosclerosis and diabetes mellitus, increased oxidative stresss has been extensively documented as a major pathogenic factor [1,2]. Although increased oxidative stress has been shown in experimental animals with extreme Mg deficiency [3], it is still not clear whether there is a direct causal link between Mg deficiency, oxidative stress and the resulting disease process or whether these are parallel unrelated phenomena.

Magnesium is also frequently used with success as an adjuvant in the treatment of conditions, such as chronic fatigue, where neither Mg deficiency nor increased oxidative stress have been established [4,5]. It is not clear whether the beneficial effect of Mg in these patients is due to the replenishment of possibly depleted Mg body stores or whether it is due to a direct pharmacological effect of Mg on energy metabolism, an effect which is independent of the state of body stores [6].

In order to address these questions we have measured the status of Mg by a parenteral retention test and some parameters of oxidative stress in a group of patients with a complaint of chronic fatigue before and after supplementation with nutritional doses of Mg for more than 3 months to ensure slow replenishment of body stores.

2. Methods and results

Apart from the routine clinical examination and laboratory tests, the 97 consecutive patients with chronic fatigue enrolled in the study (25% male, 16% smokers, with ages ranging from 14 to 73 years, with BMI and blood biochemistry values within the normal ranges) were also evaluated as follows.

Chronic fatigue syndrome (CFS), diagnosed according to the CDC criteria (7), was present in 54% of the patients. Fibromyalgia (FM) was positive (when 11 or more out of 18 trigger points were painful) in 52%. FM was significantly associated with CFS

T. Theophanides and J. Anastassopoulou (eds.), Magnesium: Current Status and New Developments, 99-104.

(p0.007). The cryptotetany test was positive in 81% but was not associated with either FM or CFS.
Mg deficiency was established when 20% or more of the Mg infused (0.2 mEq/Kg over 4 hrs) during the parenteral loading test was retained (8). 47% of the patients (30% male) were Mg deficient. There was no significant association between the presence of Mg deficiency and/or sex and CFS, FM or positive cryptotetany.
Mg concentrations in plasma, urine and red blood cells (RBC) were measured before and after the intravenous loading test by atomic absorption spectrometry (AAS). There were no significant differences between the groups being Mg deficient or not. There was no significant correlation between the % of Mg retention during the loading test and any of the Mg concentrations measured. Patients with CFS had a tendency for higher RBC Mg, those with positive cryptotetany had lower plasma Mg and those with FM had a lower increase in plasma Mg and higher increase in RBC Mg after the intravenous loading test.(Tables 1-4).
A dietary anamnesis, conducted during the first loading test, revealed that daily intakes of Mg (6.1±1.9 mg/Kg/day (range 2.6-10.4) were compatible with the RDA in this group of patients and that there were no differences in intakes of nutrients between the groups being Mg deficient or not. Total daily Mg intake correlated positively with total intake of calories, fats, cholesterol, polyunsaturated fats and fibers (r= 0.82 p< 0.0005 (55). Mg intake/Kg/day correlated positively with serum levels of HDL-Cholesterol and vitamin E and negatively with uric acid , creatinine and BMI (r= - 0.41 p<0.005 (52).
Patients with Mg deficiency had significantly lower albumine (4.27±0.37 g/dl (42) than those not deficient (4.42±0.39 (49) p<0.05). The tendency to have lower iron (with higher total iron binding capacity, TIBC), higher alkaline phosphatase and triglycerides in the Mg deficient group did not reach significance.
Global antioxidative capacity of plasma was measured using the TEAC method (9), where two effects of the antioxidant(s) were distinguished : prolongation of the lagtime of the reaction (TEACLT), inhibition of the reaction velocity measured after 6 minutes (TEAC6). TEACLT correlated with uric acid levels (r=0.72 p<0.0005 (29) and TEAC6 with albumin (ro= 0.47 p <0.05 (20) and TIBC (r= - 0.50 p<0.01 (24).
TEACs measured before and after the intravenous loading test were not significantly different, pointing out to the lack of an acute effect of Mg on this parameter. However, Mg deficient patients had significantly lower TEAC6 (2.23±0.56 mM Trolox equivalents versus 2.62±0.54 in the non-deficient group p<0.05). Moreover there was a significant negative correlation between % of Mg retention and TEAC6 only (r= - 0.43 p0.012 (28). In contrast, TEACLT did not differ between the two groups (0.31±0.10 and 0.31±0.10).
Apart from the lower albumin which would explain the lower antioxidative capacity of these patients as regards reaction velocity (TEAC6), levels of other individual antioxidants such as uric acid, glutathion (GSH), vitamins E and A, glutathion peroxidase, transferrine and ceruloplasmine did not differ between the two groups.
Susceptibility of plasma to lipid peroxidation, as measured by the formation of TBARS and of fluorescent products in non-HDL lipoproteins when incubated in vitro with copper, did not differ significantly between the 2 groups. Susceptibility to peroxidation was higher in those patients with lower TEACLT and TEAC6, lower vitamin E (r=-0.46

TABLE 1. Parameters of Mg status according to Mg deficiency (Patients with Mg retention ≥ 20% after the loading test are considered as Mg-deficient)

Parameter (units)		Retention ≥ 20%	Retention < 20%	p-value
Mg plasma (AAS) (mM)	(before test)	0.82 ± 0.10 (41)	0.82 ± 0.09 (49)	ns
	(after test)	1.04 ± 0.11 (31)	1.03 ± 0.11 (29)	ns
Mg RBC (AAS) (mM)	(before test)	1.97 ± 0.22 (41)	1.95 ± 0.20 (49)	ns
	(after test)	2.00 ± 0.21 (31)	1.95 ± 0.21 (29)	ns
Mg urine (AAS) (mM)	(before test)	4.04 ± 2.57 (25)	2.91 ± 1.41 (20)	0.067
	(after test)	3.99 ± 1.89 (19)	5.27 ± 2.48 (15)	0.090
Δ in plasma Mg due to test (mM)		0.213 ± 0.090 (31)	0.213 ± 0.063 (29)	ns
Δ in RBC Mg due to test (mM)		0.016 ± 0.097 (31)	0.040 ± 0.130 (29)	ns
Δ in Urine Mg due to test (mM)		-0.25 ± 2.97 (19)	2.50 ± 1.58 (15)	0.002

TABLE 2 . Parameters of Mg status according to the presence of Chronic Fatigue Syndrome (CFS)

Parameter (units)		CFS +	CFS -	p-value
Retention (% of infused Mg retained)		21.7 ± 33.1 (32)	22.9 ± 30.8 (28)	ns
Mg plasma (AAS) (mM)	(before test)	0.82 ± 0.09 (30)	0.81 ± 0.09 (28)	ns
	(after test)	1.04 ± 0.12 (30)	1.01 ± 0.09 (28)	ns
Mg RBC (AAS) (mM)	(before test)	2.00 ± 0.21 (30)	1.90 ± 0.21 (28)	0.07
	(after test)	2.05 ± 0.19 (30)	1.91 ± 0.21 (28)	0.01
Mg urine (AAS) (mM)	(before test)	3.55 ± 2.58 (25)	3.44 ± 1.61 (18)	ns
	(after test)	5.17 ± 2.47 (20)	3.87 ± 1.62 (1)	0.09
Δ in plasma Mg due to test (mM)		0.223 ± 0.085 (30)	0.203 ± 0.068	ns
Δ in RBC Mg due to test (mM)		0.057 ± 0.112 (30)	0.011 ± 0.120	0.14
Δ in Urine Mg due to test (mM)		1.47 ± 2.93 (20)	0.48 ± 2.74 (11)	ns

TABLE 3. Parameters of Mg status before and after Mg loading test according to the presence of a positive cryptotetany test

	crypto +	crypto -	p-value
Retention (% of infused Mg retained)	22.6 ± 32.4 (47)	20.9 ± 33.7 (11)	ns
Mg plasma (AAS) (mM) (before test)	0.81 ± 0.09 (44)	0.87 ± 0.06 (11)	0.02
(after test)	1.02 ± 0.11 (44)	1.08 ± 0.06 (11)	0.05
Mg RBC (AAS) (mM) (before test)	1.96 ± 0.23 (44)	1.91 ± 0.16 (11)	ns
(after test)	1.99 ± 0.23 (44)	1.96 ± 0.17 (11)	ns
Mg urine (AAS) (mM) (before test)	3.55 ± 2.44 (30)	3.55 ± 1.79 (10)	ns
(after test)	4.37 ± 2.02 (24)	5.86 ± 2.83 (7)	ns
Δ in plasma Mg due to test (mM)	0.216 ± 0.080 (44)	0.214 ± 0.065 (11)	ns
Δ in RBC Mg due to test (mM)	0.023 ± 0.114 (44)	0.052 ± 0.122 (11)	ns
Δ in Urine Mg due to test (mM)	0.782 ± 3.026 (24)	2.28 ± 1.92 (7)	0.14

TABLE 4. Parameters of Mg status before and after Mg loading test according to the presence of Fibromyalgia (FM)

	FM +	FM -	p-value
Retention (% of infused Mg retained)	17.8 ± 31.0 (32)	27.7 ± 32.0	0.23
Mg plasma (AAS) (mM) (before test)	0.81 ± 0.10 (30)	0.82 ± 0.07 (29)	ns
(after test)	1.00 ± 0.11 (30)	1.06 ± 0.10 (29)	0.04
Mg RBC (AAS) (mM) (before test)	1.97 ± 0.21 (30)	1.93 ± 0.21 (29)	ns
(after test)	2.04 ± 0.20 (30)	1.93 ± 0.21 (28)	0.05
Mg urine (AAS) (mM) (before test)	3.95 ± 2.75 (23)	3.13 ± 1.33 ((21)	0.21
(after test)	5.17 ± 2.24 (16)	4.19 ± 2.19 (16)	0.22
Δ in plasma Mg due to test (mM)	0.185 ± 0.064 (30)	0.240 ± 0.080	0.005
Δ in RBC Mg due to test (mM)	0.064 ± 0.135 (30)	0.001 ± 0.086 (29)	0.04
Δ in Urine Mg due to test (mM)	0.86 ± 3.30 (16)	1.18 ± 2.45	ns

p<0.001(59) with slope of fluorescence) and higher Mg in RBC. The intravenous Mg infusion was followed by an increase in TBARS formation at the saturation phase (after 150-180 min).

Of the first 34 patients diagnosed as being Mg deficient and accepting Mg supplementation , 24 returned and the above-mentioned parameters were remeasured in a second visit. Supplementation was oral in all but 3 patients (as Mg-rich drinking water, adaptation of the diet or Mg oxide gellules) and aimed at intakes of 10 mg/Kg/day for periods of more than 3 months. Although 11 patients still had 20% or more retention after supplementation (the non-responders), mean % retention decreased from 47±17% to 18±43% p<0.002). Supplementation did not affect significantly Mg concentrations in plasma and RBC, nor concentrations of uric acid, albumin,GSH, vitamin A, glutathion peroxidase, nor lipid concentrations or in vitro formation of TBARS. Non reponders were distinguished by having a significantly lower GSH, which remained lower even after supplementation (5.91±0.83 versus 6.91±1.13(11) in the responders (p <0.02).

However, vitamin E increased from 8.54±3.35 to 13.20±5.39 µg/ml after supplementation (p<0.001) and the slope of fluorescence formation decreased from 0.957±0.209 to 0.841±0.227 (p<0.02), this decrease being more pronounced in the group of responders. There was a non significant tendency for an increase in TEAC and in iron.

Patients with a positive diagnosis of CFS were distinguished in the 1st visit by having significantly lower albumin (4.18±0.43 g/dl(26) versus 4.46±0.53 (23) in the non-CFS (p<0.05), lower transferrine (241±35 mg/dl versus 273±50 (28) p<0.01) and higher cholesterol (209±44 mg/dl versus 173±44 in the non-CFS (p<0.05). This was accompanied by a significantly higher production of TBARS at the propagation phase : after 60 and 90 minutes (30.9±22.1 nmol MDA/mg non-HDL cholesterol versus 19.3±15.0 p<0.05) as well as higher glutathion peroxidase activity (155±48 U/l hemolysate versus 114±38 p<0.05) although other antioxidants and TEAC did not differ. These differences,which were due to the presence of CFS, persisted after supplementation with Mg.

3. Conclusions

The patients investigated in this study were not hospitalised and had healthy eating habits and normal biological parameters. Thus, in this study population, the link between Mg status and oxidative stress could be investigated in the absence of age- and disease-related factors such as dyslipemias, hyperglycemia and accumulation of products of oxidative cell damage, which might have distorted the results.

The results obtained confirm that Mg concentrations in plasma and RBC do not reflect the state of Mg body stores, as evaluated by the % Mg retention in the parenteral test, which is known to correlate with Mg stores in bone [10]. Indeed, Mg deficient patients (with a retention of 20% or more) did not have lower Mg in plasma or RBC. Even after slow replenishment of stores by oral supplementation, which was confirmed by the significant decrease in %Mg retention, these Mg concentrations did not change. Neither was Mg status in these patients influenced by the intake of Mg in the diet.

We could not establish a direct causal link between Mg deficiency and increased oxidative stress. For example, 1) global antioxidative capacity of plasma (TEAC) was not affected by the Mg infusion. 2) Mg deficient patients did not differ in the levels of individual antioxidants nor in TEACLT, which was dependent on uric acid levels. 3) Lipid peroxidation susceptibility did not differ in the Mg deficient patients, was higher in the patients with higher RBC Mg and even increased after the Mg infusion.
However, some observations point to a possible indirect influence of Mg status on other factors which do play a significant role on oxidative stress status. For example, 1) patients with a higher daily intake of Mg, expressed per Kg body weight had a lower BMI and higher serum levels of vitamin E and HDL-cholesterol, which are considered as protective factors against atherosclerosis. 2) Supplementation with Mg was accompanied by a significant increase in serum Vitamin E and a decrease in lipid peroxidation susceptibility as measured by the slope (velocity) of formation of fluorescent products. 3) TEAC6 was lower in Mg deficient patients who also had lower albumin levels, but both parameters remained low after Mg supplementation . 4) Mg deficient patients who did not replenish their Mg body stores after supplementation (non-responders) had persistently lower GSH before and after supplementation.
In contrast, patients with CFS were not more Mg deficient but did have a persistently (even after Mg supplementation) higher lipid peroxidation susceptibility as measured by TBARS formation in the propagation phase, which was accompanied by persistently higher glutathion peroxidase, but lower albumine and transferrine.
In conclusion, Mg status plays a role in lipid peroxidation susceptibility and antioxidant capacity of plasma only indirectly, by influencing the levels of other modulatory factors such as vitamin E and albumin.

References

1. Rude, RK. (1993) Magnesium metabolism and deficiency. *Endocrinology and Metabolism Clinics of North America,* **22(2),** 377-395.
2. Wolff SP. Diabetes mellitus and free radicals. *Br.Med.Bull.***49**, 642-652,1993.
3. Kramer, J.H., Misik, V., Weglicki, W.B. (1994). Magnesium-deficiency potentiates free radical production associated with postischemic injury to rat hearts: vitamin E affords protection. *Free Rad.Biol & Med.* **16(6)**, 713-723,.
4. Cox, I.M.,Campbell, M.J., Dowson, D. (1991) Red blood cell magnesium and chronic fatigue syndrome. *Lancet* **337**, 757-760.
5. Hinds, G., Bell, N.P., McMaster, D., McCluskey, D.R. (1994) Normal red cell magnesium concentrations and magnesium loading test in patients with chronic fatigue syndrome. *Ann.Clin.Biochem.* **31**, 459-461,.
6. Sheperd, C. Magnesium and chronic fatigue syndrome. Lancet 337 : 1095, 1991.
7. Holmes, G.P. et al. (1988) Chronic fatigue syndrome: a working case definition. *Ann.Int.Med.* **108,** 387-389.
8. Ryzen, E., Elbaum, N., Singer, F.R., Rude, R.K. (1985). Parenteral magnesium tolerance testing in the evaluation of magnesium deficiency. *Magnesium* **4**,137-147,
9. Rice-Evans, C., Miller, NJ. (1994) Total antioxidant status in plasma and body fluids. *Meths.Enzymol.* **234**, 279-293.
10. Cohen, L., Laor, A. (1990).Correlation between bone magnesium concentration and magnesium retention in the intravenous magnesium load test. *Magnes.Res.* **3**, 271-274,

SELF-CARE MANAGEMENT FOR THE PATIENT WITH NORMOMAGNESEMIA MAGNESIUM DEFICIENCY

Herbert C. Mansmann, Jr[1] and Paris T. Mansmann[2]
[1] *Jefferson Medical College of Thomas Jefferson University, Alumni Hall (Room 543), 1020 Locust Street, Philadelphia, PA. 19107-6799,* [2] *West Virginia University, Box 9167, 2178 Health Science Center North, Morgantown, WV, 26506-9237*

1. INTRODUCTION

A comprehensive review of the magnesium literature failed to disclose any articles describing self-care or self-management of normomagnesemia magnesium (Mg) deficiency (NMgD) requiring oral dosing with Mg supplication (MgS). A review of eleven books on Mg, mostly research conference proceedings, likewise showed a lack of interest in this essential modality of clinical medical care. Yet physician guidelines are available that recommend, that at every opportunity the patient should be exposed to self-management education for a partnership in comprehensive asthma care, including self-medication[1]. Many books exist on this subject for asthmatics, as well for diabetics[2].

While patients with any degree of NMgD, can benefit by learning this routine, the authors believe it is essential in the most complicated situations, in order for the patient to feel in control of the disorder, their life and thus their outcome. Moreover only the patient can live through the rapid changes on a daily bases. Even seeing a physician weekly is an inadequate substitution for this experience and can only lead to a hopeless feeling and noncompliance, with worsening of the situation. Hypomagnesemia is surely to follow, requiring more intense acute medical therapy, and then followed by oral MgS. Self-care will again become essential to prevent acute illness and care. The purpose of this paper is to assure the reader that this modality can be very successful, and to recommend patient characteristics, and methods to implement self-care management of NMgD in selected patients with multiple medical problems and hence variable daily MgS needs.

2. OPTIMAL PATIENT CHARACTERISTICS

Self-care or self-management requires that the patient have the following characteristics **(in bold type)** in order to be successful with this approach to his/her medical therapy. The patient must be **educable** in order to become **knowledgeable** about the disease and

T. Theophanides and J. Anastassopoulou (eds.), Magnesium: Current Status and New Developments, 105-108.

the treatment methods. The patient must be sufficiently **flexible** to modify most life-style behaviors that aggravate magnesium deficiency (MgD). He/she must be **motivated** to learn what must be done and when to do it. The subject must be reasonably **organized** in able to do the necessary tasks and to keep accurate daily notes, and have an **adaptable** nature in order to the meet the need to change doses in either direction on the same day, based on frequent changing symptoms. Must be **objective,** so that the structure and the intensity of the treatment does not become overwhelming, and become an **active team member** oriented to properly utilize a necessary clinical care support systems.

The elimination of the life style behaviors is most helpful in the control of NMgD. This includes the control or omission of alcohol, caffeine, high intake of glucose, nicotine, phosphate containing sodas, theobromine, theophylline, and stress.

3. DOSING RECOMMENDATIONS AND REQUIREMENTS

A fast-acting supplement is five 500mg tablets of magnesium gluconate (MgG), each with 27mg of Mg^{2+}, giving a total dose of 135 mg of elemental Mg^{2+} which is effective within an hour on an empty stomach, and a repeat dose every hour is well tolerated. The delayed acting supplements was a 250mg (elemental Mg) tablet of magnesium oxide (MgO), which has a peak effect at 5-6 hours and may require dosing every 6 hours (the most frequent dosing tested).

An important goal is to establish the maximum tolerated dose that completely controls symptoms for 24 hours. The dose is need dependent and varies from 1 tablet every 12 hours to 4 tablets every 6 hours.

4. PERSONAL EXPERIENCE

A 71 year old male obese chronic dieter with asthma and non-insulin dependent diabetes mellitus, required daily high dose steroids, 40mg one day and & 20mg the next day of methylprednisolone (MP) in order to control polymyositis. After 4 months of this high steroids, the aldolase serum level fell from 9.8U/L to 4.8U/L(normal 2.0-7.0). At this time the dose of MP was slowly reduced to 40mg on alternate days. Since the aldolase has remained normal the steroid dose is now, a year later, being reduce by ½ mg/month and is at 7.5mg q2nd day. For several months the diabetes was under very poor control, in spite of glipizide XL 10mg daily, with swings between hyper and hypoglycemia. The HgA1c fell from 9.1% to 6.8% in 4 months with diet control and qid blood sugar monitoring. A 100 lb weight loss occurred. Many nights symptoms similar to NMgD (dysphagia, fasciculations and muscle cramps), hypoglycemia, and restless leg syndrome had to be differentiated, by a blood glucose level or a clinical response to MgG (135mg stat) in one hour. At 5 months edema of the feet and legs required Dyazide and furosemide. By this time the Mg requirement had gone from 1,500 mg of elemental Mg (sMg 2.2mEq/L and rbcMg of 2.1mEq/L) initially to

3,000mg of Mg/d. Since then the Mg requirement has varied between 3,000mg/d, which resulted in constipation, and 4,000mg/d, which occasionally caused diarrhea. A MgG dose of 135mg at bedtime and at 3AM controlled the recurrent nighttime neuromuscular symptoms and assured restful sleep. This patient knew self-care for his asthma and his diabetes, and the addition of self-care of his NMgD was easily facilitated. It is worth noting that the patient's nystagmus, first noted and has been continuous since age 24 years, disappeared when the Mg dose reached 2,000mg/d. The rbcMg was 4.5mEqL. Calcium, potassium and copper supplements were a fixed dose, but the zinc (Zn) deficiency resulted in corneal edema[3] with severe blurred vision requiring frequent daily adjustment of the Zn supplement.

The patient had the following causes of urinary Mg and Zn wasting that lead to the severe NMgD: chronic dieting with starvation diarrhea, frequent use of diet soda, high coffee and chocolate intake, frequent asthma medications, and daily verapamil (causes hypomagnesemia and hypozincemia). The addition of steroids increased hyperglycosuria and along with the diuretics resulted in an increased urinary Mg wasting. The dose of Mg needed frequent adjustments as discussed below.

5. PATIENT SUGGESTIONS AND RECOMMENDATIONS

5.1. LEARN TO COPE WITH DIARRHEA AND CONSTIPATION

1. Diarrhea is a common symptom in many gastrointestinal diseases, and the development of MgD is well documented. Less well known is the diarrhea that occurs with dieting, starvation induced. Here exists a paradox, that is also often unrecognized, which is this type of diarrhea is very responsive to an adequate dose of oral Mg. Diarrhea not induced by Mg is not a contraindication to Mg supplementation.
2. The development of constipation means that the maximum tolerated dose has not been reached.

5.2. IMPORTANT OVERDOSE SYMTOMS TO RECOGNIZE

1. Diarrhea is the most common symptom of Mg overdose, but it is important to remember that diarrhea could be due to an acute viral infection or sorbitol found in many diet foods and candy. It also could be due to a dietary indiscretion such as in the case of a known food allergy or intolerance.
2. Significant multiple muscle over-relaxation.
3. Urinary bladder relaxation with significant decrease in nocturia. In the absence of 1&2 above, this occurrence is likely to be very acceptable

5.3. WHEN TO INCREASE THE DOSE

1. Constipated bowel movement, increase next MgO dose 25%.

2. With an occurrence of one or more established end point symptoms; fasciculations, muscle cramps, dysphagia, and/or restless legs take a MgG dose of 135mg Mg^{2+}.
3. An increase of the MgO dose is taken when a new medication known to cause an increase in urinary Mg^{2+} excretion causes symptoms.
4. An increase in disease specific condition(s) that causes changes in Mg^{2+} metabolism; hyperglycemia documented by self-monitoring of blood sugar, an obvious increase in stress, etc.
5. In 2-4 above the maintenance dose should be increased 25%.

5.4. WHEN AND HOW TO DECREASE THE DOSE

1. With watery diarrhea decrease only next dose 50%, since this is usually short lived, if present at the time of the next maintenance dose repeat the above lower dose. Then try to control symptoms with a 25% lower dose.
2. When a very soft bowel movement occurs, decrease the next dose 25%. May need to alternate doses regularly, such as 75% and 100% of maximum tolerated dose.
3. Comfortably soft bowel movements are a desirably objective and should be considered optimal.

5.5. WHAT TO DO IF PATIENT FORGETS A DOSE OR DOSES

1. If it has been less than ½ the time between doses, take ½ the missed dose immediately. If acute symptoms occur in 2-6 hours take 135mg of MgG.
2. If over ½ the time has pasted, treat acute symptoms with MgG every hour if symptoms are not relieved. Don't forget the next maintenance dose.
3. If after 12 hours, take 135 mg of MgG every hour until symptoms are tolerable, and at the same time restart the maintenance dose at the previous dose frequency.

6. CONCLUSION

NMgD can be controlled by self-care management techniques.

References

1. Guidelines for the diagnosis and management of asthma; National Heart, Lung, and Blood Institute of the National Institute of Health. 1997, *NIH Publication* **No. 97-4051A.**
2. Valentine V. Biermann J. Toohey B., Diabetes Type II & What To Do. Publ. Lowell House, Los Angeles, California, USA.
3. King JC, Keen CL. Zinc. In MODERN NUTRITIO IN HEALTH AND DISEASE, Ed by M.E. Shils, J.A. Olson and M. Shike , 8th Ed., Publ Lea & Febiger, Philadelphia

Key Words; asthma, corticosteroids, diabetes, magnesium, polymyositis, zinc

MAGNESIUM METABOLISM AND PHYSICAL EXERCISE IN HUMANS

M. NISHIMUTA, N. KODAMA, H. TAKEYAMA, F. TOYOOKA
The National Institute of Health and Nutrition. Tokyo (162), Japan

1. Introduction

Effects of physical exercise on mineral metabolism have been studied by many researchers during last decade. However, the results are contradictory, because intensity, duration, and sort of exercise of the papers were different each other as well as the timing of drawing blood after exercise and the dietary conditions were not the same among the experiments. So we detected the effects of physical exercise on magnesium (Mg) metabolism under several conditions.

2. Sweat Magnesium

Sweat minerals during relatively heavy exercise by bicycle ergometer (1.5kp, 50rpm, 66min, twice a day) were measured under low or moderate intake of salt, 6g/d or 10g/d, respectively. At the 6g study, sweat contents of sodium (Na) were as low as 100mg/L, while those at the 10g study were much higher. On the contrary, Sweat contents of calcium (Ca) and Mg at the 6g study, increased according to the experimental days, while those at the 10g study kept low levels (Fig. 1.)[1].

These suggested that diets poor in salt lead to a inevitable loss of Ca and Mg through the skin. Bone minerals (Ca, Na and Mg) may be absorbed at the same time when Na is deficit in the body, to utilize sodium in the bone.

3. Urine and plasma Magnesium

Urine Ca and Mg during relatively light exercise by bicycle ergometer (50rpm; intensity was controlled as to keep heart rate 120bpm, 20min) decreased during and after the exercise, accompanied with a temporal decrease in serum Mg [2]. These suggested that magnesium in the blood stream during light exercise entered into the cell, and permitted their blood level decrease[3,4]. On the contrary, 24h urine Ca and Mg as well as phosphorus during heavy training in young male athletes, gradually increased day by

T. Theophanides and J. Anastassopoulou (eds.), Magnesium: Current Status and New Developments, 109-113.

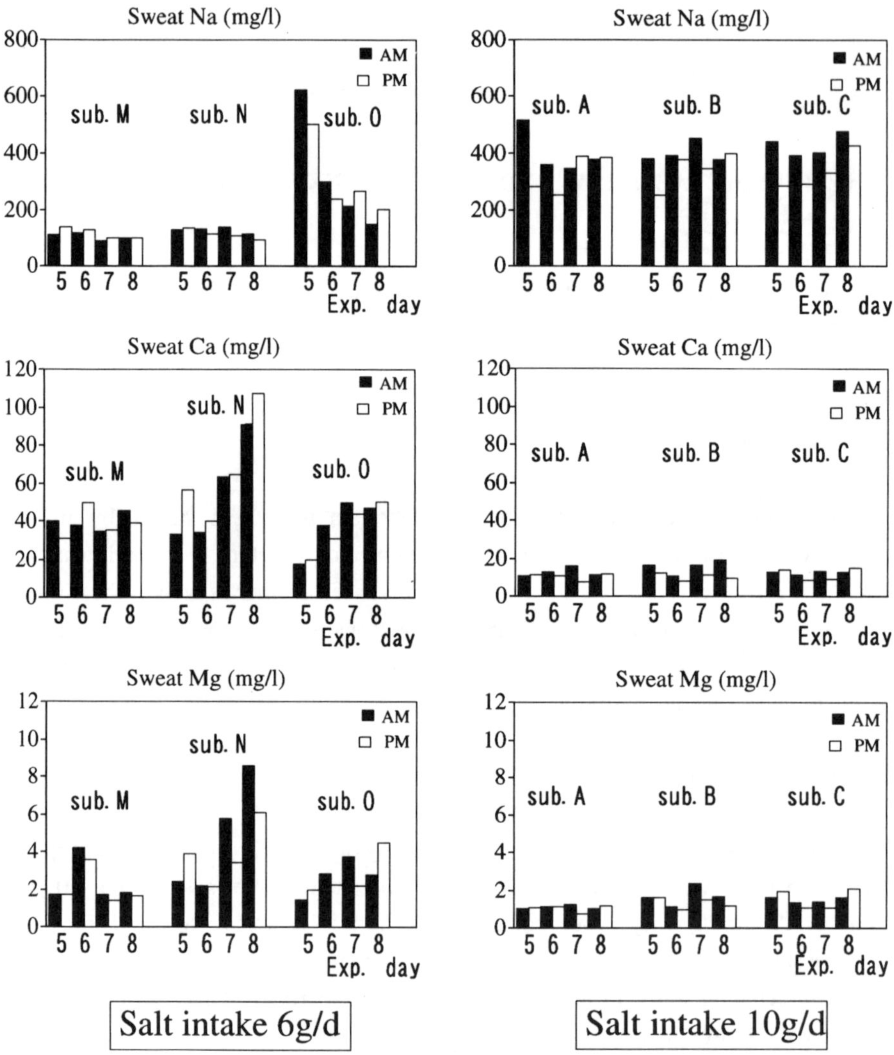

Fig. 1. Sodium(Na), calcium(Ca) and magnesium(Mg) contents of arm sweat during relatively hard exercise by bicycle ergometer (1.5kp, 50rpm, 66min twice a day) under salt intake of 6g/d (left) or 10g/d (right) in young female students. Sweat contents of Na at the 6 g/d experiment were lower, while those of Ca and Mg were higher, compared with those at the 10g/d.

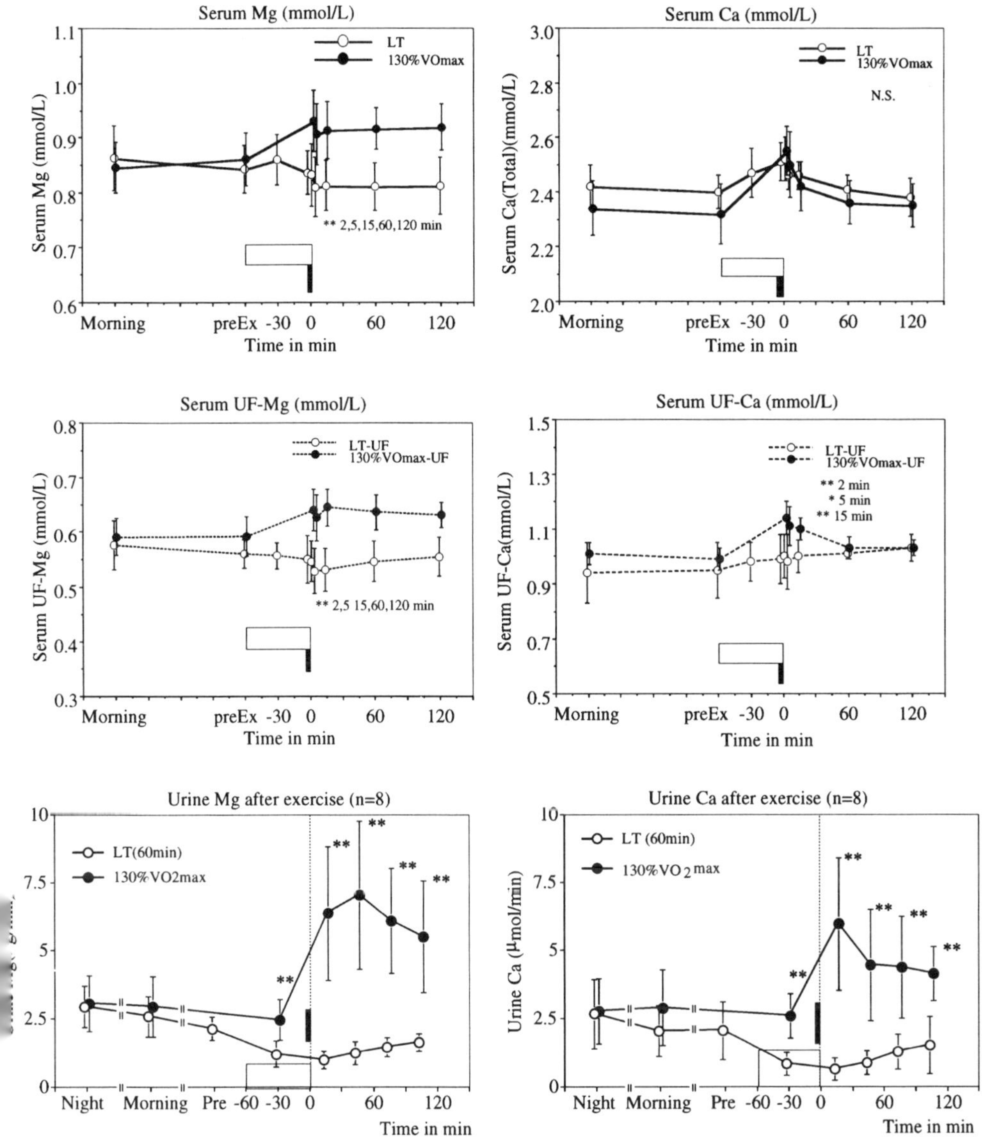

Fig. 2. Serum, ultrafiltrated-serum and urine magnesium (Mg) and calcium (Ca), during and after light (aerobic) or heavy (anaerobic) physical excrcise. Light exercise was 50rpm, 60min, lactic threshold (LT) level of intensity, and heavy exercise was 90 rpm, at most 3 min, 130 % of maximum oxygen uptake (130% VO2max) level of intensity (mean+s.d.). Eminent uresis of Mg and Ca associated with increased serum and ultrafiltrated Mg were observed at the anaerobic exercise.
*p<0.05, **p<0.01

day[5].

These results suggested that light exercise decreased, while heavy exercise increased in urine excretion of Ca and Mg.

The effects of physical exercise affecting to the metabolism of divalent cations seemed to depend on the intensity of exercise.

4. Aerobic and anaerobic exercise and Magnesium

So we compared these difference among exercise using aerobic and anaerobic exercise. Aerobic exercise we used were that by bicycle ergometer (50rpm, 60min; intensity was as hard as those of lactic threshold (LT) for the subjects). Anaerobic exercise were those with the intensity of 130% of maximum oxygen consumption(130% VO_{2max}), 90rpm, until beginning of lowering performance or at most 3 min. LT and VO_{2max} for the subjects were determined before the experiment. Eight male students took part in this study (Fig.2.)[6,7]. It is clearly demonstrated that, the anaerobic exercise increases in serum level of Mg as well as in urine excretion of Mg and Ca, without consumption of much energy, if compared with aerobic one. This may suggested that the difference of exercise on the metabolism of divalent cations was mainly due not to the quantity, but to the quality of exercise, that is, aerobic or anaerobic.

5. Conclusion

Magnesium metabolism is affected by physical activity or exercise in a different manner when its intensity is different. In the point of tissue respiration, Mg influx into or efflux from the cell may depend, at least in a part, on the style of energy consumption, aerobic or anaerobic.

To understand the effects of exercise on Mg metabolism, further studies are needed to focus on the blood level of pH, pO2 and pCO2 during exercise.

Further, because of the increase after heavy exercise, the evaluation of 'ionized Mg' or ultrafiltrated Mg for the health has not be established in this study and is still in a dense fog.

References

1. Nishimuta, M., Kodama, N., Ono, K., Kobayashi, S., and Suzuki, K.(1985) Mineral contents in arm sweat at a low mineral diet with special reference to the onset of physical exercise. JJSMgR:4;13-21.
2. Nishimuta, M., Yamada, T., Kobayashi, S. and Suzuki, K.(1983) Post exercise depression of serum magnesium in man. JJSMgR:2;49-56.
3. Nishimuta, M., Kodama, N., Hitachi, Y., Ozawa, M., SAhmed, S.M. and Oomori,T.(1991) A mineral

balance study in male long distance runner. JJSMgR :10;243-253

4. Nishimuta, M., Kodama, N., Takeyama, H. and Suzuki, K.(1996) Uresis of calcium and magnesium after diets followed by physical exercise -increased intestinal absorption after physical exercise-. JJSMgR :15;133-139.
5. Nishimuta, M. and Kobayashi, S. (1988) Mineral requirements for athletes estimated by the magnesium levels in diet, supplement, urine, serum and ultrfiltrated serum. Sports Science Report VIII Sports Science Committee. JAAA. pp. 46-47.
6. Takeyama, H., Fuchi, T., Kodama, N., Yoshitake, Y. and Nishimuta, M.(In submission)
7. Nishimuta, M., Fuchi, T., Takeyama, H., Kodama, N. and Yoshitake, Y. (In submission)

IN VITRO EFFECT OF GLUCOSE ON INTRALYMPHOCYTE FREE MAGNESIUM

P. DELVA , M. DEGAN, C. PASTORI AND A. LECHI
Department of Internal Medicine, University of Verona, Verona, Italy

1. INTRODUCTION

It has been known from some time that diabetes mellitus and the plurimetabolic syndrome are characterized by a decrease of intracellular magnesium [1,2]. Despite the in vivo important role of magnesium on glucose metabolism, very little is known on the mechanisms linking glucose and magnesium intracellular homeostasis. It has been reported that in vitro incubation of non-nucleated cells (erythrocytes) in the presence of glucose was followed by a decrease of ionized magnesium as measured by a NMR technique [3]. The recent synthesis of a new fluorescent magnesium-sensitive dye has allowed the measurement of intracellular free magnesium in nucleated cells. We therefore measured lymphocythe free magnesium in nucleated cells, i.e. human lymphocytes, incubated in the presence or in the absence of glucose.

2. METHODS

Human lymphocytes were isolated by sedimantation on Ficoll , obtaining cell preparations with a percentage of lymphocytes higher than 97 % (Trypan blue). Three separate aliquots of lymphocytes (6.10^6 each) were treated with furaptra (10 μmol/l) to measure free intracellular magnesium as previously described [2]. Fluorescence emission at 510 nm was measured with alternate excitation at 335 and 370 nm within a thermostatically controlled cuvette holder (31 °C) in a Hitachi F-2000 fluorescence spectrophotometer. The intracellular cation concentration considered was the mean of the concentrations of three separate experiments. The transmembrane leak of fluorescence probe was reduced by limiting the temperature to 31 °C . Interference with the reading by residual probe leakage was eliminated by addition of EGTA/EDTA. Calibration was performed after cell lysis with Triton-X. Intracellular free calcium (Ca_i) was assassed by means of analogue procedure based on the dye fura-2 [2]. The experiments were conducted in the absence or in the presence of D- or L-glucose 10 mm/l. Intralymphocyte ATP concentration was assessed by means

T. Theophanides and J. Anastassopoulou (eds.), Magnesium: Current Status and New Developments, 115-11'

of a bioluminescence technique [4,5] after cell lysis by freezing. Data were statistically analyzed using paired t-test.

3. RESULTS

Incubation of lymphocytes with 10 mM glucose is followed by a statistically significant decrease in Mg_i (M ± SD without glucose, n=4: 316 ± 45 μmol/l; with glucose, n=4: 245 ± 31 μmol/l ; P=0.003). The effect of glucose is specific as the Mg_i variation could be detected only by incubating the cells with 10 mM D-glucose no effects being detected with L-glucose (M ± SD, D-glucose, n=3: 280 ± 42 μmol/l ; L-glucose, n=3: 366 ± 27 μmol/l ; P=0.009). Glucose incubation of lymphocytes did not seem to modify Ca_i (M ± SD; without glucose, n=3: 62 ± 13 nmol/l ; with glucose, n=3: 57 ± 4 nmol/l). Glucose deprivation did not induce an intracellular ATP (ATP_i) variation (ATP_i, M±SD: without glucose, n=3: 21 ± 4 ng/10^6 cells; with glucose, N=3, : 21 ± 3 ng/10^6 cells). By incubating the cells in a Na-free medium we found a significant decrease of Mg_i in the presence of glucose as compared to cells without it (M±SD, without glucose, n=3: 608 ± 45μmol/l; with glucose, n=3: 487 ± 40 μmol/l; P<0.01). The furaptra emission spectra obtained from lymphocytes in the presence and the absence of glucose are identical.

4. DISCUSSION

Resnick et al. described a significant reduction of intraerythrocyte magnesium *in vitro* by incubating erythrocytes in a glucose-rich medium. The present data confirm this previous finding extending the observation to metabolic-complete cells, i.e. lymphocytes. Furthermore, these observations confirms previous *in vivo* data describing alterations of intraerythrocyte magnesium concentration following an oral glucose load.

In our study we can rule out the hypothesis that the decrease of Mg_i induced by glucose may be due to a substantial decrease of Ca_i or a glucose effect on cellular enzymes capable of de-esterifying the permeable form of furaptra. In fact, Ca_i is not modified by incubation of lymphocytes in glucose medium. Furthermore, the emission spectra of intracellular furaptra are the same either in the absence and in the presence of glucose. The glucose effect on Mg_i seems specific as no effect on Mg_i could be induced by the stereoisomer L-glucose. If considering the experimental conditions used in the present study, in the absence of insulin, it may be hypothized that glucose itself may act on intracellular magnesium homeostasis. The observation that intralymphocyte ATP concentration did not vary in our experimental conditions whether or not glucose is present in the incubation media, suggests that the glucose-induced Mg_i decrease is independent from ATP as suggested previously [3]. We have tested the hypothesis that the magnesium decrease induced by glucose may be due to the activity of the Na-Mg exchanger. The latter is the only magnesium membrane transport system known so far and catalizes the exchange of extracellular Na (Na_o) for Mg_i. Thus, in experiments conducted in the absence of Na_o, the glucose-induced Mg_i decrease is still present suggesting that the action of this transport is not crucial. In conclusion, from this preliminar set of experiments, we confirm in nucleated cells, that glucose may exert a direct effect on intracellular magnesium metabolism that is independent from intracellular ATP variations. Regardless of the cellular mechanisms involved, these data may contribute to explain some aspects of Mg_i homeostasis in hyperglycemic states such as non-insulin-dependent diabetes-mellitus.

References

1. Mather HM, Nishet JA, Burton GH, Poston G, Bland JM, Bailey PA, Pilkington TR. (1979) Hypomagnesemia in diabetes. *Clin. Chim. Acta.*; **95**:235-242.
2. Delva P, Pastori C, Degan M, Montesi G, Lechi A. (1996) Intralymphocyte free magnesium in a group of patients with essential hypertension. *Hypertension*; **28**:433-439.
3. Resnick LM, Barbagallo M, Gupta RK, Laragh JH. (1993) Ionic basis of hypertension in diabetes mellitus. role of hyperglicemia. *Am J Hypertens*, **6**:413-417.
4. Soderling E, Le Bell J, Laikko I, Larmas M. (1981) Conditions for assay of ATPase from biological materials using a bioluminescence technique. *Clinica Chimica Acta* , , **111**:33-37.
5. Resnick LM, Gupta RK, Gruenspan HL, Laragh JH. (1988) Intracellular free magnesium in hypertension: relation to peripheral insulin resistance. *J. Hypertension*; **6** (suppl. 4):S199-S201.

SERUM MAGNESIUM AND LIPID PROFILE IN CHILDREN WITH CONTROLLED INSULIN DEPENDED DIABETES MELLITUS

CH. GHAWAR, L. ALOUANE, CH. BEN RAYANA, K. NAGATI
Ecole Superieure des Siences et Techniques de la Sante. Institut National de Nutrition de Tunis

1. Introduction

Magnesium deficiency has been frequently described in both adults and children with diabets[1]. There have been contradictory findings regarding whether there is any relationship between the degree of diabetic control and the changes in magnesium content[2]. The role of Mg depletion in lipid peroxydation has recently been demonstrated[3] and hypomagnesmea is estimated a risk factor in the occurrence of cardiovascular lesions[4].

The purpose of our study was to investigate serum Mg concentration and the lipid profile in children with insulin depended diabets mellitus (IDDM) and relate it to the diabetes metabolic control as assessed by fructosamine levels.

2. Subjects and Methods

Fifty eight children with IDDM (25 males and 33 females) with age ranging from 8 to 16 years, served as the study group and were compered to fifty eight healthy children matched for age and sex. The diabetics were selected among children in a holiday camp sponsored by the Tunisian Diabetic Association, and the control group from a Tunisian SOS village.

The study was done during the third week of the diabetes surveillance. For the age matched control group, biochemical test, chest X - ray and clinical check up were performed and no renal, pulmonary or metabolic affection were observed. Diabetic control was assessed on the basis of fasting glycaemia and fructosamine. Serum magnesium, cholesterol, trigliceride, Apo A_1, Apo B_{100} were assessed for both patients and healthy controls. Duration of the disease, age and Body Mass Index were recorded.

Fructosamine was estimated by colorimetric method based on Tetrozolium Nitroblue (Boerhinger/Germany). Apo A_1 , Apo B_{100} were estimated by immunoturbidimetry and serum Mg by atomic absorption spectrophotometry.
The Student's t - test and regression analysis were used to assess statistical significancy.

T. Theophanides and J. Anastassopoulou (eds.), Magnesium: Current Status and New Developments, 119-121.

3. Results

In our diabetic group only 6 children were having a poor control as defined by a fructosamine level superior to the upper limit of 270 µmol/l. These children were not included in this study.

Table 1 gives anthropometric parameters and laboratory data on the 52 diabetic children with good control (fructosamine < 270 µmol/l) and 52 non diabetic children who agreed to cooperate in the study. Results were expressed as mean ± SD.

TABLE 1: Anthropometric parameters and laboratory data in children with iddm and controls

	IDDM group (n = 52)	Controls group (n = 52)
Age (years)	11.4 ± 2.3 (8 - 16)	11.5 ± 2.2 (8 - 16)
Duration (years)	3.0 ± 2.8(4 months - 13 yr.)	
BMI Kg/m^2	17.4 ± 2.2(12.6 - 23.4)	16.7 ± 2.0(13.5 - 21.9)
Magnesium (mmol/l)	0.97 ± 0.23(0.57 - 1.28)	0.89 ± 0.15(0.62 - 1.23)
Cholesterol (mmol/l)	3.90 ± 0.60[1](2.64 - 5.01)	3.54 ± 0.75(2.11 - 5.87)
Triglycerides (mmol/l)	0.79 ± 0.24[2](0.42 - 1.31)	0.69 ± 0.24(0.40 - 1.49)
Apo A_1 (g/l)	1.68 ± 0.28[3](1.15 - 2.33)	1.18 ± 0.22(0.68 - 1.63)
Apo B_{100} (g/l)	0.95 ± 0.13[3](0.72 - 1.36)	0.61 ± 0.16(0.40 - 0.87)
Apo A / Apo B	1.81 ± 0.40[3](1.20 - 2.70)	2.14 ± 0.69(0.81 - 3.88)
Fructosamine (µmol/l)	182 ± 30(111 - 260)	
Glycaemia (mmol/l)	7.74 ± 2.28(2.40 - 12.8)	

[1]p<0.007; [2]p<0.03; [3]p<0.001

Fructosamine was significantly related to glycaemia in our diabetics (r = 0.361; p<0.008). We found no significant difference in serum magnesium concentrations between non diabetic controls and diabetic subjects.
Hypercholesterolemia in diabetics is well known[5]. A significant increase was observed between patients and healthy controls as concerns serum cholesterol concentration (p<0.007).

A significant increase in diabetic Apo A_1 and Apo B_{100} versus control subjects (p<0.001) was observed. A corresponding increase in serum triglyceride was also found (p<0.03). Serum magnesium did not correlate with age, BMI, fructosamine, blood glucose and diabetes duration. There was no difference in serum Mg for both sexes.

4. Discussion

Optimal glycaemic control is a major aim in the treatment of diabetes mellitus and fructosamine is recognised as a reliable index of short term glycaemic control[6].

By the end of the third week of diabetes surveillance, 52 children with IDDM were having a good control as assessed by fructosamine levels.In our study fructosamine was significantly correlated to glycaemia. However, a lack of association between serum Mg and fructosamine was observed.

Insulin - depended diabetes mellitus is often associated with hypomagneseamia[7]. Mg deficiency produced hyperlipaemia characterised by increase plasma levels of triglycerides[8]. From our findings, despite the good control with normalised serum Mg levels, a significant increase in serum triglyceride ($p<0.03$) and total cholesterol ($p<0.007$) was observed in diabetic versus healthy children.

The persistent elevation in lipid fractions indicates that short term improvement in glycaemic control did not seem to ameliorate the lipidic status in children with IDDM. These alterations may be the result of several mechanisms. Apo A_1 /Apo B_{100} ratio in diabetic children was significantly decreased ($p<0.001$), when compered to healthy controls. This fact suggests that lipid metabolism alterations appear early in children with IDDM. It is therefore necessary to have a contineous diabetes control, involving adjusted diet with a sufficient Mg supply, adequat physical activity and medical follow up to prevent magnesium depletion and atherogenesis.

References

1. Rohn RD., Pleban P., Jenkins LL. (1993): Magnesium, zinc and copper in plasma and blood cellular components in children with IDDM. Clin. Chim. Acta. **215**; 21 - 28.
2. Djurhuus S. (1994): Magnesium status in patients with insulin - depended diabetes mellitus. Magnes. Res. 7, suppl. 1, 22 - 23.
3. Durlach J., Durlach V., Bac P., Rayssiguier Y., Bara M. & Guiet - Bara A. (1993): Magnesium and aging. II. Clinical data: aetiological mechanisms and physiopathological consequences of magnesium deficit in the elderly. Magnes. Res. **6**, 379 - 394.
4. Singh R.B., Niaz M.A., Ghosh S. & Rastogi V. (1996): Epidemiological study of magnesium status and risk of coronary artery disease in elderly rural and urban populations of North India. *Magnes. Res.* **9**, 165 - 172.
5. Srivastava V.K., Chauhan A. K. & Lahiri V.L. (1993): The significance of serum magnesium in diabetes mellitus. *Ind. J. Med. Sciences.* **47**, 119 - 123.
6. Lederer J. (1992): Le regime du diabetique. *Manuel de dietetique. Ed. Maloine.* **VIII**, 71 -88.
7. Kobbah A.M., Hellsing K., Tuvemo T. (1988): Early changes of some serum proteins and metals in diabetic children. *Acta. Paediatr. Scand.* **77**, 734 - 740.
8. Corica F., Allegra A., Di Benedetto A., Giacobbe M.S., Romano G., Cucinotta D., Buemi M. & Ceruso D. (1994): Effects of oral magnesium supplementation on plasma lipid concentrations in patients with non - insulin - depended diabetes mellitus. *Magnes. Res.* **7**, 43 - 47.

MAGNESIUM IN THERMOREGULATION

STENDIG-LINDBERG G., LRCPI. FRSM.[1], MORAN D[2], SHAPIRO Y.[1,2]
[1]*Department of Physiology and Pharmacology, Sackler Faculty of Medicine, Tel-Aviv University,* [2]*Heller Institute of Medical Research, Sheba Medical Center, Tel Hashomer*

1. ABSTRACT

Nine apparently healthy male subjects, aged 19-25, free from disease, medication or addiction, who gave informed consent underwent 10 days acclimatization process in a heat chamber. They were exposed for 2 h. daily to a temperature of 40°C and 40% relative humidity during low work load. Serum, mononuclear and erythrocyte magnesium (S-Mg, M-Mg and E-Mg respectively) were monitored on day 1, 5 and 10 before and after the heat exposure. S-Mg decreased while M-Mg increased during heat exposures suggesting a shift of S-Mg to mononuclear cells. At the end of the acclimatization process S-Mg was significantly decreased ($p<0.003$). E-Mg increased up to day 5 and decreased gradually approximating the baseline by day 10. Mg appears to play a significant role in heat acclimatization.

2. INTRODUCTION

Body temperature may be regulated by Mg in two ways. One is through its central sedative effect on the hypothalamus and the second through its peripheral effect achieved by reducing the neuromuscular excitibility [1]. Mg is lowered during hyperthermia due to its loss via sweat and Mg diuresis [2,3]. Mg deficiency may constitute an important factor in the pathogenesis of malignant hyperthermia e.g. heat stroke [3]. This made us particularly interested in the possible role of Mg in the etiology of exertional heat stroke [4,5].

We studied 8 unacclimated apparently healthy young males exposed, during 2h, on two consecutive days, in a heat chamber to a temperature of 40°C, and 50% relative humidity, while stepping up and down a 30 cm high bench at the rate of 12 steps/min [6] and found a 12% decrease of S-Mg after the first exposure and a significant ($p<0.01$) decrease of 19% after the second.

We have established that Mg deficiency occurs as a sequel of strenous effort [7-9]. Our findings were confirmed by Deuster et al.[10] and reviewed by Rayssiguier et al., [11], Stendig-Lindberg [12] and others. During our studies we observed that the effect of

T. Theophanides and J. Anastassopoulou (eds.), Magnesium: Current Status and New Developments, 123-126.

strenous effort on Mg was enhanced when the probands were exposed to high ambient temperatures.

In the present experiment we explored the intracellular as well as the extracellular Mg during the heat acclimatization process in order to observe the dynamics of Mg changes during heat exposure.

3. MATERIAL AND METHODS

Nine apparently healthy male subjects, aged 19-25 free from serious disease, intercurrent infection, ongoing medication, or addiction, who gave informed consent, underwent a 10 day long acclimization procedure consisting of exposure to hot climate (40^{0}C, 40% relative humidity) in a heat chamber, for 2 h daily. During the exposure the probands walked on a treadmill elevated by 3% at a speed of 5 km/h. Their extrapolated daily mean food Mg content was about 400 mg.

Estimations of serum magnesium concentration (S-Mg), mononuclear magnesium content (M-Mg), and erythrocyte magnesium content (E-Mg), were made before and after the heat exposures on days 1,5 and 10, using methods previously described [13-15].

4. RESULTS

S-Mg - The baseline mean S-Mg was at the lower border of the national reference interval (estimated by substracting only 1 SD from the mean, which is much closer to the reference range found at steady state, than the spurious value obtained after the conventional substraction of 2 SD [15]. The baseline mean S-Mg decreased significantly at the end of the acclimatization procedure (p=0.014, Student's dependent t-test). The fluctuations of S-Mg, i.e. a decrease after a heat exposure with an attempt at a rebound on its termination, were statistically significant on days 5 and 10 (P=0.005 and P=0.046, respectively) and inverse to those of M-Mg. The analysis of variance with repeated masures showed a statistically significant difference between S-Mg values before and after the heat exposures (P=0.019).

M-Mg - The baseline mean M-Mg was significantly below the national mean found in apparently healthy Israeli population (x 164.8, SD 28.3, n=20, reference interval 108.2 - 221.4, % Cl 152.4 - 177.2 fg/cell,p=0.000, Students independent t-test [15]. M-Mg increased during a heat exposure and fell on its termination. Its fluctuations were inverse to those of S-Mg. The baseline value did not differ significantly from the end value, but on analysis of variance with repeated measures a trend towards lower border line significance (p=0.102) was observed.

E-Mg - E-Mg showed a bell shaped curve which peaked after the heat exposure on day 5, followed by a steady decrease which became statistically significant on day 10 (p=0.046, Student's t-test). There was no statistically significant difference between the baseline E-Mg and the end value (NS, Student's dependent t-test).

5. DISCUSSION

The initial mean S-Mg and M-Mg were significantly lower than those reported earlier in apparently healthy Israeli population samples [15,16]. Since the probands were carefully screened to exclude possible presence of concomitant disease, medications or addiction, which may lower Mg, the decrease of the mean S-Mg and M-Mg confirms earlier findings of a relatively high prevalence of Mg deficiency in Israeli apparently healthy population [16] and points to a recent trend to an increased prevalance of Mg deficiency. There was a significant decrease of S-Mg and a trend towards lower border line significant decrease of M-Mg after termination of the experiment which indicates the presence of Mg deficiency. The rise of the M-Mg content occured simultaneously with a decrease of S-Mg after heat exposures indicating a shift of Mg from the extracellular space to the mononuclear cells. The E-Mg content increased up to the fifth day of exposure and thereafter decreased, coming down to lie on the tenth day close to the baseline value. Our findings indicate that Mg plays a significant role in the heat acclimatization process.

The decrease of S-Mg and M-Mg at the end of 10 brief exposures shows a cumulative effect; the longer the duration of heat exposure, the more severe the Mg deficit.

We hypothesize that the energy depletion which characterizes heat stroke [17-19] may be due to the presence of profound Mg deficiency. Since Mg plays a pivotal role in energy metabolism, Mg deficiency should be looked for in conditions of failure of thermoregulation, especially in exertional heat stroke. The latter would appear to constitute the result of the combined effects of 1) the Mg deficiency of strenous effort 2) of the Mg deficiency resulting from its loss via sweat and Mg diuresis in response to high ambient temperature 3) the increased requirement for Mg caused by the increased energy metabolism of hyperthermia.

Acknowledgement We wish to thank Professor WEC Wacker of Harvard University Medical School for his encouragement and support.

References

1. Heagy, F.C. and Burton, A.C. (1948) Effect of intravenous injection of magnesium chloride on the body temperature of the unanaesthesized dog, with some observations on magnesium levels and body temperature in man, *Am J Physiol* **152**, 407-416.
2. Joy, R.Y.T., Beisel, W.R. and Goldman, R.F. (1964) Metabolic responses to artificial hyperthermia in man, *The Physiologist* 7, 170.
3. Durlach, J. (1988) Magnesium in clinical practice John Libbey London, Paris, pp.143-145.
4. Epstein, Y., Sohar, E. and Shapiro, Y. (1995) Exertional heat stroke. A preventable condition, *Isr J Med Sci* **31**, 454-462.
5. Shapiro, Y. and Seidman, D.S. (1990) Field and clinical observations of exertional heat stroke patients, *Med Sci Sports Exerc* **22**, 6-14.
6. Epstein, Y., Stendig-Lindberg, G., Shapiro, Y., Rosenblum, J. and Galun, E. (1983) The effect of exercise under severe heat load on serum magnesium. Proceedings of the First Joint Meeting of the Israeli Societies for Life Sciences, Jerusalem, A155.
7. Stendig-Lindberg, G., Shapiro, Y., Epstein, Y., Galun, E., Schonberger, E., Graff, E., Wacker, W.E.C. (1987) Changes in serum magnesium concentration after strenuous exercise, *J Am Coll Nutr* **6**, 35-40.
8. Stendig-Lindberg, G., Shapiro, Y., Graff, E., Schonberger, E., Wacker, W.E.C. (1989) Delayed metabolic changes after strenuous exertion in trained young men, *Magnes Res* **2**, 211-216.

9. Stendig-Lindberg, G., Wacker, W.E.C., Shapiro, Y. (1991) Long-term effects of peak strenuous effort on serum magnesium, lipids and blood sugar in apparently healthy young men, *Magnes Res* **4**:59-65.
10. Deuster, P.A., Dolev, E., Kyle, S.B., Anderson, R.A., Schoomaker, E.B.(1987) Magnesium homeostasis during high-intensity aerobic exercise in men, *J. Appl Physiol* **62**(2),545-550.
11. Rayssiguier, Y., Guezennec, C.Y., Durlach, J. (1990) New experimental and clinical data on the relationship between magnesium and sport, *Magnes Res* **3**, 93-102.
12. Stendig-Lindberg G. Magnesium deficiency and physical performance (1995) In Magnesium and Physical Activity (Edited by Vecchiet L,) pp. 183-198, The Parthenon Publishing Group. New York, London.
13. Elin, R.J., Hosseini, J.M. (1985) Magnesium content of mononuclear blood cells, *Clin Chem* **31**, 377-380.
14. Stendig-Lindberg, G, Penciner, J, Rudy, N, Wacker, WEC. (1984) Comparison of diluents for serum magnesium estimation by atomic absorption spectrophotometry, *Magnes* **3**, 50-56.
15. Stendig-Lindberg, G., Harsat, A., Graff, E. (1991) Mononuclear and erythrocyte magnesium in apparently healthy Israelis, *J Eur Clin Chem Clin Biochem* **29**, 833-836.
16. Stendig-Lindberg, G, Rudy, N, Penciner, J, Chayne, M, Katcharow, O. (1983) Serum magnesium pattern in apparently healthy Israeli population,*Magnes* **2**, 26-35.
17. Hales, J.R.S. (1987) Proposed mechanisms underlying heatstroke,in Hales, J.R.S. and Richards, D.A.B Ed.s, pp. 85-99, Elsevier Science Publishers, Amsterdam.
18. Hubbard, R.W. (1990) Heatstroke pathophysiology: The energy depletion model, *Med Sci Sports Exerc* **22**, 19-28.
19. Kozlowski, S., Brzezinska, Z., Kruk, B., Kaciuba-Uscilko, H., Greenleaf, J.E. and Nazar, K. (1985) Exercise hyperthermia as a factor limiting physical performance. Temperature effect on muscle metabolism, *J Appl Physiol* **59**, 766-773.

INTERRELATIONS BETWEEN MAGNESIUM AND TBARS IN DIFFERENT RAT TISSUES, AFTER ADMINISTRATION OF GINSENG

L. PEREZ GALLARDO, J. ROMERA, R. DE MIGUEL and A. SANCHEZ-MAYORAL
Area de Bioquimica y Biologia Da Molecular. Univ.of Valladolid E. U. Fisioterapia. Nicolas Rabal 17, 42003 SORIA. SPAIN.

1. INTRODUCTION

Ginseng has been shown to have antioxidant properties, ginsenoside can increase body resistance to many harmful factors and protect tissues from damage when an organism is in stress [1,2]. On the other hand, magnesium may be a factor that enhances the antioxidative mechanisms as in magnesium deficiency conditions there is a decrease of anti-oxidative defenses [3]. To elucidate whether interrelation exist between magnesium or calcium and the antioxidative effect of ginseng, a study was carried out on magnesium, calcium and lipid peroxidation levels in heart, skeletal muscle, intestine, kidney, spleen and liver after administration of two kinds of ginseng.

2. MATERIAL AND METHODS

We performed a study on 17 normal wistar male rats weighing 200-300 g with the aim of evaluating the effect of Panax ginseng or Eleutherococcus senticosus after 6 weeks=92 administration. Six rats received an 0.8 % solution of root powder from P. ginseng (group P), another group (n=3D6) received an 0.8 % solution of root powder from E. senticosus (group E), and finally the control group (group C), consisting of 5 rats received tap water. All groups received food containing 7100 mg/Kg Ca and 2100 mg/Kg Mg and drink and libitum. At the end of the 6 weeks the rats were anaesthetized and blood was extracted. Then, the heart, skeletal muscle, intestine, kidney, spleen and liver were separated. A fraction of these organs were subjected to the action of a basic digestion and the calcium and magnesium concentrations were analized by espectrophotometry of atomic absortion. Lipid peroxidation was assessed by measurements of thiobarbituric acid reactive substance (TBARS) on all tissues studied according to the Uchiyama method [4]. The results obtained from all tissues in the thre groups were analized by one-way analysis of variance and when the analysis was significant ($P<0.05$) results were compared by Student t-test.

3. RESULTS

Results of calcium, magnesium and TBARS on heart, skeletal muscle, intestine, kidney, spleen and liver are presented in Table 1. Heart, skeletal muscle and intestine TBARS in groups P and E decreased while spleen and liver TBARS increased when compared with control group. There was a significant decrease in heart, skeletal muscle and liver Mg values and heart, kidney, spleen and liver Ca concentrations in group P with respect to C group. Simultaneously, there was a significant increase in kidney and spleen Mg concentration and skeletal muscle Ca values when comparing these results between P and C groups. With the exception of kidney TBARS levels the results obtained in group E follow the same tendency as were observed in group P. With respect to the interrelations between Mg and TBARS, these results show a significant positive

T. Theophanides and J. Anastassopoulou (eds.), Magnesium: Current Status and New Developments, 127-128.

correlation in heart and intestine and a non significant negative correlation (r =3D - 0.5110) only in liver.

TABLE 1. Effect of Panax ginseng and Eleutherococcus senticosus on magnesium, calcium and TBARS levels in different rat tissues. Values means =B1 SD. *P< 0.05, **P< 0.01 and P< 0.001 vs control group.

	Mg (mg/g)	**Ca (mg/g)**	**TBARS(nmol/l)**
heart	7.76±0.05	1.03±0.02	26.86±0.76
Control (n=5)	6.44±0.04***6.	0.62±0.02***0.	11.85±0.93***
P. ginseng (n=6)	67±0.12***	66±0.03***	8.20±0.68***
E.senticosus n=6)			
skeletal muscle	8.18±0.16	1,28±0.03	68.51±3.10
Control (n=5)	6.70 ±0.09***	1.94±0.04***	27.35±2.14***
P.ginseng (n=6)	8.09±0.17	2.06±0.17***	29.57±1.88***
E. senticosus(n=6)			
intestine	6.12±0.07	1.58±0.03	16.43±1.18
Control (n=5)	5.90±0.12	1.49±0.08	9.99± 0.29***
P.ginseng (n=6)	5.67±0.21	1.10±0.03***	7.84±0.50***
E.senticosus(n=6)			
kidney	6.60±0.09	1.18±0.06	37.03±1.36
Control (n=5)	7.50±0.08***	1.03±0.04**	35.29±2.76**
P. ginseng (n=6)	6.72±0.11	0.96±0.02**	45.88±1.32**
E. senticosus(n=6)			
spleen	6,73±04=	1.11±0.02	2.80±0.08
Control (n=5)	7.35±0.10***	0.90±0.03**	7.47±0.30***
P. ginseng (n=6)	7.10±0.08***	1.15±0.06	11.17±0.30***
E. senticosus(n=6)			
liver	6.47±16	0.86±0.03	1.56±0.07
Control (n=5)	4.77±0.24***	0.56±0.02***	2.28 ±0.12**
P. ginseng (n=6)	5.67±0.21***	0.84±0.02	2.89±0.35**
E. senticosus(n=6)			

References

1. Kim, H., Lee, Y.H, Kim, S.I and Jin, S.H. (1988) Effect of Polyacetylene Compoundsfrom Korean Ginseng on Lipid Peroxidation. *5th S.* pp. 81-86.
2. Deng Hui-ling and Zhang Jun-Tian. (1991), Anti-lipid peroxidative effect of ginsenoside RB1 RG1. Chinese *Medical J.* 104 395-398.
3. Freedman A, Cassidy M and Weglicki W. (1991),: Magnesium-deficient myocardium demonstrates an increased susceptibility to an in vivo oxidative stress.*Magnesium Res.* **4**, 185-189.
4. Uchiyama M and Mihara M. (1978), Determination of malonaldehyde precursor in tissues by thiobarbituric acid test. *Analytical Biochemistry* **66**, 271-278.

CHOLESTEROL METABOLISM IN HUMAN UMPILICAL ARTERIAL ENDOTHELIAL CELLS CULTURED IN LOW MAGNESIUM MEDIA

FRED A. KUMMEROW,
Burnsides Research Laboratory, University of Illinois, 1208 W. Pennsylvania Ave., Urbane, IL 61801 USA

In our previous studies we found that abnormal lipid metabolism in magnesium-deficient rabbits is reflected by hypercholesterolemia, increased phospholipid concentration and elevated cholesterol oxides. In *in vitro* studies we observed that incorporation of [^{14}C] acetic acid into phospholipid was inhibited, its incorporation into triglyceride was stimulated, but its incorporation into cholesterol was not affected by magnesium deficiency in cultured human umbilical arterial endothelial cells.

The cells grown in medium containing 94 μM of magnesium decreased their cellular uptake of free cholesterol and the reduced uptake was significant from 6 h to 24 h of the treatment. When magnesium concentrations in medium were enhanced from 94 μM to 188 μM or more, no affected uptake was observed until 48 h of the exposure. Cholesterol synthesis in control groups was increased progressively with time during 48 h of the experimental period. The incorporation of the labeled mevalonolactone into free cholesterol in the cells exposed to the medium containing 94 μM of magnesium was obviously lower than in control cells after 48 h of the treatment. When the effect of different magnesium concentrations in the medium was investigated, magnesium levels at both 94 μM and 188 μM in the medium had an inhibiting role in the incorporation of mevalonolactone into free cholesterol.

A significantly enhanced incorporation of labeled precursors, [^{3}H]mevalonolactone or [^{3}H]oleate, into cholesteryl esters was observed in the cells treated with low magnesium at 12 h and 24 h of the exposure. However, at 48 h of the treatment, no obvious alternations of cholesterol esterification were seen in the cells exposed to magnesium at any low levels. When [^{3}H]cholesterol was used as a precursor, a significant stimulation of cholesterol esterification was only detected in the cells treated with low magnesium for 24 h, compared with ester formation in the control group. This study indicated that cholesterol esterification was stimulated in the endothelial cells exposed to low magnesium media.

It is known that the glutathione redox cycle and catalase are important antioxidant defense systems. The oxidation of reduced glutathione (GSH) by hydrogen peroxide

T. Theophanides and J. Anastassopoulou (eds.), Magnesium: Current Status and New Developments, 129-130.

(H_2O_2) is catalyzed by glutathione peroxidase (GSH-Px). The oxidized glutathione is then reconverted to reduced glutathione by glutathione reductase (GSHRed). In experiments performed with cultured arterial endothelial cells (ECs), it was found that the ability of the cells resistant to lysis by exogenously generated H_2O_2 was mainly from the glutathione redox cycle because disruption of the cycle resulted in the cells lysis by H_2O_2.

The intact endothelial layer of arteries is an important barrier for preventing the development of cardiovascular disease. Inflammation may play a leading role in the disease process. Chronic inflammation may arise following infections, smoking or long-term stressful conditions which also may produce an activation of phagocytic leukocytes. An increase in reactive oxygen species released by activated leukocytes is also produced by other risk factors such as diabetes, hypertension and hypercholesterolemia. Since the endothelial layer is continually exposed to free radicals, this layer is easily damaged. An increase in H_2O_2 in the culture medium could immediately kill the cultured cells. Endothelial cells exposed to a low concentration of H_2O_2 were still killed and the proliferation of surviving cells was inhibited. Moreover, H_2O_2 release from ECs was detected in several laboratories. ECs are, therefore, both targets and sources of free radicals, such as superoxide anion, H_2O_2 and nitric oxide.

The effects of magnesium deficiency on glutathione redox cycle and catalase activities, which play the main antioxidant role, were investigated in cultured human arterial endothelial cells. The concentrations of released hydrogen peroxide (H_2O_2) and intracellular glutathione, and the activities of glutathione reductase (GSHRed), glutathione peroxidase (GSH-Px) and catalase were measured after the cells were exposed to low magnesium (569, 380, 190 and 95 μM) medium, respectively, until 72 h. Magnesium at a level of 949 μM was used as a control. The results showed that under magnesium deficient condition (1) the decrease of total glutathione level was form the diminishment of reduced glutathione, (2) H_2O_2 release was not affected, (3) consumption of exogenous H_2O_2 was enhanced, (4) the activities of GSH-Px was stimulated, the activities of GSHRed was shown to be stimulated first and then inhibited, and the activities of catalase was not detected to have any significant alternation. In conclusion, glutathione redox cycle plays the main antioxidant role in the presence of magnesium deficient condition.

CHILDHOOD OBESITY, INSULIN RESISTANCE, AND MAGNESIUM METABOLISM

Akio Kobayashi, MD, FACN, Hitoshi Okabe, MD, Tomio Nozue, MD, Akiko Sawa, MD, Yasuo Takagi, MD, Fujio Uemasu, MD, Pei Hui Lan, MD

Department of Pediatrics, University of Showa/Toyosu Hospital, Toyosu 4-1-18, Koto-ku, Tokyo 135, Japan

Glucose metabolism and insulin resistance are known to be deeply associated with magnesium metabolism. Obesity is occasionally complicated both with impaired glucose tolerance and with insulin resistance. We studied magnesium metabolism in obese children with insulin sensitiveness and those with insulin resistance.

Subjects and methods

The subjects were 11 children with obesity, of which 4 were insulin-sensitive and 7 insulin-resistant. The diagnosis of obesity was based on the body weight over 20% of the ideal body weight. Insulin sensitiveness was based on 50% reduction of blood glucose or more after IV insulin (0.1U/kg), and insulin-resistance on the reduction of blood glucose less than 50% of the pre-infusion values.
Before and 15 min. after IV insulin (0.1U/kg) serum Mg and K, whole blood ionized Mg (Mg^{++}) and Ca (Ca^{++}), and blood glucose were determined.
Clinical characteristics are shown in the Table.

Table Clinical characteristics of the subjects

	Insulin-sensitive	Insulin-resistant
No.	4	7
Sex(M/F)	3/1	6/1
Obesity index(%)	53.8± 9.5	46.5±18.7*
Blood pH	7.46±0.04	7.44±0.02
Serum Mg(mg/dl)	2.30±0.41	2.23±0.23
Whole blood Mg^{++}(mmol/l)	0.53±0.01	0.54±0.04
Whole blood Ca^{++}(mmol/l)	1.28±0.01	1.29±0.03*
Serum K (mEq/l)	4.00±0.18	4.09±0.23*

* $p<0.05$

Results

Blood glucose levels were reduced after IV insulin in both insulin-sensitive and -resistant subjects and the lowest values were obtained at 15 min. after IV insulin. The values returned to the preinfusion values at 60 min. after IV insulin (Fig. 1).

T. Theophanides and J. Anastassopoulou (eds.), Magnesium: Current Status and New Developments, 131-134.

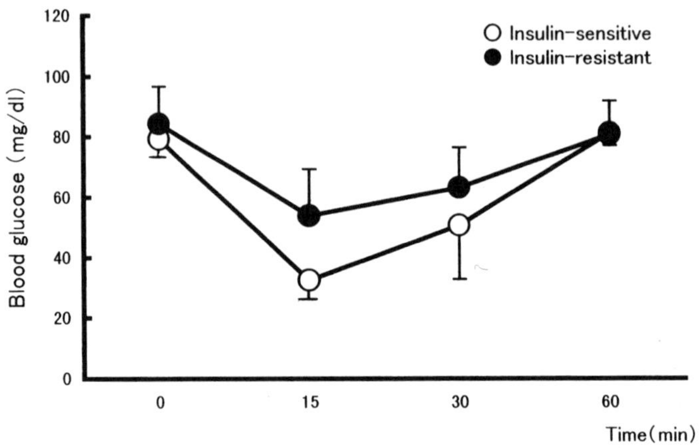

Fig. 1 Blood glucose levels after iv insulin: insulin-sensitive (○), insulin-resistant (●).

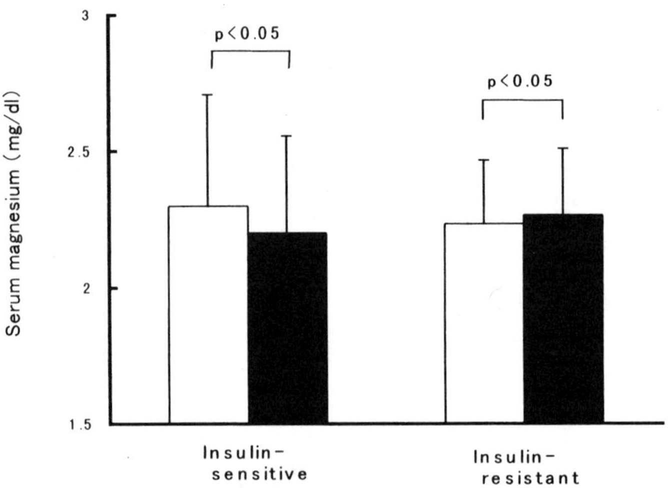

Fig. 2 Serum magnesium levels before (□) and 15 min. after iv insulin (■).

Serum Mg and whole blood Mg^{++} levels were reduced in the insulin-sensitive group (2.30 ± 0.41 to 2.20 ± 0.36 mg/dl, and 0.533±0.013 to 0.523± 0.010 mmol/l, both $p<0.05$), and raised in the insulin-resistant group (2.23 ± 0.23 to 2.27± 0.24 mg/dl, and 0.539 ± 0.041 to 0.543 ± 0.035 mmol/l, both $p<0.05$) at 15 min. after IV insulin (Fig. 2 and 3).

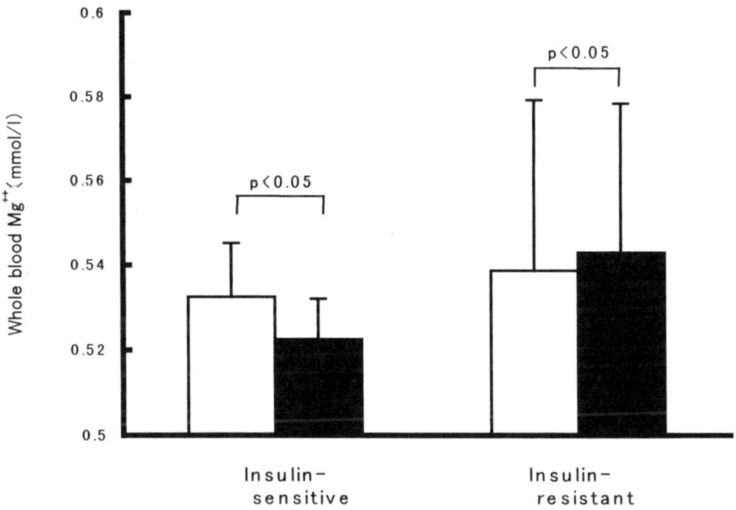

Fig. 3 Whole blood ionized Mg^{++} levels before (□) and 15 min. after iv insulin (■).

Whole blood Ca^{++} levels were raised in both insulin-sensitive (1.28± 0.01 to 1.30 ± 0.04 mmol/l, $p<0.05$) and –resistant group (1.29 ± 0.03 to 1.31 ± 0.05 mmol/l, $p<0.05$) at 15 min. after IV insulin (Fig. 4).
Serum K levels were reduced in both insulin-sensitive (4.00± 0.18 to 3.58± 0.27 mEq/l, $p<0.05$) and -resistant group (4.09± 0.23 to 3.72 ±0.29 mEq/l, $p<0.05$) (Fig. 5).

Summary and Conclusion

Serum magnesium and whole blood Mg^{++} levels were reduced after iv insulin in obese children with insulin sensitiveness and raised in those with insulin resistance. These results suggest that insulin resistance decreases a cellular uptake of magnesium in childhood obesity.

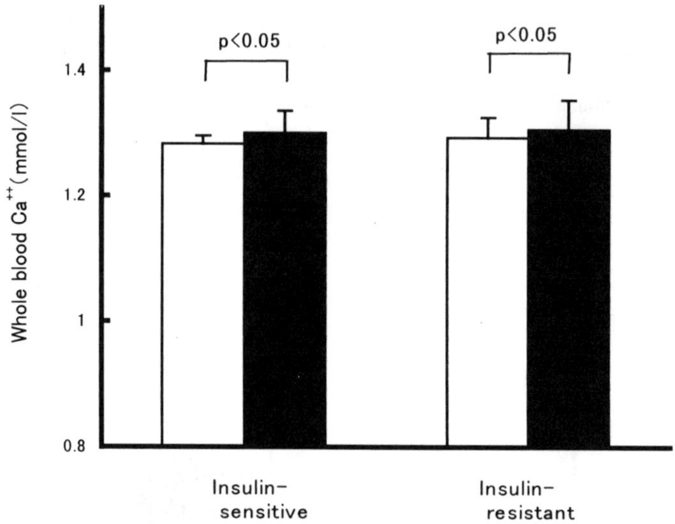

Fig 4 Whole blood Ca^{++} levels before (□) and 15 min. after iv insulin (■).

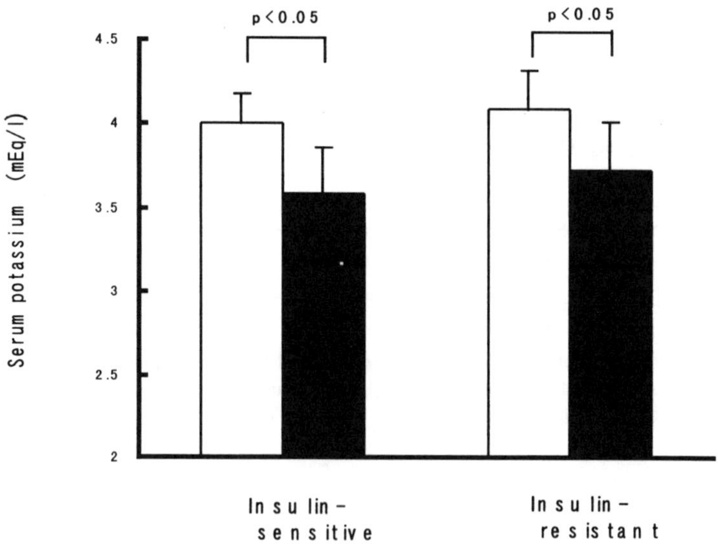

Fig. 5 Serum potassium levels before (□) and 15 min. after iv insulin (■).

EFFECT OF NUTRIENTS ON CALCIUM AND MAGNESIUM IN BONE

YOSHINORI ITOKAWA
Emeritus Professor, Kyoto University, Kyoto, Japan

1. INTRODUCTION

Approximately 99% of the calcium and 70% of the magnesium in the body are in the bones and teeth and the store of these minerals in the bones is utilized in the body. In this paper, the author intend to summarize the effect of various nutrients on levels of calcium and magnesium in bones. The calcium and magnesium levels in bones represent directly the nutritional status of these minerals in the body.

2. INTER-RELATIONSHIP BETWEEN CALCIUM AND MAGNESIUM (1-3)

Male Wistar rats were separated into two groups. To one group calcium deficient diet was given and to another group normal diet was given. After 9 weeks of these dietary regimens, all rats were sacrificed and concentrations of mineral in various tissues determined. Calcium level in bone decreased in calcium deficient rats. In addition, calcium levels in blood, brain, heart and kidney decreased in calcium deficient rats. Magnesium levels in bone decreased significantly in calcium deficient rats. In contrast, magnesium levels in blood, liver and spleen increased in calcium deficient rats (Fig 1).

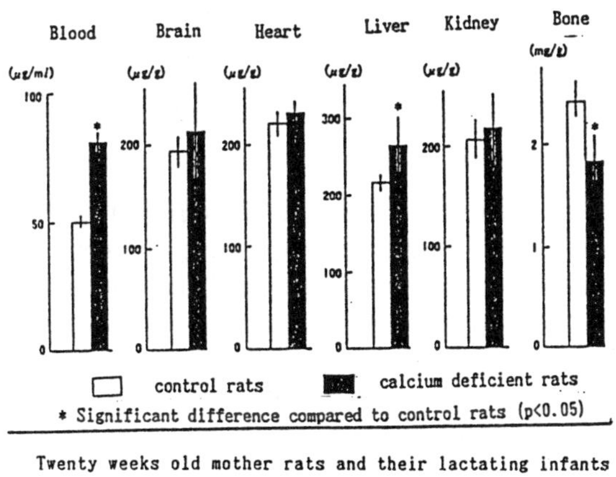

Figure 1 Magnesium Concentrations in Bone and Other Tissues

T. Theophanides and J. Anastassopoulou (eds.), Magnesium: Current Status and New Developments, 135-138.

Twenty weeks old mother rats and their lactating infants were raised in calcium deficient diets for 3 weeks. In calcium deficiency, calcium levels in bone were decreased significantly in both mother rats and lactating infants. Magnesium levels in bone were also decreased in mother rats, while calcium levels in bone in lactating infants were not decreases. Thus the effects of calcium deficiency were remarkable for the mother rats rather than lactating infants.
To male wistar rats, magnesium deficient diet was given for 4 weeks and magnesium and calcium levels in various tissues were compared to control rats. Magnesium levels in bone decreased significantly in magnesium deficient rats. In addition, magnesium levels in serum, heart and muscles were decreased in magnesium deficient rats, however, significant difference was not observed in magnesium levels of brain, liver, spleen, kidneys and testicles. Calcium levels in bone decreased significantly in magnesium deficient rats. In contrast, calcium levels increased in serum, heart, kidneys and muscles (Fig 2). Thus, it is feasible that calcium and magnesium are released from the bones to other tissues thereby promoting a homeostatic adaptation.

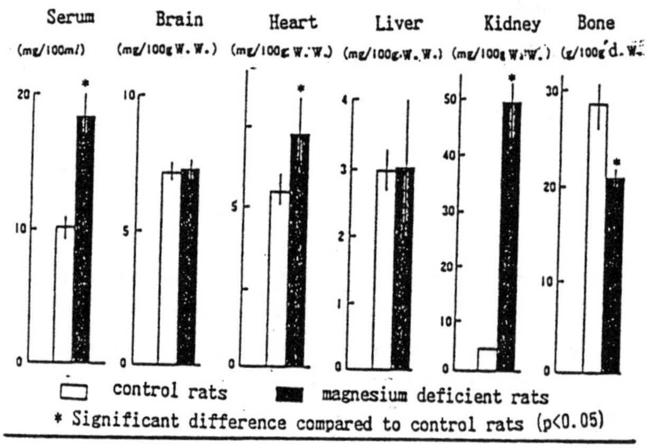

Figure 2 Calcium Concentrations in Bone and Other Tissues

3. INFLUENCE OF LOW PROTEIN AND CALCIUM DEFICIENCY ON BONE MINERALS (4)

Female Wistar rats were separated into 9 groups, and 9 different synthetic diets (each diet contains different level of protein and calcium) were given to each group. After 5 weeks of these dietary regimens, all rats were sacrificed and calcium and magnesium levels in bones and serum were determined. In calcium deficient groups, calcium and magnesium levels in bone decreased and plasma calcium levels decreased. When comparison was made among the calcium deficient groups, calcium and magnesium levels in bone were higher in low protein-calcium deficient groups than higher protein-calcium deficient groups (Fig 3). It is probable that protein deficiency inhibits calcium depletion and consequently the influence of calcium deficiency is less significant in the condition of deficiency in both protein and calcium.

4. CHANGES IN MAGNESIUM AND CALCIUM LEVELS IN BONES INN MAGNESIUM DEFICIENT RATS IN RELATION TO THIAMIN EXCESS OR DEFICIENCY (3)

To clarify the influence of thiamin deficiency or a dietary excess of thiamin on magnesium deficiency, six different synthetic diets (each diet contains different level of thiamin and magnesium) were fed to six groups of Wistar rats and determined magnesium and calcium levels in bones and various tissues after 4 weeks of these dietary regimens. In magnesium deficient groups, magnesium and calcium levels in bones were decreased. In addition, magnesium levels in serum, muscle and heart decreased and calcium levels in serum, muscles, heart and kidneys increases.

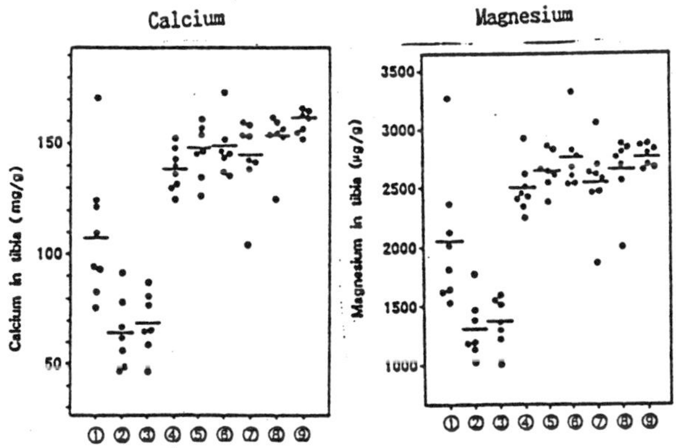

Group	Diet Protein (%)	Diet Calcium (mg/kg)	Group	Diet Protein (%)	Diet Calcium (mg/kg)	Group	Diet Protein (%)	Diet Calcium (mg/kg)
①	5	0	④	5	250	⑦	5	500
②	15	0	⑤	15	250	⑧	15	500
③	20	0	⑥	20	250	⑨	20	500

Figure 3 Calcium and magnesium Concentration in Tibia (Individual Rats)

All changes in magnesium and calcium concentrations seen in magnesium deficient animals including magnesium and calcium levels in bones were significantly alleviated when the animals were also deficient in thiamin. It is assumed that thiamin deficiency inhibits magnesium depletion and consequently the influence of magnesium deficiency is less significant in the thiamin and magnesium deficient animals. In contrast, dietary excess thiamin may accelerates the magnesium deficiency. As thiamin requires magnesium for the binding of thiamin and protein, magnesium deficiency in magnesium deficient rats may be accelerated by excess thiamin.

5. EFFECT OF DIETARY TIN DEFICIENCY OC MAGNESIUM AND CALCIUM LEVELS IN BONES (5)

To investigate the essentiality of tin for animals, two different synthetic diets were fed to male Wistar rats for 3 weeks, Group 1 fed a diet containing 1,990 ng tin/g (normal diet) and group 2 fed a diet containing 17 ng/g (tin deficient diet). The rats in group 2 showed poor growth lowered response to sound, and alopecia, with decreases food efficiency compared with rats in group 1. In the changes of mineral concentrations in tissues, calcium concentration increased and magnesium concentration decreased in lung compared with group 1. However, no difference was observed in magnesium and calcium concentrations in bone between these two groups. Although these results suggest that tin may be essential trace element for rat, magnesium metabolism may not be involved in the essentiality.

6. EFFECT OF A LOW RUBIDIUM DIET ON MAGNESIUM AND CALCIUM LEVELS IN RAT TISSUES (6)

To investigate the essentiality of rubidium for animals, two different synthetic diets were fed to male Wistar rats for 12 weeks. Group 1 fed a diet containing 8,120 ng rubidium/g (normal diet) and group 2 fed a diet containing 540 ng/g (rubidium deficient diet). Although growth retardation was not observed, rats in group 2 had a higher urea nitrogen level in plasma. Magnesium levels in bones decreased significantly in group 2 as compared to group 1, while no difference was observed on calcium levels in bones between these groups. Further studies are needed to stipulate the essentiality of rubidium for animals but it is probable that magnesium metabolism may be involved in rubidium metabolism.

References

1. Iwabuchi A., Notani C., Tsutsui Y., Kimura M., Itokawa Y. (1985): Changes in tissue mineral levels in calcium deficient rats. *J. Jap. Soc. Mg. Res.* **4**: 45-51.
2. Yagi N., Morioka M., Maruyama Y., Yamada K., Takeda K., Itokawa Y. (1986): Effect of magnesium and calcium deficiency on mineral metabolism of mother rats and their lactating infants. *J. Jap. Soc. Mg. Res.* **5**: 9-14.
3. Itokawa Y., Fujiwara M. (1973): Changes in tissue magnesium, calcium and phosphorus levels in magnesium deficient rats in relation to thiamin excess and deficiency. *J. Nutr.* **103**: 438-443.
4. Takeda T., Kimura M., Itokawa Y. (1993): Imbalance of calcium, magnesium and phosphorus in bone and other tissues of rats induced by low protein and calcium deficiency. *J. Nutr. Sci. Vitaminol.* **39:** 355-363.
5. Yokoi K., Kimura M., Itokawa Y. (1990): Effect of dietary tin deficiency on growth and mineral status in rats. *Biol. Trace Elem. Res.* **24**: 223-231.
6. Yokoi K., Kimura M., Itokawa Y. (1994): Effect of low rubidium diet on macro-mineral levels in rat tissues. *J. Jpn. Soc. Nutr. Food Sci.* **47**: 295-299.

MAGNESIUM AND CALCIUM CONTENTS IN REFORM CEREAL FOODS AND UPTAKE OF THESE ELEMENTS

ANNA CSIKKEL-SZOLNOKI[1] AND SÁNDOR A. KISS[2]
[1]Department of Inorganic and Analytical Chemistry, Attila József University, Szeged, H-6720, Dóm tér 7, Hungary, [2]Hungarian Magnesium Society, Szeged, H-6726, Fõ fasor 73 A/2, Hungary

1. ABSTRACT

There were investigated the total and uptaking (soluble) magnesium and calcium contents in different wheat, oat and barley foods.
The total element determination was performed after the digestion with cc. HNO_3 in microwave apparatus. Uptake of two elements was investigated at 37°C for 1 hour both in distilled water and in synthetical gastric juice (0.1 M HCl and pepsin).
Spectrochemical analysis by atomic absorption was used for measurement of all elements.
It was found, that total magnesium content was higher in the wheat bran (466mg/100 g) than in the wheat sprout (291 mg/100 g) and in the oat bran (233 mg/100 g). Hulled products (flakes) made from seeds contain same magnesium amounts (150-170 mg/100 g). The uptake (soluble) of magnesium is higher (3-8 fold) in HCl-pepsin mixture than in distilled water from different cereal foods, which means 50-90 per cent amount of the total magnesium.
The total calcium contents in different cereal products are in about 10 per cent amount of total magnesium contents, which result small Ca/Mg ratio (5-20 $\times 10^{-2}$). The uptake (solubility) of calcium is smaller than the uptake of magnesium, because the calcium-complexes are more stable (for example Ca-fitase) than the magnesium-complexes. This is very important for the uptake of magnesium from different foods.

Keywords: cereal, magnesium, calcium, solubility (uptake)

2. INTRODUCTION

The ionequilibrium is very important in the living cells. According to Liebig-law, the ratio of ions is constant in the organism (for example K : Mg = 6 : 1, Ca : Mg = 8 : 1). It follows from this, if one's amount of them increases or decreases, then the amount of the other ion also changes. For these reasons the uptake of different ions takes place with required ratio from foods, which is very importent for the organism.

T. Theophanides and J. Anastassopoulou (eds.), Magnesium: Current Status and New Developments, 139-142.

Compensation of ions may accomplish with drinking water or with foods. Sós [1] established already in 1942, that there are wrong alimentary customs in Hungary. For example the ingestion of papilionaceous plants (soya, pea, lentil with high magnesium contents) is not sufficient. Brans, made from different seeds, are very good magnesium sources.

Magnesium to be found as magnesium-fitase in brans, which dissolves slowly, so its uptake is wrong. From this reason the bran to be found as not good magnesium source for some researcher.

The uptake of magnesium and calcium was investigated both in distilled water and in synthetical gastric juice (0.1 M HCl and pepsin) from different reform cereal foods.

3. METHODS

3.1. DIGESTION [2]

Berghof B type microwave apparatus was used for digestions.

Approximately 0.2 g of dried powdered sample was measuremed and then it was transferred in a teflon bomb of the microwave apparatus. After the addition of 5 cm^3 of concentrated nitric acid, the sample was digested by the following way:

Warm up:	*Digestion:*	*Cooling:*
Power supplied: 80% (850 W)	Power supplied: 80%	Power supplied: 0%
Temperature: 140°C	Temperature: 190°C	Temperature: 100°C
Time: 2 min. 30 sec.	Time: 7 min. 30 sec.	Time 10 min.
P-range: 5°C	P-range: 20°C	P-range: 5°C

3.2. Instrumentation

The elemental analyses were performed by atomic spectroscopy. UNICAM 919 atomic absorption spectrometer was used.

Experimental parameters for measurements:

Wavelengths: Ca 422.7 nm, Mg 285.2 nm Bandpass: 0.5 nm

Acetylene pressure: 100 kPa Lamp current: 15 mA

Air pressure: 30 bar Burner heigh: 12.5 mm

4. RESULTS AND DISCUSSION

Table 1 lists the mean magnesium, calcium contents and the Ca/Mg ratio in different reform foods dissolved them both in distilled water and in gastric juice.

It was found, that the total magnesium content was higher (466 mg/100 g) in wheat bran than in oat bran (233 mg/100 g). The uptake of magnesium from wheat bran is better (in water 23 per cent in gastric juice 90 per cent) than from oat bran (in water 9.4 per cent, in gastric juice 60 per cent). There are the smallest difference in the total magnesium contents of different husked seeds (wheat 149 mg/kg, oat 167 mg/kg, barley 162 mg/kg).

TABLE 1 Magnesium and calcium contents in reform cereal foods and uptake of these elements

Cereale	Magnesium					Calcium					Ca/Mg x 10^{-2}		
	total	solubility in water	solubility in gastric juice			total	solubility in water		solubility in gastric juice		total	solubility in water	solubility in gastric juice
	mg/100 g	mg/100 g	%	mg/100 g	%	mg/100 g	mg/100 g	%	mg/100 g	%			
Wheat													
husked	149	16.0	11	102	69	14.4	2.00	13.9	10.4	72	9.7	2.0	10
bran	466	105	23	419	90	25.1	2.11	8.4	14.9	60	5.4	2.0	3.6
sprout	291	104	36	263	91	23.6	2.42	10	16.5	70	8.1	2.3	6.3
Oat													
husked	167	14.0	8.4	102	61	33.6	1.73	5.1	14.6	44	20	12	14
bran	233	22.0	9.4	141	60	38.0	1.21	3.2	15.3	40	16	5.4	11
Barley													
husked	162	31.0	19	81,4	50	7.95	2.21	28	6.75	85	4.9	7.1	8.7

The total calcium contents in different cereal products are about 10% of the total magnesium contents, which result small Ca/Mg ratio. The Table 1 contains also the Ca/Mg ratio both in the wheat and in the oat. The Ca/Mg ratio is smaller in the bran (wheat $3.6 \cdot 10^{-2}$, oat $11 \cdot 10^{-2}$) than in the husked products (wheat $10 \cdot 10^{-2}$, oat $14 \cdot 10^{-2}$) in the gastric juice, which means higher magnesium contents.

The experimental data agree with previous other plant investigations (Csikkel, Kiss [3]), namely there are higher total magnesium contents in the bran than in the other parts of seeds. According to the ionequilibrium-law the wheat products are better magnesium sources than for example the oat ones, because the wheat and barley products have smaller Ca/Mg ratio ($5.4\text{-}9.7 \cdot 10^{-2}$ and $4.9 \cdot 10^{-2}$) than the oat products ($16 - 20 \cdot 10^{-2}$).

The uptake (solubility) of magnesium shows a significant difference in the different cereal products. The soluble magnesium amounts are higher (in water 14.0-105 mg/100 g, in gastric juice 81.1-419 mg/100 g) than the soluble calcium (in water 1.20-2.42 mg/100 g and in gastric juice 6.75 - 16.5 mg/100 g), because the magnesium and the calcium are as antagonists.

The bran is very important food with high fiber content, too.

5. CONCLUSIONS

Magnesium and calcium dispersions are not equal in the different parts of seeds. The bran has higher magnesium and calcium contents than the inner starchy part.

Solubility of the magnesium and calcium is 3-12 fold in gastric juice than in distilled water. Solubility of calcium salts (fitase) is worse than the solubility of magnesium salts. This fact is also very importent in respect of magnesium source considering the calcium-magnesium antagonism.

Sequence of "reform" cereal foods is, investigated by us: wheat bran, wheat sprout, oat bran, husked wheat and oat and husked barley, from aspect of magnesium source.

References

1. Sós, J. /1942/: Magyar Táplálkozástan, Hungarian *Food science*, Rózsavölgyi Kiadó, Budapest, pp. 257.
2. Csikkel-Szolnoki A.(1995): Comparable investigations of different plant digestions XXXVIII. *Hungarian Spectroscopic Congress*, Paks, Hungary
3. Csikkel-Szolnoki, A. and Kiss A.S. (1995): *Magnesium, calcium and aluminium contents in different varieties of soya seeds*, Lucrari Stiintifice of Univ. Agricult. Banatului, Timisoara, Romania, pp. 182.

MAGNESIUM IN DIGESTA OF HORSES FED DIETS CONTAINING DIFFERENT AMOUNTS OF PHYTATE

T. MATSUI, Y. MURAKAMI and H. YANO

Division of Applied Biosciences, Graduate School of Agriculture,
Kyoto University
Sakyo-ku, Kyoto, 606-01, Japan.

1. Introduction

Approximately 70% of total plant phosphorus exists as inositol hexaphosphate (phytate). Phosphorus in phytate is generally considered to be poorly available in many mammals because 95% of phytate was recovered in the ileum of humans and dietary phytase improved phosphorus availability in pigs fed diets consisting of plant proteins rich in phytate. Phytate is also suggested to suppress magnesium absorption through making insoluble complex with magnesium because dephytinization of diets increased magnesium solubility in the small intestine and magnesium absorption (Hirabayashi, et al., 1995). On the other hand, Hintz et al. (1973) suggested that phytate phosphorus was available in equine because there was phytase activity in the lower gut. Thus, phytate may not affect magnesium absorption in horses. This experiment was conducted to study apparent magnesium movement in the gastrointestinal tract of horses fed different contents of phytate.

2. Materials and Methods

Seven Thoroughbred horses were fed diets containing low phytate diet (phytic acid; 5.6 g/kg) or high phytate diet (phytic acid; 16 g/kg) for 5 days. The diets were added Cr2O3 as an unabsorbable marker. They were then killed 3 hours after the last feeding, and digesta was collected from distal mid jejunum, ileum, descending colon and rectum within 30 min after slaughter. Magnesium concentrations were measured using an atomic absorption spectrophotometer. Phytate in diets and digesta was determined by the method of AOAC.

T. Theophanides and J. Anastassopoulou (eds.), Magnesium: Current Status and New Developments, 143-144.

3. Results and Discussion

Magnesium was apparently absorbed in the upper small intestine of each group. However, the high phytate group tended to absorb less magnesium than did the low phytate group. Magnesium was also absorbed in the lower small intestine of the low phytate group but magnesium absorption was not clear in the high phytate group. In the large intestine, magnesium was secreted in the both groups but was more in the low phytate group than in the high phytate group. Phytate was degraded in the large intestine of each group. The degradation of phytate was more remarkable in the high phytate group. These results suggested that phytate suppressed magnesium absorption in the small intestine. The less magnesium secretion in the large intestine of the high phytate group may be resulted from the adaptation to less absorption of magnesium in the small intestine. Magnesium was mainly absorbed prior to the degradation of phytate and thus phytate would interfere with magnesium absorption in horses. Otherwise the degradation of phytate possibly increased available magnesium and increased magnesium absorption in the large intestine, which might partly lower apparent secretion of magnesium in the large intestine of the high phytate group although the primary site of magnesium absorption has been reported to be the small intestine in horses (Hintz and Schryver, 1972) and apparent magnesium absorption in the whole digestive tract tended to be lower in the high phytate group than in the low phytate group.

TABLE 1. Movements of magnesium and phytic acid in the digestive tract of horses fed diets containing different amounts of phytate

Segments	Low phytate group	High phytate group	Low phytate group	High phytate group
	Magnesium (g/day)		Phytic acid (g/day)	
Upper small intestine[1]	8.37±1.54	4.72±2.96	0.31±0.35	0.58±3.80
Lower small intestine[2]	1.22±0.07	3.40±5.16	0.25±1.11	0.11± 4.32
Large intestine[3]	-6.24±0.96	-2.04±0.58**	2.17±0.59	9.01±4.01*
Whole gut[4]	7.35±3.22	3.97±1.37	2.54±1.54	10.61±0.52**

Each value was mean±SD. *, P<0.05; **, P<0.01; Significantly different from the low phytate group.
1, Stomach to distal mid jejunum 2, Distal mid jejunum to ileum 3, Ileum to descending colon
4, Stomach to rectum

4. References

Hintz, H.F., and Schryver, H.F. (1972) Magnesium metabolism in the horse. Journal of Animal Science 35,755-759.

Hintz, H.F., Williams, A.J., Rogoff, J., and Schryver, H.F. (1973) Availability of phosphorus in wheat bran when fed to ponies. Journal of .Animal Science 36,522-525.

Hirabayashi, M., Matsui, T., and Yano, H. (1995) Fermentation of soybean by Aspergillus Usamii increases magnesium availability in rats. Japanese Journal of Magnesium Research 14,45-53.

INFLUENCE OF DIFFERENT DIETARY MAGNESIUM AND CALCIUM INTAKE LEVELS ON THE BEHAVIOR OF ^{28}Mg AND MAGNESIUM IN MICE

MIEKO KAWAMURA[1], TATSUO IDO[2], REN IWATA[2], SHUICHI KIMURA[3] AND TAKAKO IKEDA[3]
1)Department of Applied Biochemistry, Faculty of Agriculture, Tohoku University, Sendai 981, 2)Radiopharmaceutical Chemistry, Cyclotron and Radioisotope Center, Tohoku University, Sendai 980-77, JAPAN 3) Syowa Women's University, Tokyo 154, JAPAN

Magnesium, the second most important intracellular cation is found in all tissues and may affect many function of the body. Magnesium deficiency can affect the behavior of other trace elements and minerals. Magnesium uptake in normal mice were investigated with ^{28}Mg by tissue biodistribution and autoradiography[1-3]. The present report describes the effect of varying quantities of dietary magnesium and calcium on the behavior of magnesium and ^{28}Mg in mice organs, in particular the myocardium.

1. EXPERIMENTAL PROCEDURE

Forty eight ddY male mice(4 weeks old) were divided into 4 groups. Each group was administered an experimental diet containing magnesium or calcium for 6 days. The control (group 1) received a normal diet containing standard quantities of magnesium and calcium. Group 2 received the same diet except that the daily calcium intake was twice that of the control. Group 3 and 4 were administered a magnesium deficient diet As in the control and group 2, the daily calcium intake of group 4 was twice that of group 3. The concentration of magnesium and calcium in the 4 experimental diets were measured; Group 1 (Mg:Ca=40 mg:503g), Group 2 (Mg:Ca=40 mg:965 mg), Group 3 (Mg:Ca=3mg:459 mg), and Group 4 (Mg:Ca=3mg:946mg).

2. RESULTS AND DISCUSSION

The autoradiograms of mice fed the magnesium deficient diet (group 3 and 4) revealed a higher accumulation of ^{28}Mg in the heart and small intestine when compared to the other groups. The biodistribution of ^{28}Mg in the sampled tissues showed that group 3 had the highest uptake levels in the heart , liver, and small intestine while the lowest uptake level in bone. The biodistribution of ^{45}Ca in the four groups did not show significant differences. When compared to group 1, the magnesium levels in the heart and bone tissues of levels in the heart and bone tissues of group 4 showed the lowest

T. Theophanides and J. Anastassopoulou (eds.), Magnesium: Current Status and New Developments, 145-146.

values group 4 showed the lowest values. Similarly, the magnesium concen-tration in the livers of groups 2,3, and 4 showed substantially lower. The calcium concentration in the heart of group 3 and 4 were higher compared to the remaining groups. The results indicated that the experimental group fed the magnesium deficient diet showed a decrease both in the concentration of magnesium and the uptake of ^{28}Mg in bone tissue. Tissue analysis of the heart, and liver showed a decrease in the concentration of magne sium and a increase in the uptake of ^{28}Mg. None of the administered calcium to magnesium ratios molar ratio 20-190) (affected the magnesium transfer to the organs.

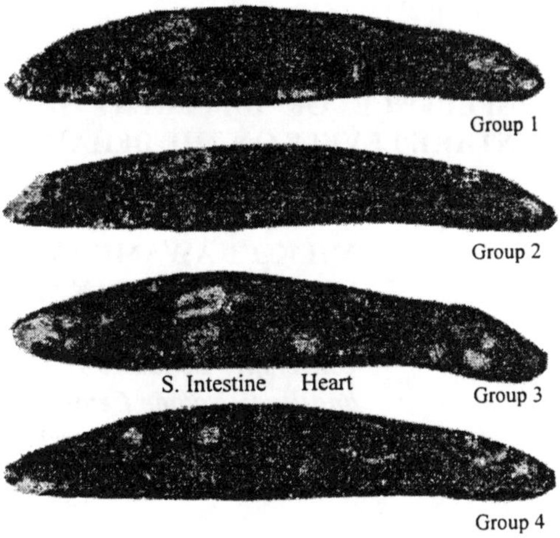

Autoradiograms of ^{28}Mg in mice.

TABLE 1. Biodistribution of ^{28}Mg in mice fed the experimental Ca-Mg diets .

	120 min (DAR)			
Tissue	Group 1	Group 2	Group 3	Group 4
Blood	0.43±0.30	0.59±0.02	0.39±0.10	0.69±0.44
Heart	2.06±0.19	2.18±0.39	2.79±0.22*	2.59±0.47
Lung	0.78±0.12	0.79±0.17	0.83±0.14	0.83±0.04
Liver	1.27±0.18	1.21±0.08	2.22±0.07*	1.81±0.44
Spleen	0.92±0.03	0.98±0.02	0.92±0.12	1.04±0.20
S. Intestine	1.55±0.18	1.88±0.13	2.35±0.35*	2.13±0.14
Kidney	2.30±0.27	2.36±0.36	2.49±0.06	2.63±0.54
Muscle	0.41±0.08	0.41±0.01	0.42±0.05	0.64±0.17
Bone	2.55±0.48	2.47±0.35	1.21±0.03*	1.47±0.19*
Brain	0.19±0.01	0.20±0.03	0.29±0.02	0.30±0.04

Data are means±SE values from three or four mice. *P<0.05 compared with group 1,

TABLE 2. Magnesium and calcium levels in the heart.

Group	Mg (mg/g)	Ca (mg/g)	Ca / Mg
1	0.178±0.007	0.075±0.010	0.42
2	0.171±0.005	0.078±0.009	0.45
3	0.165±0.005	2.119±1.252	12.81
4	0.160±0.003*	0.750±0.263**	4.68

Data are means±SE values from six mice. *P<0.05, **P<0.01 compared with group 1.

References

1. Iwata R. and Ido T. et al. (1992), *J. Radioanal. Chem. Article,* **159,** 233.
2. Kawamura M., Ido T. and Iwata R. et al. (1993)., *CYRIC Annual Report*, 136.
3. Kawamura M., Ido T. and Iwata R. et al. (1993), *Biomed. Res. Trace Elements* **4** 127

DIETARY MAGNESIUM INTAKE IN FRENCH ADULT POPULATION

PILAR GALAN[1], PAUL PREZIOSI[1], VINCENT DURLACH[2], LOUNDRES RIBAS[3], DJAMILA BOUZID[4], BERNANDETTE FIEUX[1], ALAIN FAVIER[5], SERGE HERCBERG[1]

[1]Institute Scientifique et Technique de la Nutrition et l'Alimentation (Conservatoire Nationale des Arts et Metiers 2 rue Contè, F-75003 Paris, France), [2]CHU de Reims, [3]Universite de Barcelone, [4]Sanofi Winthrop (Gentilly), [5]Laboratoire de Biochimie, CHU de Grenoble, France

1. Introduction

In the last few generations, in developed countries, the reduction of energy intake and the increase in consumption of foods containing only energy without minerals or vitamins, have contributed to decrease the total mineral and vitamin density per unit of energy in diet, and particularly magnesium density. For these reasons, the usual diets in developed countries may no longer meet the magnesium recommended dietary allowances for numerous people. Unfortunately, little information exists on the dietary magnesium intake of large representative samples of populations, particularly in European countries. Using data issued from the SU.VI.MAX study developed in France since 1994, the present study was carried out to assess dietary magnesium intake of a national sample of adults living in France.

2. SUBJECTS AND METHODS

This study was realized on a national sample of adults living in France and participating in the **SU.VI.MAX** cohort. The "**SU**pplementation en **VI**tamines et **M**ineraux **A**ntio**X**ydants" Study is a randomized double-blind, placebo-controlled, primary-prevention trial designed to test the efficacy of a daily supplementation with antioxidant vitamins and minerals, at nutritional doses, on reducing the major health problems in industrialized countries, especially main causes of premature death; 12,735 eligible subjects (women aged 35 to 60 years; and men aged 45 to 60 years) were included in 1994 and followed-up for 8 years.
Assessment of magnesium dietary intake was realized on a subsample of 5,448 subjects (3,111 women and 2,337 men) who reported six 24-h dietary records during a one-year period. For this, they received a free, tiny, central processing unit, specifically developed for the study, loaded with ad hoc software. This material enabled subjects to

T. Theophanides and J. Anastassopoulou (eds.), Magnesium: Current Status and New Developments, 147-149.

fill out computerized questionnaires off line and to transmit data during a brief telephone connection. Via the Minitel Telematic Network, they were connected to the main SU.VI.MAX computer server. The Minitel is a small terminal widely used in France as an adjunct to the telephone. Subjects were aided in this task by conversational facilities of the software, and an instruction manual for codification of foods, including photographs for estimations of portion size. The energy and magnesium content of the diet was estimated using a computerized food composition table (data from the CIQUAL table completed by Mac Cance and Widowson). Daily magnesium intake was compared to the French Recommended Dietary Allowances (Dupin et al, 1992).
Data were compiled on an Alpha-VMS system using SAS and a specific database developed for handling the data by SAS.

3. Results

Mean energy intake (± SD) was 1,811 ± 482 Kcal/ (or 7,570 ± 2,015 KJ/d) in women and 2,428 ± 623 Kcal/ (or 10,149 ± 2,604 KJ/d) in men. Total daily magnesium intake from dietary sources for the two sexes and different age groups are presented in Table 1.

TABLE 1. Daily magnesium intake (mean + SD)

Age (years)	Men (*mg/d*)	Women (*mg/d*)
35-40	-	278 ± 129
44-45	-	286 ±136
50-55	365±161	278±130
55-60	362±154	277±129
	373±160	276±126

Magnesium intakes were higher in men than in women. In each sex, there was no difference according to age. The percentage of subjects with dietary magnesium intakes below the recommended dietary allowances (RDA) according to French recommendations is presented in Table 2.

TABLE 1. Percentage of subjects with dietary magnesium intakes below the recommended dietary allowances (RDA)

	Men	Women
>RDA	27.6%	22.8%
>2/3 RDA &< RDA	54.5%	54.3%
> 1/3RDA&< RDA	17.2%	22.4%
< 1/3RDA	0.7%	0.5%

Seventy-seven percent of women and 72% of men had dietary magnesium intakes lower than recommended dietary allowances (330 mg/d for women; 420 mg/d for men); 23% of women and 18% of men consumed less than 2/3 of these RDA.
Differences were found between sex groups in the relative contribution of the different foods to magnesium intakes. Ceral products represented the main contribution in men (21%) and in women (19.8%).

Acknowledgments
This work has been supported by a grant of Sanofi Winthrop

References

Dupin H. (1992) Apports Nutritionnels Conseillès pour population Francaise. Rapport du CNERNA. Editions Technique et. Documentation Lavoisier, Paris, 1992, 146 pages.

GROWTH CHECK AND MAGNESIUM IMBALANCE ON YOUNG RATS BY OVER INTAKES OF CALCIUM.

M.Kimura, K.Matumura, N.Hatsuda, T.Takeda, S.Noumi*, Y.Yokoyama* Department of Social & Preventive Medicine Graduate School of Medicine Kyoto University, Kyoto JAPAN, *Akou Chemical Co. Ltd. Akou JAPAN

1. INTRODUCTION

Osteoporosis in women, that is age-associated bone loss 1) and calcium intake (537 mg/day in average of Japanese) by national nutritional survey less than the current RDA for calcium of 600 mg/day are focused in Japan 2). Therefore recently a significant increase in supply of calcium supplements and the introduction of many kinds of calcium-fortified foods is widely resulted. To Prevent bone loss, public or research attention for increasing of calcium intake was reported 3-5). Effects of low- or high- calcium intakes on zinc balance in humans were also reported 6,7). On the other hand, protein and lipid intakes are increasing with westernized food habits of Japanese 2). In our previous, it was studied that high protein intake with lower calcium intake induced a turn for the worse of osteoporosis 8). Some reports were shown as following that studies using isolated protein sources had shown an increase in urinary calcium excretion and a negative calcium balance in subjects fed a high-protein diet compared with a moderate-protein diet 9), and when phosphate was allowed to rise naturally as the protein intake was increased, no hypercalciuria nor fall in calcium balance was observed 10). Effect of high sodium intake on calcium retention in postmenopausal women was also demonstrated 11,12). But high intake of calcium induced low zinc absorption 13-17). In this report, lesions by over intake of calcium on health and mineral status in rats are studied.

T. Theophanides and J. Anastassopoulou (eds.), Magnesium: Current Status and New Developments, 151-168.

2. MATERIALS AND METHODS

Fifty 3-weeks-old male Wistar rats were purchased from Japan SLC (Shizuoka, Japan), respectively divided into 5 groups, and housed individually in a stainless steel cage with raised wire bottoms. The rats were given diets and water ad libitum. Drinking water used was deionized double-distilled in quartz (Quartz Distiller Zj-IDS, Fujiwara Factory, Tokyo, Japan) and provided in polycarbonate containers. The environmental temperature was maintained at 23_1 C.

Four kinds of diets containing the different level of calcium (Ca1:0.6%:control, Ca2:1.2%:2 times, Ca5:3%:5 times, Ca10:6%:10 times) were prepared as shown in table 1. Animals were weighed weekly.

The rats were fed experimental diets for 4 weeks. After the dietary regimens, the animals fasted overnight and anesthetized with sodium pentobarbital. Blood was collected fromthe abdominal aorta with a heparinized plastic syringe free from mineral concentration. The tissues (brain, heart, aorta, lung, liver, spleen, kidney, testis, muscle, bone) were removed and weighed.

The tissues, blood, and plasma were taken into borosilicate test tubes. A mixture of nitric acid and perchloric acid was added to the concentrations. The contents were heated for 24 hours at 85 C, followed by heating for 5 hours at 110 C in hot block bath (Model TPB-62, Advantec Toyo Ltd., Tokyo, Japan). Digestion was completed by the appearance of nearly colorless solution. Calcium, magnesium, phosphorus, sulfur, iron, zinc, copper and manganese concentrations in the samples were measured with inductively coupled plasma emission spectrometry (ICPS-1000 II, Shimadzu, Kyoto, Japan) by external calibration with an internal standard of yttrium. For the quality control, NIST bovine liver 1577a (National Institute of Standards and Technology, Gaithersburg, MD) was analyzed.

All chemicals used for analyses were the best grade obtained fro Nacalai Tesque, Kyoto, Japan.

Data were treated by type III two-way analysis of variance with general linear model using the computer software SYSTAT, Version 5.2 (SYSTAT, Evabston, IL). Means of the individual groups were compared by Tukey's multiple comparison with a per-experiment rate of 0.05.

3. RESULTS

Figure 1 shows the body weight curve, amounts of food intake and water drank in the rats during the experimental period. The body weight gain in groups Ca5 and Ca10 were significantly low compared with these in group Ca1 or Ca2 (control). The food intakes in groups were lower in group Ca1 or Ca2 (control). Feed efficiency was lower in groups Ca5, Ca10 than in groups Ca1, Ca2, and lowest in group Ca10. But the amounts of drinking water were high in groups Ca5, Ca10 compared wit these in groups Ca1, Ca2. The growth parameter shows in table 2. All growth parameters in groups Ca5, Ca10 were lower than in Ca1, Ca2.

Weights in tissues of the rats were shown in table 3. Weights of brain, heart, lung, spleen, kidney, and bone in groups Ca5, Ca10 rats were significantly low compared with these in groups Ca1, Ca2 rats. Weights of liver in groups Ca2, Ca5, Ca10 rats were significantly low compared with these in groups Ca1 rats, and testis in group Ca5, Ca10 rats were significantly low compared with these in groups Ca1, Ca2 rats.

Calcium concentrations in tissues of the rats are shown in table 4. Calcium concentrations in spleen, kidney and testis of group Ca5, Ca10 rats were lower than in groups Ca1, Ca2, and in heart in group Ca10 rats were lower than in other groups.

Magnesium concentrations in tissues of the rats are shown in table 5. Magnesium concentrations in bone of groups Ca2, Ca5, Ca10 rats were significantly low compared to these of groups Ca1. Magnesium concentrations in blood of groups Ca2, Ca5, Ca10 rats were lower than in groups Ca1 and in kidney of groups Ca5, Ca10 rats were lower than in groups Ca1, Ca2. Magnesium concentrations in testis of group Ca10 rats were lower than in other groups.

Phosphorus concentrations in tissues of the rats are shown in table 6. Phosphorus concentrations in kidney and testis of groups Ca5, Ca10 rats were lower than in groups Ca1, Ca2.

Table 7 shows sulfur concentrations in tissues of the rats. Sulfur concentrations in kidney of groups Ca5, Ca10 rats were lower than in groups Ca1, Ca2, and in liver, testis of group Ca10 rat was lower than in groups other groups.

Table 8 shows iron concentrations in tissues of the rats. Iron concentrations in lung of group Ca5 rats were higher than in group Ca1.

Zinc concentrations in tissues of the rats were shown in table 9. Zinc concentrations in kidney of groups Ca5, Ca10 rats were lower than in groups Ca1, Ca2, and in heart, bone of group Ca10 rat was higher than in other groups.

Copper concentrations in tissues of the rats were shown in table 10. Copper concentrations in lung, liver, testis of group Ca10 rats were higher than in groups Ca1, Ca2, and in kidney, muscle of groups Ca5, Ca10 rat were lower than in group Ca1..

4. DISCUSSION

Our findings clearly suggest that high calcium intake induce the lesion in growth and mineral imbalance of young rats. Calcium concentration was no change in blood of each group, but decreased in bone, and increased in heart, lung, liver, spleen, testis, brain & muscle, especially much increased in artery (2-3 times) & kidney (30-40 times) by over intake of calcium (Ca5 or Ca10) compared to control. Magnesium and phosphorus concentrations were decreased in bone, kidney etc. and increased in testis by over intake of calcium. Iron concentration decreased in liver, spleen and increased in lung, on the other hand copper and sulfur concentrations increased in liver etc. and decreased in kidney by calcium over intake. Zinc concentration increased in bone & blood in rats fed high calcium diet.

Effects oı nutritional factors in osteoporosis in women, that is age-associated bone losses were reviewed by Heabey 1) . To Prevent bone loss, public or research attention for increasing of calcium intake was reported 3-5). Recently a significant increase in supply of calcium supplements and the introduction of many kinds of calcium-fortified foods is widely resulted. Effects of low- or high- calcium intakes on zinc balance in humans were also reported 6,7). On the other hand, protein and lipid intakes are increasing with westernized food habits of Japanese 2). In our previous report, it was studied that high protein intake with lower calcium intake induced a turn for the worse of osteoporosis 8). Some reports were shown as following that studies using isolated protein sources had shown an increase in urinary calcium excretion and a negative calcium balance in subjects fed a high-protein diet compared with a moderate-protein diet 9), and when phosphate was allowed to rise naturally as the protein intake was increased, no hypercalciuria nor fall in calcium balance was observed 10).

Some of animal studies have shown that high calcium consumption resulted in decreased absorption and retention of zinc, retarded growth and decreased serum alkaline phosphatase activity14-16).

The effect of the amount of calcium fed on growth was greater in animals than in human. In this study of young rats, the growth reduced remarkably, but zinc concentrations of tissues were not decrease, total amount of zinc in the body was low, the efficiency of diet was inferior in the rats fed high calcium diets. These facts suggest that balance of mineral or other nutrients were broken down by high calcium intake, metabolism of zinc was inferior and the growth was prevented. That is, over calcium absorbed from high calcium diets must discharge and other mineral for example magnesium, phosphorus or zinc is excreted with calcium at same time, therefore deficiency, imbalance or lesion of utilization in other mineral was resulted.

In summary, our study indicates that long-term high calcium supplementation or besides dietary calcium with calcium of over 3-5 times of RDA calcium induced lesion of growth or health based mineral imbalance.

We thank to Prof. Y.Itokawa to help for the study.

5. SUMMARY

Recently, osteoporosis in women and low calcium intake are focused in Japan, and calcium supplement is generally used. On the other hand, protein and lipid intakes are increasing with westernized food habits of Japanese. It was reported in our previous report that high protein intake with lower calcium intake induced a turn for the worse of osteoporosis. In this report, lesions by over intake of calcium (Ca) on health and mineral status are studied.

Forty male Wistar rats were divided into 4 groups and the rats fed 4 kinds of diets contained 4 different Ca contents (1:normal, 2:2 times, 3:5 times, 4:10 times) were kept for 4 weeks. Body weights were under half in group 3 and 4 compared to these in group 1 (control). After wet digestion of the various tissues of the rats in mineral concentrations (calcium, magnesium, phosphorus, sulfurs, iron, zinc) were determined by inductively coupled plasma emission spectrometry.

Ca concentration was no change in blood of each group, decreased in bone, and increased in heart, lung, liver, spleen, testis, brain & muscle, especially much increased in artery (2-3 times) & kidney (30-40 times) of group 3 or 4 compared to group 1 (control). Magnesium concentrations were decreased in blood, kidney & muscle, especially under half in bone, and increased in testis & artery of group 3 or 4 compared to group 1 (control). Phosphorus concentration decreased in bone & kidney, and increased in heart & testis of group 3 or 4 compared to group 1 (control). Suffer concentrations decreased in kidney and increased in liver of group 3 or 4 compared to group 1 (control). Iron concentrations decreased in liver & spleen and increased in lung of group 3 or 4 compared to group 1 (control). Zinc concentrations increased in bone & blood of group 3 or 4 compared to group 1 (control). These results showed that over intake of calcium supplement induced the mineral unbalance of the severest kind and consequently health lesion was resulted.

6. REFERENCES

1. Heaney, R.: Nutritional factors in osteoporosis, Annu Trv Nutr **328** (1993), 287-316

2. National nutritional survey in Japan 1996 (1997)

3. Reid, I.R., Ames, R.W., Evans, M.C., Gamble, G.D., Sharpe, S.j.: Effect of calcium supplementation on bone loss in postmenopausal women, N Engl J Med **81** (1993), 460-464

4. awson-Hughes, B.D., Dallal, G., Krall, E.A., Sasowski, L., Sahyoun, N., Tannenbaum, S.: A controlled trial of the effect of calcium supplementation on bone density in postmenopausal women, N Engl J Med **323** (1990) 878-883

5. Aloia, J.F., Vaswani, A., Yeh, J.K., Ross, P.L., Flaster, E., Dilmanian, F.A.: Calcium supplementation with and without hormone replacement therapy to prevent postmenopausal bone loss, Ann Intrtn Med **120** (1994) 97-10

6. Wood, R.J., Zheng, J.J.: High dietary calcium intakes reduce zinc absorption and balance in humans, Am J Clin Nutr **65** (1997) 1803-1809

7. McKenna, A.A., Ilich, J.Z., Andon, M.B., Wang, C., Matkovic, V.: Zinc balance in adolescent females cosuming a low- or high-calcium diet, Am J Clin Nutr **65** (1997) 1460-1464

8. Takeda, T., Kimura, M., Itokawa, Y.: Imbalance of calcium, magnesium and phosphorus in bone and other tissues of rats

induced by low protein and calcium deficiency, J Nutr Sci Vitaminol **39** (1993) 355-363

9. Orwoll, E.S.: The effects of dietary protein insufficiency and excess on skeletal health, Bone **13** (1992) 343-350

10. Spencer, H., Kramer, L., Osis, D.: Do protein and phosphorus cause calcium loss? J Nutr **118** (1988) 657-560

11. Zarkadas, M., Gougeon-Reyburn, R., Marliss, E.B., Block, E., Alton-Mackey, M.: Sodium chloride supplementation and urinary calcium excretion in postmenopausal women.,Am J Clin Nutr **50** (1989) 1088-1094

12. Devine, A., Criddle, R.A., Dick, I.M., Prince, R.L.: A longitudial study of the effect of sodium and calcium intakes on regional bone density in postmenopausal women, Am J Clin Nutr **62** (1995) 740-745

13. Forbes, R.M.: Nutritional interactions of zinc and calcium, Am J Clin Nutr **55** (1992) 992-996

14. Hockstra, W.G., Lewis, P.K., Phillips, P.H., Grmmer, R.H.: The relationship of parakeratosisi, suuplemental calcium and zinc to the content of certain body components of swine, J Anim Sci **15** (1956) 572-564

15. Luecke, R.W., Hoeter, J.A., Brammell, W.S., Schmidt, D.A.: Calcium and zinc in parakeratosisi of swine, J Anim Sci **16** (1957) 3-11

16. Dursun, N., Aydogan, S.: Comparative effects of calcium deficiency and supplements on the intestinal absorption of zinc in rats, Jpn J Physiol **44** (1993) 157-166

17. Argiratos, V., Samman, S.: The effect of calcium carbonate and calcium citrate on the absorption of zinc in healthy female subjects, Bur J Clin Nutr 45 (1994) 198-204

18. Snedeker, S.M., Smith, S.A., Greger, J.L.: Effect of dietary calcium and phosphorus levels on the utilization of iron, copper, and zinc by adult males, J Nutr **112** (1982) 136-143

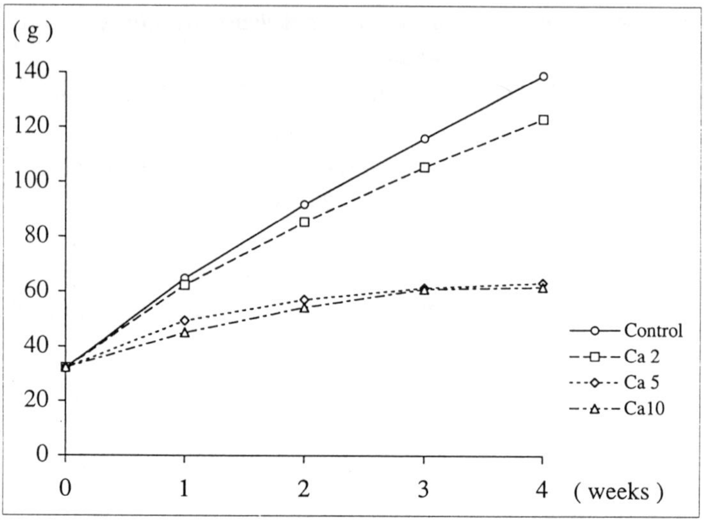

Growth curve

Composition of synthetic diets (g/100g)

	Control	Ca 2	Ca 5	Ca10
Calcium Carbonate	0.0	1.5	5.9	13.3
Sucrose	33.0	31.5	27.1	19.7
Potato Starch	30.0	30.0	30.0	30.0
Casein	20.0	20.0	20.0	20.0
Olive Oil	10.0	10.0	10.0	10.0
Cellulose powder	2.0	2.0	2.0	2.0
Mineral mixture	4.0	4.0	4.0	4.0
Vitamin mixture	0.5	0.5	0.5	0.5
DL-Methionine	0.3	0.3	0.3	0.3
Choline Chloride	0.2	0.2	0.2	0.2

Growth Parameters in Rats Fed High Calcium Diets

	Control	Ca 2	Ca 5	Ca10
Body weight gain (g)	107±13	91±10	31±11 **	30±18 **
Food intake (g)	289±30	270±31	170±31 **	182±58 **
Food efficiency	0.37±0.02	0.34±0.01	0.18±0.04 **	0.16±0.04 **
Water intake (g)	357±36	397±56	408±57	435±90

Data are mean ± SD

* significant difference ($p < 0.05$)

** significant difference ($p < 0.01$)

Weight of Various Tissues in Rats Fed High Calcium Diets
(per 100g Body Weight)

	Control	Ca 2	Ca 5	Ca10
Brain (g)	1.24±0.08	1.39±0.14	2.41±0.31 **	2.54±0.47 **
Heart (g)	0.39±0.02	0.43±0.05	0.42±0.04	0.41±0.03
Lung (g)	0.49±0.04	0.49±0.05	0.65±0.07 **	0.62±0.06 **
Liver (g)	4.40±0.29	4.08±0.33	3.81±0.61 *	3.68±0.27 **
Spleen (g)	0.29±0.03	0.27±0.03	0.23±0.05 *	0.23±0.04 *
Kidney (g)	0.97±0.06	0.96±0.06	1.53±0.14 **	1.70±0.48 **
Adrenal (mg)	26±5	31±6	53±7 **	50±9 **
Testis (g)	1.44±0.10	1.49±0.25	1.96±0.28 **	1.55±0.51
Tibia (g)	0.36±0.03	0.39±0.06	0.56±0.08 **	0.57±0.08 **

Data are mean ± SD

* significant difference ($p < 0.05$)

** significant difference ($p < 0.01$)

Calcium Concentrations in Blood and Various Tissues

		Control	Ca 2	Ca 5	Ca10
Blood	(μg/mℓ)	101±30	104±14	91±19	116±23
Heart	(μg/g)	41.3±1.5	53.2±14.7	55.2±8.5	71.1±18.7 **
Lung	(μg/g)	77.2±8.8	84.3±10.8	82.9±11.6	92.9±12.2
Liver	(μg/g)	43.0±3.4	44.3±6.4	54.0±11.4	50.5±12.3
Spleen	(μg/g)	45.1±3.4	42.2±1.7	94.4±42.5 **	70.4±21.3
Kidney	(μg/g)	75.5±14.7	72.6±8.6	2787±2468 **	1669±386
Testis	(μg/g)	38.7±1.5	40.7±2.4	47.6±6.5	55.7±9.1 **
Muscle	(μg/g)	70.6±13.9	69.1±19.8	80.8±18.7	112±41 *
Tibia	(mg/g)	120±6	123±12	117±20	111±20

Data are mean ± SD

* significant difference ($p < 0.05$)

** significant difference ($p < 0.01$)

Magnesium Concentrations in Blood and Various Tissues

		Control	Ca 2	Ca 5	Ca10
Blood	(μg/mℓ)	44.6±5.7	36.3±3.3 *	35.0±8.2 **	35.1±4.0 **
Heart	(μg/g)	204±12	200±14	200±9	216±10
Lung	(μg/g)	138±9	136±10	127±13	131±12
Liver	(μg/g)	232±11	232±14	226±28	232±8
Spleen	(μg/g)	214±14	205±16	199±19	194±24
Kidney	(μg/g)	206±18	193±13	167±10 **	175±14 **
Testis	(μg/g)	152±2	151±3	162±10	169±9 **
Muscle	(μg/g)	282±15	281±16	269±15	266±17
Tibia	(mg/g)	2.38±0.13	1.42±0.13 **	1.18±0.28 **	1.09±0.22 **

Data are mean ± SD

* significant difference ($p < 0.05$)

** significant difference ($p < 0.01$)

Phosphorus Concentrations in Blood and Various Tissues

		Control	Ca 2	Ca 5	Ca10
Blood	(μg/mℓ)	463±68	451±54	470±101	518±95
Heart	(mg/g)	2.31±0.14	2.34±0.12	2.31±0.15	2.52±0.11 *
Lung	(mg/g)	2.55±0.15	2.60±0.16	2.52±0.26	2.64±0.23
Liver	(mg/g)	3.52±0.15	3.61±0.13	3.38±0.32	3.39±0.14
Spleen	(mg/g)	3.56±0.21	3.44±0.28	3.31±0.31	3.42±0.41
Kidney	(mg/g)	2.96±0.13	2.96±0.18	2.47±0.25 **	2.61±0.29 *
Testis	(mg/g)	2.07±0.05	2.08±0.05	2.27±0.13 *	2.50±0.20 **
Muscle	(mg/g)	2.35±0.15	2.38±0.15	2.30±0.20	2.31±0.23
Tibia	(mg/g)	62.8±3.5	58.7±4.3	55.4±9.5	54.4±11.3

Data are mean ± SD

* significant difference ($p < 0.05$)

** significant difference ($p < 0.01$)

Sulfur Concentrations in Blood and Various Tissues

		Control	Ca 2	Ca 5	Ca10
Blood	(mg/mℓ)	1.77±0.22	1.80±0.16	1.88±0.36	1.93±0.36
Heart	(mg/g)	2.57±0.13	2.57±0.06	2.48±0.13	2.66±0.08
Lung	(mg/g)	1.81±0.13	1.83±0.12	1.75±0.17	1.84±0.18
Liver	(mg/g)	2.46±0.22	2.69±0.14	2.64±0.19	2.73±0.10 *
Spleen	(mg/g)	2.09±0.12	2.09±0.09	2.05±0.20	2.11±0.27
Kidney	(mg/g)	2.31±0.14	2.30±0.14	1.98±0.16 **	2.02±0.17 *
Testis	(mg/g)	1.13±0.01	1.13±0.02	1.18±0.05	1.25±0.06 **
Muscle	(mg/g)	2.45±0.17	2.48±0.15	2.33±0.17	2.34±0.22
Tibia	(mg/g)	2.63±0.10	2.73±0.15	2.72±0.26	2.62±0.38

Data are mean ± SD

* significant difference ($p < 0.05$)

** significant difference ($p < 0.01$)

Iron Concentrations in Blood and Various Tissues

		Control	Ca 2	Ca 5	Ca10
Blood	(μg/mℓ)	497±64	501±53	546±105	544±96
Heart	(μg/g)	82.3±14.1	87.7±19.0	92.9±27.6	70.7±9.8
Lung	(μg/g)	69.1±7.9	71.2±7.6	88.4±17.6 *	81.4±13.3
Liver	(μg/g)	132±20	107±30	89±14 **	70±13 **
Spleen	(μg/g)	278±25	253±29	158±37 **	140±34 **
Kidney	(μg/g)	61.3±3.9	56.3±9.5	66.0±12.0	54.7±17.3
Testis	(μg/g)	15.6±1.5	14.5±1.5	16.1±2.3	17.8±3.2
Muscle	(μg/g)	13.9±1.0	13.8±1.9	14.4±3.0	14.7±1.2
Tibia	(μg/g)	54.5±4.7	49.1±2.6	50.1±4.9	43.9±9.6

Data are mean ± SD

* significant difference ($p < 0.05$)

** significant difference ($p < 0.01$)

Zinc Concentrations in Blood and Various Tissues

		Control	Ca 2	Ca 5	Ca10
Blood	(μg/mℓ)	7.47±0.54	7.53±1.13	7.20±1.23	8.91±1.74
Heart	(μg/g)	15.7±0.6	15.9±1.1	15.5±0.7	17.8±2.0 *
Lung	(μg/g)	18.1±1.2	18.7±1.5	16.6±1.4	17.8±1.4
Liver	(μg/g)	24.5±2.2	26.2±2.0	30.2±8.9	29.4±2.9
Spleen	(μg/g)	18.7±0.7	18.2±0.6	17.1±1.3	18.0±2.4
Kidney	(μg/g)	19.7±1.1	19.0±1.0	17.1±2.7	17.6±1.5
Testis	(μg/g)	22.0±0.8	22.1±0.7	22.0±2.0	22.1±2.1
Muscle	(μg/g)	12.0±1.6	10.9±2.0	11.8±2.2	12.3±2.7
Tibia	(μg/g)	66.1±9.7	64.8±7.0	78.1±11.8	85.6±12.1 *

Data are mean ± SD

* significant difference ($p < 0.05$)

** significant difference ($p < 0.01$)

Copper Concentrations in Blood and Various Tissues

		Control	Ca 2	Ca 5	Ca10
Blood	(μg/mℓ)	1.27±0.11	1.44±0.21	1.46±0.13	1.65±0.41
Heart	(μg/g)	5.05±0.31	5.12±0.44	5.00±0.27	5.22±0.30
Lung	(μg/g)	1.32±0.10	1.39±0.15	1.50±0.30	1.63±0.20 *
Liver	(μg/g)	4.63±0.46	4.59±0.29	6.53±2.15	9.34±3.65 **
Spleen	(μg/g)	1.26±0.06	1.20±0.08	1.59±0.29	1.59±0.31
Kidney	(μg/g)	5.25±0.40	3.81±0.28 **	2.87±0.70 **	2.90±0.44 **
Testis	(μg/g)	1.41±0.03	1.39±0.05	1.58±0.12	1.75±0.17 **
Muscle	(μg/g)	1.11±0.09	1.05±0.08	0.87±0.08 **	0.83±0.09 **

Data are mean ± SD

* significant difference ($p < 0.05$)

** significant difference ($p < 0.01$)

NEUROPEPTIDES, FREE RADICAL STRESS AND ANTIOXIDANTS IN MODELS OF MG-DEFICIENT CARDIOMYOPATHY

W.B. WEGLICKI, I.T. MAK, B.F. DICKENS, R.E. STAFFORD, A.M. KOMAROV, B. GIBSON, M.M. CASSIDY, T.M. PHILLIPS,[1] AND J.H. KRAMER

Department of Physiology and Experimental Medicine, and The Immunochemistry Laboratory,[1] Department of Medicine, The George Washington University Medical Center, Washington, D.C. 20037 USA

Summary: Dietary Mg-deficiency in rodents induces a pro-oxidant/proinflammatory condition *in vivo* characterized by enhanced free radical (lipid radicals and nitric oxide) production, accumulation of oxidation products and pro-oxidant metals, depletion of endogenous antioxidants, and elevated circulating levels of inflammatory mediators. Untreated, this condition leads to cardiomyopathy, as well as reduced myocardial tolerance to imposed oxidative stresses (lipid hydroperoxide exposure; ischemia/reperfusion). The early detection of elevated circulating substance P levels (by dietary day 3) in this model led to the hypothesis that this neuropeptide may trigger subsequent inflammatory and pro-oxidant events *in vivo*. Treatment *in vivo* with substance P receptor antagonists (or with select antioxidants) provided substantial protection against pro-oxidant stress, evidenced by the attenuation of blood glutathione loss and cardiac lesion formation *in vivo*, and postischemic injury *in vitro*. The protection provided by the receptor antagonists demonstrates a direct contribution of early substance P-mediated inflammation in the subsequent development of the Mg-deficient cardiomyopathy and reduced tolerance to postischemic oxidative stress.

1. Introduction

Free radicals [1, 2] may be important contributors to the development of myocardial and vascular lesions [3-6] known to occur in animal models of dietary Mg-deficiency. Mg-deficient rats have elevated plasma levels of thiobarbituric acid reactive substances [7], indicative of increased lipid peroxidation. Moreover, Mg-deficient animals have

T. Theophanides and J. Anastassopoulou (eds.), Magnesium: Current Status and New Developments, 169-178.

increased sympathetic activity [8], and catecholamine-induced cardiomyopathy may occur partly through a free radical mechanism [9]. Catecholamine injection potentiated Mg-deficient myocardial lesion formation [1], and this amplification was attenuated by antioxidant therapies [1] suggesting a critical role for free radicals in lesion development. Mg-deficiency is also associated with significant accumulation of tissue and plasma iron [10]; this pro-oxidant metal is key in mechanisms of free radical production and lipid and protein oxidation, and also induces cardiomyopathy [11]. The similarities between the vascular lesions associated with Mg-deficiency and those observed in pathobiologic conditions involving either over-production of free radicals or depleted endogenous antioxidant defenses (vitamin E or selenium deficiencies) suggest that Mg-deficiency may induce an antioxidant deficit which predispose animal tissues to exogenously- or endogenously-induced oxidative injury

Mg-deficiency in rats is characterized by temporal changes in circulating levels (peak detection after day 10 of the diet) of inflammatory mediators [12, 13], such as histamine, interleukin-1 (IL-1), interleukin-6 (IL-6), and tumor necrosis factor-alpha (TNFα), which precede leukocyte tissue infiltration ($\geq$ 3 weeks) and cardiac necrosis (after 4 weeks) [14, 15]. Moreover, Mg-deficiency also elicited a neurogenic response during the first week of the diet [13, 16, 17]; this observation supports the hypothesis [18] that neuropeptides trigger the cascade of later inflammatory/pro-oxidant events in this model. We provide evidence which suggests that: (1) dietary Mg-deficiency induces an inflammatory/pro-oxidant condition which predisposes the myocardium to further injury following imposed oxidative stresses; and (2) these inflammatory/pro-oxidant events may be initiated by specific neurogenic mediators.

2. Methods

Golden Syrian hamsters (~100 g) and Sprague-Dawley rats (~150 g) were placed on Mg-deficient (~10% of National Research Council's recommendation) or Mg-sufficient diets for 2 and 3 weeks, respectively, as described previously [15, 19]. Antioxidant intervention procedures using hamsters (probucol as dietary supplement; vitamin E, *d,l*-propranolol and *d*-propranolol as time-release subcutaneous pellets) [1, 19, 20] and rats (substance P receptor antagonist L-703,606 as subcutaneous pellets) [15] were described previously.

Oxidized myocardial protein content was determined by incorporation of dinitrophenyl hydrazine (DNPH) in soluble proteins according to Oliver *et al* [21]. Total red blood cell (RBC) glutathione was measured by the DTNB (5,5'-dithiobis-2-nitrobenzoic acid)-GSSG reductase 'recycling' assay [22]. Nitrosyl hemoglobin (HbNO), an indicator of nitric oxide production *in vivo*, was measured by electron spin resonance (ESR) spectroscopy (77 K) in packed RBCs as previously described [23]. Cardiovascular inflammatory lesions [19, 20, 24] were analyzed by light microscopy and morphometry to determine lesion density (number/area) and size (area/mm^2) in tissue sections fixed in 4% formalin and stained by hematoxylin-eosin. Changes in plasma Mg levels were measured by atomic absorption spectroscopy. Circulating levels of substance P (Sub P), calcitonin gene-related peptide (CGRP), vasointestinal peptide (VIP), interferon-gamma, histamine and prostaglandin E_2 (PGE_2) were measured using antigen capture ELISA and chemiluminescence detection as described [13, 16].

Isolated working hearts from Mg-deficient and Mg-sufficient rats were perfused with Krebs-Henseleit buffer under physiologic conditions [15]. After 30 minutes of stabilization, hearts were exposed to different severities of *exogenous oxidative stress* by adjusting the infusion rate (0.10, 0.70, 1.75 ml/min) of 0.88 μM tert-butyl hydroperoxide (t-BuOOH)/iso-osmotic saline. Hearts were exposed to three intervals (10 minute for each infusion rate), and each interval was followed by t-BuOOH-free perfusion for measurement of cardiac pressure-volume work [15]. Additive percent inhibition of cardiac work was plotted against t-BuOOH perfusate concentration. *Endogenous oxidative stress* was induced using the isolated 30 min global ischemia/15 min reperfusion model [15, 25]. Effluent lipid hydroperoxide (LOOH) content and electron spin resonance (ESR) spin trapping using α–phenyl-t-butylnitrone (PBN) for detection of lipid peroxidation-derived free radicals, were measured during reperfusion [15, 25].

3. Results And Discussion

Serum magnesium levels in the hamster dropped 70% after two weeks on the Mg-deficient diet. Serum Mg in the rat decreased 70% by 1 week of the diet and greater than 80% by 3 weeks; this finding agrees with those of others [26] that 3 weeks is sufficient to induce severe hypomagnesemia in the rat. Mg deficiency in rodents results in the development of inflammatory cardiovascular lesions which can be quantified by microscopic techniques [19, 20, 24]. *In vivo* intervention studies using classic and novel antioxidants were conducted to determine if free radical toxicity contributed to lesion development in the Mg-deficient model (Fig. 1). We reported that vitamin E treatment concurrent with the diet provided substantial dose-dependent protection against cardiovascular lesion formation in hearts from Mg-deficient hamsters [19] and rats [27]. Since this lipophilic antioxidant does not alter the development of hypomagnesemia (serum Mg content), vitamin E must be protecting the myocardial membranes, which are vulnerable to free radical attack [28]. The anti-hyperlipidemic agent, probucol, also provided substantial protection against cardiomyopathic lesion formation in the hamster (Fig. 1) at concentrations which had no effect on the hyperlipidemia [1]. This protection may be due largely to the antioxidant properties of probucol, since it is known that probucol can inhibit the oxidation of low density lipoproteins [29] and cardiac sarcolemmal membranes [1]. Select cardioactive drugs have been shown to possess secondary antioxidant properties [30, 31], and those examined were found to be protective against lesion development [20, 32]. Both the

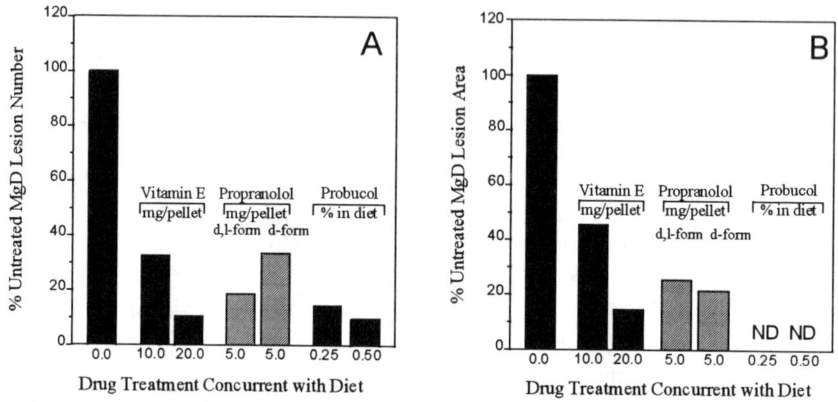

Figure 1: Protective effects of vitamin E, propranolol, and probucol against Mg-deficiency-induced myocardial lesion formation in Golden Syrian hamsters. Values are expressed as % of untreated Mg-deficient control (A) lesion number or (B) lesion area, and were derived from reference #s 1, 19, 20 & 24. ND is not determined.

pharmacologically active (*d,l-*) and nonactive (*d-*) forms of propranolol, a β–adrenergic receptor antagonist, protected equally against Mg-deficiency-induced lesion formation (Fig. 1); this suggests that protection was not associated with β–adrenergic receptor blockade, but rather with the agent's antioxidant properties [33]. Further evidence for this mechanism of protection comes from our findings that *d*-propranolol effectively reduced lipid peroxidation-derived carbon-centered and alkoxyl radical formation in oxidatively-stressed sarcolemmal membranes [34] and the postischemic rat heart model [31], respectively.

Dietary Mg-deficiency leads to a pro-oxidant state [2, 19, 20, 24] which could predispose the heart to heightened injury when an additional stress is imposed [25]. Others have shown that Mg-deficient animals exhibit reduced levels of blood and tissue antioxidants (GSH, vitamin E and C) [2, 35, 36] and elevated plasma levels of pro-oxidants (iron and copper) [10] which can participate in free radical generating pathways. That Mg-deficient animals experience oxidative stress was further suggested by the accumulation of protein oxidation products in myocardial (Fig. 2B), brain and kidney tissues [37]; and that Mg-deficiency resulted in about 50% loss of RBC glutathione in hamsters and rats (Fig. 2A). This loss of glutathione was probably due to enhanced oxidative consumption during Mg-deficiency, since it was attenuated by vitamin E supplementation [22]. Mg-deficient rats also exhibited elevated plasma malondialdehyde concentrations [2, 7, 10], and a 60% increase in blood lipid hydroperoxide levels [12, 37], which is consistent with the pro-oxidant nature of this model.

Increased levels of lipid hydroperoxides may impose an additional toxic stress to the Mg-deficient myocardium, since they can react with transition metals to form potent oxidants such as alkoxyl radicals (LO•) [31]. In this regard, isolated perfused Mg-sufficient and Mg-deficient hearts exposed to increasing t-BuOOH concentrations

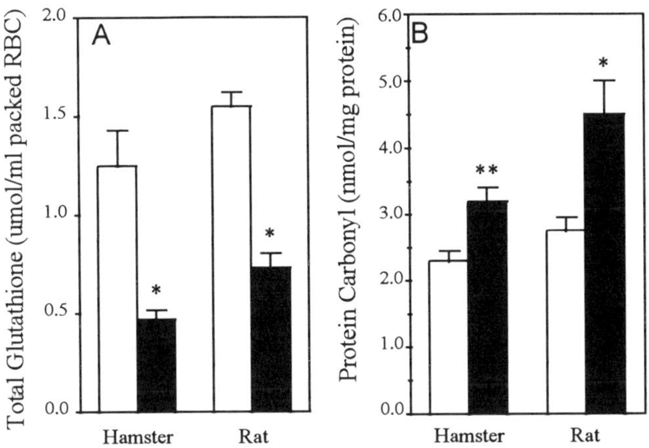

Figure. 2: Effect of dietary Mg restriction on (A) RBC glutathione and (B) cardiac tissue protein oxidation product (carbonyl) levels in 2 week hamster and 3 week rat dietary groups. Values are means ± SE of 4 Mg-deficient (closed bars) or Mg-sufficient (open bars) animals/group. **$p<0.05$ and * $p<0.01$ vs Mg-sufficient

experienced progressively greater losses of contractile function(Fig. 3); however, Mg-deficiency notably increased myocardial sensitivity at each t-BuOOH concentration. The reason(s) for this differential sensitivity is unclear, since several possible mechanisms (or combination thereof) may lead to the further reduction of contractile function in Mg-deficient hearts, including: (i) a pre-existing depression of endogenous antioxidant defenses [2, 35], reducing the heart's ability to detoxify hydroperoxides or free radicals; (ii) diet-induced alterations in calcium homeostasis due to impaired Na/K-ATPase activity [4]; or (iii) diet-mediated reduction in ATP synthesis, resulting from partial inactivation of tissue creatine phosphokinase activity (Starke-Reed, Kramer & Weglicki, unpublished observation).

We have also demonstrated that Mg-deficient hearts are more susceptible than their Mg-sufficient counterparts to endogenous oxidative stresses like ischemia/reperfusion [25]. Postischemic Mg-deficient hearts *in vitro* displayed greater levels of total alkoxyl radical production (1.7 to 2 time higher) and lower functional recovery (42 to 48% less) than Mg-sufficient. Vitamin E treatment (1.2 mg/day) concurrent with the 3 week Mg-deficient diet, provided significant protection with respect to these injury parameters, and also decreased myocardial LDH release, suggesting less tissue injury.

Our recent findings suggest that the pro-oxidant condition which develops during dietary Mg-deficiency and predisposes the myocardium to reperfusion injury may be triggered by earlier neurogenic events [18]. Mg-deficiency increases circulating plasma levels of the neuropeptides substance P [13], CGRP [16], and VIP (Phillips & Weglicki, unpublished observation); peak increases occurred in the rat during the first week (days 5-7) of the Mg-restricted diet, well before that of histamine and inflammatory cytokines [13, 16, 17]. Moreover, the earliest significant rise in plasma levels of these neuropeptides occurred within three days of initiating the diet (Fig. 4); interestingly,

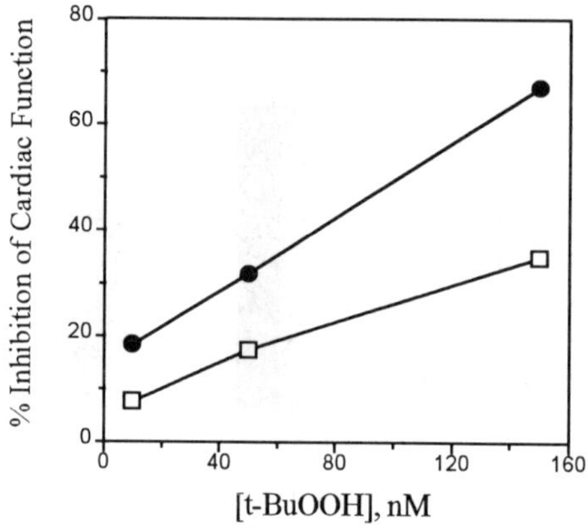

Figure 3: Effect of t-butyl hydroperoxide (t-BuOOH) on function of perfused Mg-sufficient (square) and Mg-deficient (circle) rat hearts (n=3 group). Hearts were subjected to three intervals of t-BuOOH exposure (10 min for each infusion rate) and each exposure was followed by a 10 min recovery period. Percent inhibition of cardiac pressure-volume work was determined, and means of the additive % inhibition were plotted against perfusate concentration.

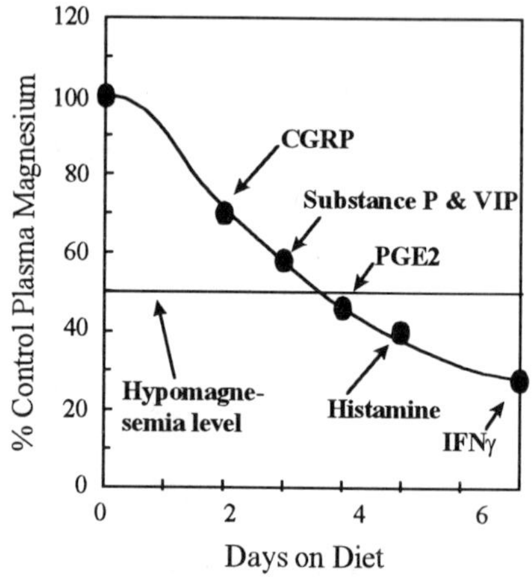

Figure 4: Earliest detection of significant increases in circulating neuropeptide and inflammatory mediators during the period of dietary Mg-deficiency in the rat. The earliest dietary day exhibiting a significant elevation ($\geq$ 25 % above baseline) in the level of each mediator was plotted against plasma magnesium content (% of normal). Data was obtained from reference #s 12, 13, & 16.

these early changes were also detected prior to achieving hypomagnesemia (50% of normal value ~ day 4), which approaches a clinically-obtainable level. In view of the early detection of substance P and its wide range of proinflammatory actions [15], we investigated its contribution to events during Mg-deficiency which promote *in vivo* production of nitric oxide (NO), the loss of RBC glutathione, and the enhanced susceptibility of the Mg-deficient myocardium to postischemic stress.

Substance P can induce production of NO and other free radicals (superoxide anion) from endothelial and inflammatory cells [38]. Indeed, recent studies have demonstrated significant increases of plasma NO products (nitrate+nitrite) in the Mg-deficient rodent [23, 39], which is consistent with observations of heightened substance P levels in this model. Moreover, we also detected pronounced nitrosyl-hemoglobin (HbNO) ESR signals in RBCs of Mg-deficient rats (Fig. 5A); this signal was completely abolished by L-NAME (a nitric oxide synthase inhibitor), confirming that the increased HbNO signal was derived from excessive NO synthesis during the dietary period. L-NAME treatment also effectively attenuated the loss of RBC glutathione (Fig. 5B), suggesting a role for NO leading to the consumption of this antioxidant. Circulating neutrophils obtained from Mg-deficient rats also displayed a 5-7 fold higher superoxide anion production [40]. When considered together, these findings suggest that during Mg-deficiency, substance P enhanced NO and superoxide production leads to the formation

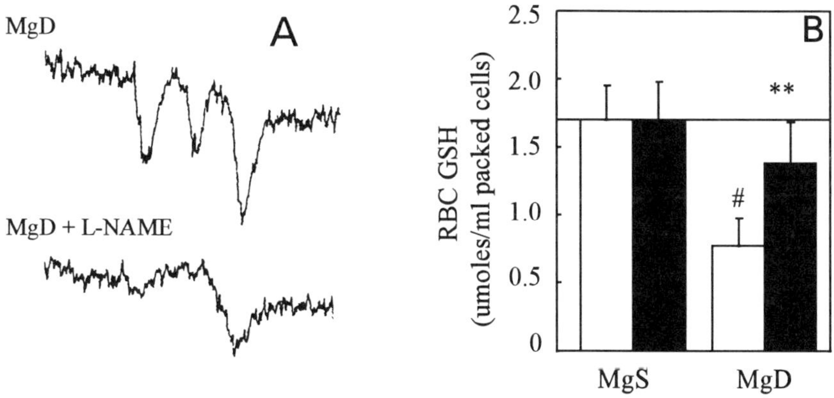

Figure 5 Effects of L-NAME (0.5 mg/ml in drinking water) on (A) nitric oxide production and (B) glutathione loss in RBCs from 3 week Mg-deficient rats. *(A):* Electron spin resonance (ESR) spectra (at 77 K) from rat RBCs. The upper spectrum displays a nitrogen hyperfine triplet structure (nitrogen hyperfine splitting constant = 17.5 gauss, centered at g = 2.012) typical of HbNO. In the lower spectrum L-NAME treatment abolished the HbNO; this spectrum displayed only the free radical signal naturally associated with RBCs, which was masked in the upper spectrum by high field lines of nitrogen triplet. (B) RBC glutathione levels in rats (n=4) with (solid bar) or without (open bar) L-NAME treatment. # $p<0.01$ and ** $p<0.02$ vs untreated MgS and MgD, respectively. Other conditions were described in reference #23.

of the highly toxic peroxynitrite, which directly mediates glutathione loss and tissue injury [40].

Since many of the *in vivo* actions of substance P are mediated through neurokinin-1 (NK-1) receptors [41], agents which selectively inhibit substance P's interaction with this receptor may protect against the pro-oxidant events associated with Mg-deficiency. Treatment with the NK-1 receptor antagonist CP-96,345 effectively prevented (90%) blood glutathione loss and reduced formation of perivascular inflammatory lesions in 3 week Mg-deficient rats [17]. Moreover, the NK-1 receptor antagonist L-703,606 [42], was shown to inhibit plasma NO product formation by 70-75% (Mak & Weglicki, unpublished observation). Thus, inhibiting the receptor-mediated actions of substance P attenuated the cellular events and subsequent peroxynitrite formation, which may induce the loss of endogenous glutathione and may contribute to the pro-oxidant stress and cardiac lesion development associated with Mg-deficiency.

The effect of treatment during the 3 week dietary period with L-703,606 was also examined in postischemic rat hearts [15]. NK-1 receptor blockade *in vivo* significantly reduced postischemic reperfusion injury to Mg-deficient hearts *in vitro* [15]. L-703,606 treatment (3.5 mg/pellet) substantially improved postischemic functional recovery (2 x higher) and myocardial tissue preservation (59% lower LDH release) [15], and reduced the associated oxidative injury, as evidenced by substantial decreases in lipid alkoxyl radical (Fig. 6A) and lipid hydroperoxide (Fig. 6B) formation. In contrast, postischemic hearts from Mg-sufficient rats did not benefit from *in vivo* or acute treatment

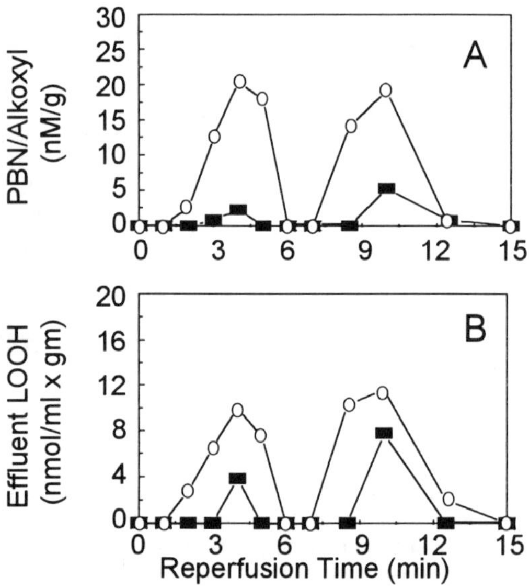

Figure 6: Representative time-courses of (A) lipid alkoxyl free radical and (B) lipid hydroperoxide (LOOH) detection in postischemic myocardial reperfusion effluents from untreated (circle) or L-703,606-treated (square; 3.5 mg/pellet) Mg-deficient rats. Hearts were subjected to 30 min ischemia followed by ESR spin trapping with PBN for alkoxyl radical detection and a colorimetric assay for LOOH.

(supplemented in perfusate) with L-703,606; this reflects the fact that these animals do not experience a neurogenic response *in vivo*.

These findings with the NK-1 receptor antagonist demonstrate a direct connection between substance P-mediated inflammation *in vivo* and reduced tolerance of Mg-deficient rat hearts to postischemic stress *in vitro* [15]. The early occurrence (dietary week 1) of a neurogenic response in this model is consistent with the hypothesis [18] that specific neuropeptides may trigger a series of subsequent inflammatory and pro-oxidant events *in vivo* (after dietary day 9), which ultimately induces cardiomyopathy or a predisposition to imposed oxidative stresses.

4. Acknowledgment

This work was supported by U.S.P.H.S. NIH RO1-HL-36418 and RO1-HL-49232.

5. References

1. Atrakchi, A.H., Bloom, S., Dickens, B.F., Mak, I.T. & Weglicki, W.B. (1992): Hypomagnesemia and isoproterenol cardiomyopathies: Protection by probucol. *J. Cardiovas. Pathol.* **1,** 155-160.

2. Günther, T., Vormann, J., Hollriegl, V., Disch, G. & Classen, H.-G. (1992): Role of lipid peroxidation and vitamin E in magnesium deficiency. *Magnesium Bull.* **14,** 57-66.
3. Altura, B.M. & Altura, B.T. (1985): New perspective on the role of Mg in the pathophysiology of the cardiovascular system. 1. Clinical aspects. *Magnesium* **4,** 226-244.
4. Ahmad, A. & Bloom, S. (1989): Sodium pump and calcium channel modulation of Mg-deficiency cardiomyopathy. *Am. J. Cardiovasc. Pathol.* **2,** 277-283.
5. Riggs, J.E., Klingberg, W.G., Flink, E.B., Schochet, S.S., Jr., Balian, A.A. & Jenkins, J.J., III. (1992): Cardioskeletal mitochondrial myopathy associated with chronic magnesium deficiency. *Neurology* **42,** 128-130.
6. Bloom, S. (1988): Magnesium deficiency cardiomopathies. *Am. J. Cardiovasc. Pathol.* **2,** 7-17.
7. Mahfouz, M.M. & Kummerow, F.A. (1989): Effect of magnesium deficiency on delta 6 desaturase activity and fatty acid composition of rat liver microsomes. *Lipids* **24,** 727-732.
8. Caddell, J.L., Kupiecki, R., Proxmire, D.L., Satoh, P. & Hutchinson, B. (1986): Plasma catecholamines in acute magnesium deficiency in weanling rats. *J. Nutr.* **116,** 1896-1901.
9. Singal, P., Kapur, N., Dhillon, K., Beamish, R. & Dhalla, N. (1982): Role of free radicals in catecholamine-induced cardiomyopathy. *Can. J. Physiol. Pharmacol.* **60,** 1390-1397.
10. Itokawa, Y. (1987): Tissue minerals of magnesium-deficient rats with thiamine deficiency and excess. *Magnesium* **6,** 48-54.
11. Rahko, R.S., Salerni, R. & Uretsky, B.F. (1986): Successful reversal by chelation therapy of congestive cardiomyopathy due to iron overload. *J. Am. Coll. Cardiol.* **8,** 436-440.
12. Weglicki, W.B., Phillips, T.M., Mak, I.T., Cassidy, M.M., Dickens, B.F., Stafford, R.E. & Kramer, J.H. (1994): Cytokines, neuropeptides, and reperfusion injury during magnesium deficiency. *Ann. N. Y. Acad. Scien.* **723,** 246-257.
13. Weglicki, W.B., Phillips, T.M., Cassidy, M.M., Mak, I.T., Dickens, B.F., Stafford, R.E., & Kramer, J.H. (1995): Pro-oxidant stress in Mg-deficiency: Role of neuropeptides and cytokines. In: *The Oxygen Paradox*, eds. K.J.A. Davis & F. Ursini, pp. 773-782. Padova, Italy: University of Padova Press.
14. Bruce, A., Bruce, V., Kurantsin-Mills, J., Cassidy, M.M., Stafford, R.E. & Weglicki, W.B. (1995): Myocardial myeloperoxidase and cardiomyopathic injury during magnesium deficiency. *FASEB J.* **9,** A708.
15. Kramer, J.H., Phillips, T.M. & Weglicki, W.B. (1997): Magnesium-deficiency enhanced postischemic myocardial injury is reduced by substance P receptor blockade. *J. Mol. Cell. Cardiol.* **29,** 97-110.
16. Weglicki, W.B., Mak, I.T., Stafford, R.E., Dickens, B.F., Cassidy, M.M. & Phillips, T.M. (1994): Neurogenic peptides and the cardiomyopathy of Mg-deficiency: Effects of substance P-receptor inhibition. *Mol. Cell. Biochem.* **130,** 103-109.
17. Weglicki, W.B., Mak, I.T. & Phillips, T.M. (1994): Blockade of cardiac inflammation in Mg-deficiency by substance P receptor inhibition. *Circ. Res.* **24,** 1009-1013.
18. Weglicki, W.B. & Phillips, T.M. (1992): Pathobiology of magnesium deficiency: a cytokine/neurogenic inflammation hypothesis. *Am. J. Physiol.* **263,** R734-R737.
19. Freedman, A.M., Atrakchi, A.H., Cassidy, M.M. & Weglicki, W.B. (1990): Magnesium deficiency-induced cardiomyopathy: Protection by vitamin E. *Biochem. Biophys. Res. Commun.* **170,** 1102-1106.
20. Freedman, A.M., Cassidy, M.M. & Weglicki, W.B. (1992): Propranolol reduces cardiomyopathic injury induced by magnesium deficiency. *Magnes. Trace Elem.* **10,** 348-354.
21. Oliver, C.N., Starke-Reed, P.E., Stadtman, E.R., Liu, G.J., Carney, J.M. & Floyd, R.A. (1990): Oxidative damage to brain proteins, loss of glutamine synthetase activity, and production of free radicals during ischemia/reperfusion-induced injury to gerbil brain. *Proc. Natl. Acad. Sci. USA* **87,** 5144-5147.
22. Mak, I.T., Stafford, R.E. & Weglicki, W.B. (1994): Loss of red cell glutathione during Mg deficiency: prevention by vitamin E, D-propranolol, and chloroquine. *Am. J. Physiol.* **267,** C1366-C1370.
23. Mak, I.T., Komarov, A.M., Wagner, T.L., Stafford, R.E., Dickens, B.F. & Weglicki, W.B. (1996): Enhanced nitric oxide production during Mg-deficiency and its role in mediating red cell glutathione loss. *Am. J. Physiol.* **271,** C385-C390.
24. Weglicki, W.B., Freedman, A.M., Bloom, S., Atrakchi, A.H., Cassidy, M.M., Dickens, B.F. & Mak, I.T. (1992): Antioxidants and the cardiomyopathy of Mg-deficiency. *Am. J. Cardiovasc. Pathol.* **4,** 210-215.
25. Kramer, J.H., Misík, V. & Weglicki, W.B. (1994): Magnesium-deficiency potentiates free radical production associated with postischemic injury to rat hearts: vitamin E affords protection. *Free Radical Biol. Med.* **16(6),** 713-723.

26. Rayssiguier, Y., Gueux, E. & Weiser, D. (1981): Effect of magnesium deficiency on lipid metabolism in rats fed a high carbohydrate diet. *J. Nutr.* **111,** 1876-1883.
27. Weglicki, W.B., Stafford, R.E., Cassidy, M.M. & Phillips, T.M. (1993): Modulation of cytokines and cardiomyopathic lesions in a magnesium-deficient rodent model: effects of vitamin E and chloroquine. *Am. J. Physiol.* **264,** C723-C726.
28. Kramer, J.H., Mak, I.T. & Weglicki, W.B. (1984): Differential sensitivity of canine cardiac sarcolemmal and microsomal enzymes to inhibition by free radical-induced lipid peroxidation. *Circ. Res.* **55,** 120-124.
29. Steinberg, D. (1986): Studies on the mechanism of action of probucol. *Am. J. Cardiol.* **57,** 16H-21H.
30. Weglicki, W.B., Mak, I.T. & Simic, M.G. (1990): Mechanisms of cardiovascular drugs as antioxidants. *J. Mol. Cell. Cardiol.* **220,** 1199-1208.
31. Kramer, J.H., Misík, V. & Weglicki, W.B. (1994): Lipid peroxidation-derived free radical production and post-ischemic myocardial reperfusion injury. *Ann. NY Acad. Sci.* **723,** 180-196.
32. Freedman, A.M., Cassidy, M.M. & Weglicki, W.B. (1991): Captopril protects against myocardial injury by magnesium deficiency. *Hypertension* **18,** 142-147.
33. Mak, I.T. & Weglicki, W.B. (1988): Protection by beta-blocking agents against free radical-mediated sarcolemmal lipid peroxidation. *Circ. Res.* **63,** 262-266.
34. Mak, I.T., Arroyo, C.M. & Weglicki, W.B. (1989): Inhibition of sarcolemmal carbon-centered free radical formation by propranolol. *Circ. Res.* **65,** 1151-1156.
35. Mills, B.J., Lindeman, R.D. & Lang, C.A. (1986): Magnesium deficiency inhibits biosynthesis of blood glutathione and tumor growth in the rat. *Proc. Soc. Exp. Biol. Med.* **181,** 326-332.
36. Hsu, J.M., Smith, J.C., Jr., Yunice, A.A. & Kepford, S. (1983): Impairment of ascorbic acid synthesis in liver extracts of magnesium-deficient rats. *J. Nutr.* **113,** 2041-2047.
37. Stafford, R.E., Mak, I.T., Kramer, J.H. & Weglicki, W.B. (1993): Protein oxidation in magnesium-deficient rat brains and kidneys. *Biochem. Biophys. Res. Commun.* **28,** 569-600.
38. Mantyh, P.W. (1991): Substance P and the inflammatory and immune response. *Ann. NY Acad. Sci.* **632,** 263-271.
39. Rock, E., Astier, C., Lab, C., Malpuech, C., Nowacki, W., Mazur, A. & Rayssiguier, Y. (1995): Magnesium deficiency enhances plasma nitric oxide level in the rat. *Magnes. Res.* **8,** 237-242.
40. Mak IT, Dickens BF, Komarov AM, Phillips TM, Weglicki WB. (1997): Activation of the neutrophil and loss of plasma glutathione during Mg-deficiency---modulation effect by NOS inhibition. *Mol. Cell. Biochem.* In Press.
41. DeRose, V., Robbins, R.A., Snider, R.M., Spurzem, J.R., Thiele, G.M., Rennard, S.I. & Rubinstein, I. (1994): Substance P increases neutrophil adhesion to bronchial epithelial cells. *J. Immunol.* **152,** 1339-1345.
42. Cascieri, M.A., Ber, E., Fong, T.M., Sadowski, S., Bansal, A., Swain, C., Seward, E., Frances, B., Burns, D. & Strader, C.D. (1992): Characterization of the binding of a potent, selective, radioiodinated antagonist to the human neurokinin-1 receptor. *Mol. Pharmacol.* **42,** 458-463.

THE DIURETIC-MAGNESIUM QUESTION IN HEART FAILURE

L. ALCOCER and A.J. REYES*
National University of Mexico and Mexico City General Hospital, Tuxpan
*16, Mexico, D.F. 06760, Mexico; *Institute of Cardiovascular Theory, Sotelo 3908, Montevideo 11700, Uruguay*

1. Summary

Magnesium deficit must be prevented and treated with particular zeast in chronic congestive heart failure. Besides its usual general consequences, in the case of chronic congestive heart failure under chronic therapy with loop or thiazide-type diuretics magnesium deficit predisposes to (digitalis-induced or not) cardiac arrhythmias , causes resistance to potassium replenishment in patients with hypokalemia, could impair cardiac pump function directly at the cardiac level, and could facilitate (exaggerate) vasoconstriction.

Loop and thiazide-type diuretics increase the renal excretion of magnesium and may conduce to magnesium deficit upon prolonged administration; this possibility is enhanced by the coexistence of factors other than diuretic therapy which may concur to the development of this condition. However, the use of diuretics is mandatory in chronic congestive heart failure. Copharmacotherapy with potassium-retaining diuretics should not be instituted as a routine. Oral magnesium suplementation, correction of concomitant factors conducing to magnesium deficit, and the use of appropriate doses of diuretics appear as the best measures to prevent and/or treat diuretic-induced deficit in patients with chronic congestive heart failure and normal renal function..

2. Introduction

Congestive Heart Failure (CHF) is a syndrome that may result from various cardiovascular diseases, (hypertension, ischemic heart disease, idiopathic cardiomyopathy, valvular lesions, viral infection , alcohol abuse). It is characterized, by a constellation of symptoms and signs produced by cardiac systolic or diastolic dysfunction and associate complex circulatory and nerurohumoral adaptation and maladaptation processes.

T. Theophanides and J. Anastassopoulou (eds.), Magnesium: Current Status and New Developments, 179-187.

The importance of CHF resides in its disabling character, ominous prognosis and high incidence and prevalence. CHF is the most common medical diagnosis in patients aged above 65 in the USA, where 400 000 new cases are diagnosed each year[1]; 1ts crude prevalence ranges from 3 to 20 individuals per 1000, and its prevalence among those aged above 65 ranges between 30 and 130 individuals per 1000[20]. CHF is highly lethal (37 percent of men and 38 percent of the women died within 2 years and 82 and 67 percent respectively died within 6 years after the diagnosis was made, in the Framingham study[3]). CHF has an important socioeconomic impact (the SOLVD study estimated that 35 percent of all CHF patients were admitted to hospital each year[4]).

The objectives of the treatment of CHF are: i) to reduce mortality, ii) to relieve symptoms and improve quality of life, and iii) to reduce the socioeconomic impact.

Diuretics constitute the cornerstone of the treatment of CHF. These drugs reduce edema, breathlessness and fatigue, and constitute a life-saving tool in different types of acute pulmonary edema. The beneficial effects of prolonged treatment with diuretics occur in both the supine and non-supine positions, at rest and during everyday physical effort[5]. The New York Heart Association functional class improves [6],[7], [8], [9], [10], [11], [12], [13], and tolerance to effort during stress tests is increased [14], [15]. These beneficial actions of diuretics in chronic CHF are secondary to their renal excretory effect, i.e. they take place only when these substances relieve fluid retention[16], [17], [18], [19], [20], [21]. Cessation of treatment with diuretics in chronic CHF results in clinical deterioration mirroring the improvement that follows the institution of this therapy[22].

Loop and thiazidelike diuretics may cause hyponatremia, hypokalemia, metabolic alkalosis, magnesium deficit, adverse renal wastage (loop diuretics) or retention (thiazidelike diuretics) of calcium, decreased glucose tolerance and hyperglycaemia, and may raise total plasma cholesterol. Potassium-and-hydrogen-retaining diuretics may cause hyperkalemia and metabolic acidosis. Loop diuretics are ototoxic at high doses, hydrochlorothiazide may cause allergic pulmonary edema, and triamterene may precipitate in the tubular structures of the kidney [23]. One of the main concerns about the use of diuretics in heart failure is the potential occurrence of cardiac arrhythmias that may increase the risk of sudden death. Electrical instability is frequent in CHF, particularly in patients on long-term diuretic therapy. This instability seems to be multifactorial in origin; magnesium deficit, which can be a relevant factor, is frequently associated with hypokalemia[24]. magnesium deficit has been implicated as a risk factor for sudden death. The prevalence of cellular magnesium deficit and hypomagnesemia in CHF is higher than in controls [25], and it is increased by the use of diuretics.

3. Bodily magnesium in chronic congestive heart failure

Magnesium is an ubiquitous divalent cation. It plays an essential role in a wide range of fundamental cellular reactions, is an essential cofactor for the activity of many enzymes involved in energy generation in the cardiac and arteriolar myocites, and it seems to participate to a relevant extent in ionic equilibria across the sarcolemma (including that of calcium) [26] via ATP-dependent enzymes, and as a consequence a bearing upon the tension in cardiac muscle and in the smooth muscle of the vasa.

Total serum magnesium[27], erythrocyte magnesium[28], lymphocyte magnesium[29], [30], mononuclear blood cell magnesium [31] and skeletal muscle magnesium have been found to be decreased in patients with chronic CHF under prolonged treatment with diuretics . Unfortunately none of these variables appears to correlate linearly with intramyocardial magnesium. Intramyocardial magnesium, which is highly compartimentalized within cardiomyocytes, acts *inter alia* as a veritable cardioprotective element. In magnesium deficit, important deleterious micromorphological changes take place in the heart. Biochemically, a decrease in the cardiomyocyte magnesium content results in an increase in free cytosolic calcium and therefore in a propensity for the occurrence of cardiac arrhythmias and, in a decrease in the threshold (in terms of serum digoxin concentrations) for digitalis-induced arrhythmias [32].

The prevalence of hypomagnesaemia in CHF has been reported as 10-37 percent in patients with heart failure [33], [34], [35]. However, no hypomagnesaemia, defined as a serum magnesium level below 0.7 mmol/l, was found in any of 57 patients with chronic CHF[36]. The inconsistency of these data can be explained by the fact that total serum magnesium does not correlate well with the intracellular magnesium contents; a better correlation exists between intracellular magnesium and ionized serum magnesium[37]. New techniques for less complicated measurements of ionized serum magnesium in clinical practice are being validated[38].

The prognostic significance of hypomagnesemia, if any, is not known. In a study of 199 patients with heart failure, the relation between serum magnesium concentration and the long-term outcome was evaluated. In patients with low serum magnesium, premature ventricular contractions and episodes of ventricular tachycardia were more frequent . These patients showed a significantly higher one-year mortality (55 percent vs. 29 percent); however, there was no difference in the frequency of sudden death. These results were established after adjustment for differences in the severity of CHF and in neurohumoral alterations [33]. However, in a sub-study of the PROMISE trial, 1068 patients with NYHA class III or IV CHF presented hypomagnesemia, and in these only a relatively small percentage of patients the frequency of dysrrhythmiouplets, and 6-month survival was similar to that of patients with normal or augmented levels of serum magnesium[39]. There is not clear explanation for the differences between the results of these two studies; nonetheless, in the PROMISE trial patients with serum potassium < 3.5 mmol/l were excluded. and it is important to consider that magnesium deficit and hypokalemia appear to act as synergistic arrhythmogenic factors[40].

Severe magnesium deficit in the heart increases the calcium content of the mitochondria, to the extent of impairing the function of these organelles and therefore the energy production for cardiac contraction. The role of mitochondrial damage in the process of remodeling is intriguing and deserves profound investigation.

Magnesium deficit in chronic CHF could enhance the sensitivity of resistance vessels to vasoconstriction and/or decrease their sensitivity to vasodilatation, and thus facilitate the exaggerate increase in the afterload to the left ventricle that occurs in cardiac insufficiency, to further detriment of the pump function of the heart. Besides its actual and/or potential consequences on the circulation, magnesium deficit should be prevented and treated in CHF because some of the frequent clinical expressions of both syndromes are common. Fatigue is one such clinical expressions. This symptom is present in practically all patients with chronic CHF, and it impairs quality of life as much as breathlessness does.

Lastly, all metabolically active tissues are affected in chronic CHF and also in magnesium deficit. The coexistence of both syndromes is potentially deleterious to the entire body.

4. Causes of magnesium deficit in chronic congestive failure treated with diuretics

In theory, many factors may theoretically contribute to the development of magnesium deficit in chronic congestive heart failure [41], [42],[43]. Magnesium intake may be reduced because of decreased food intake, which is frequent in chronic CHF, particularly during treatment with digitalis. Magnesium might be poorly absorbed in decompensated and even in compensated patients, and vomiting and diarrhea may result in excess magnesium losses. Hyperaldosteronemia, which may augment the renal excretion of magnesium, develops during decompensation and is present in nearly all patients in the last haemodynamic stages of chronic CHF. Whatever other factors might contribute to the development of magnesium deficit in chronic CHF, prolonged therapy with loop and/or thiazide-type diuretics appears as the most important determinant, and it is regarded as capable of causing magnesium deficit in the absence of relevant concurrent factors[44]. Loop diuretics reduce magnesium reabsorption in the thick ascending limb of the loop of Henle and thereby increase magnesiuresis. The fall in plasma calcium and the rise in plasma parathyroid hormone subsequent to the elevation in calciuresis caused by loop diuretics could partly oppose the magnesiuretic effect of these agents, since both deviations increase magnesium reabsorption from the preurine in Henle´s loop. Early distal tubular diuretics (e.g. the thiazides) also increase the renal urinary excretion of magnesium in man [45], perhaps through an interference with the reabsorption of this cation in the distal convoluted tubule. Thiazidelike diuretics decrease the renal output of calcium, and thus tend to elevate plasma calcium and to reduce the plasma concentration of parathyroid hormone. Both these changes in the perinephronal fluid decrease the reabsorption of magnesium in the loop of Henle, and may thus contribute to the increase in urinary magnesium output caused by early distal tubular diuretics. The excess magnesium that passes from the loop of Henle to the distal convoluted tubule in these circumstances would not be reabsorbed, because the local reabsorption processes would have a limited capacity and /or because the early distal tubular diuretic would impede such reabsorption.

The increase in aldosterone which accompanies treatment with diuretics appears to contribute to the magnesiuretic action of loop and of early distal tubular diuretics. The elevation in plasma pH produced by thiazidelike substances could also partly counter for their magnesiuretic effect.

The anti-aldosterone diuretic spironolactone does not affect the urinary output of magnesium when it does not exert a natriuretic effect, as is the case when it is given to normal subjects in whom the activity of the renin-angiotensin system has been depressed through a high-sodium diet. However, when spironolactone is coadministered with loop or thiazidelike diuretics it reduces the magnesiuretic action of the latter, given that diuretics increase plasma aldosterone concentration[46].

The potassium-retaining diuretic amiloride tends to retain magnesium, but to a less extent than potassium[47]. The reason is not clear; the decrease in plasma pH caused by amiloride is known to enhance magnesiuresis. A combination of 5 mg amiloride and 50 mg hydrochlorothiazide was however found to reduce the mean serum concentration of magnesium in hypertensive patients in whom it was used once daily for several months [48]. If an amount of amiloride sufficient to counter the hypermagnesiuresis caused by a certain dose of a thiazide were combined with the latter, excess renal retention of potassium and hydrogen could occur and hyperkalemia and acidosis develop therefrom. A combination of 5 mg amiloride and 40 mg. furosemide has been found not to increase 24-hr magnesiuresis after the first dose in healthy subjects[49]; however it tended to decrease the mean serum concentration of magnesium in patients with heart failure during prolonged administration[50].

Amiloride reduces the urinary output of calcium. The reduction in calciuresis caused by the thiazide and by amiloride may be additive in provoking hypercalcemia and a rise in plasma parathyroid hormone, thus leading to increased magnesiuresis. In the case of combinations of amiloride and loop diuretics, the excess magnesiuresis caused by the loop diuretic would be offset by the potassium retaining-diuretic, due to the fact that the opposing effects of amiloride and e.g. furosemide on urinary calcium output results in isocalciuresis. Thus, calcemia and the level of parathyroid hormone would not be affected. Therefore, it appears more rational to coadminister amiloride with a loop diuretic than with a thiazide when a reduction in renal magnesium wastage is sought.

The 24-hour magnesiuretic action of the first once-daily dose of a loop or of an early distal tubular diuretic may persist with little change during prolonged treatment, or it may be attenuated to variable extent or even disappear. This happens because regulatory processes reduce the urinary excretion of magnesium when the bodily content of this ion is decreased. Therefore, magnesiuresis is not a reliable indicator of the bodily magnesium state or of the action of diuretics on the renal handling of magnesium during prolonged treatment with these drugs in CHF. Recent investigations[51] of the bodily magnesium state using the intravenous magnesium loading test have established the range of normality for the result of this test when it is carried out as per a certain technique, and have established that the repeatability of the test is acceptable. The intravenous magnesium loading test as carried out in these investigations is currently considered the most reliable procedure to evaluate bodily magnesium. Fifty-one percent of patients with CHF (not all of them treated with diuretics) who did not present with renal insufficiency were found to have magnesium deficit as diagnosed by this test [52]. Whether such patients were receiving a potassium-retaining diuretic in combination with a kaliuretic diuretic is not disclosed; if such had been the case , as many of 80 percent of patients with chronic CHF under prolonged treatment with loop and/or thiazide-type diuretics whose renal function is normal or affected minimally could present magnesium deficit. A "normal" level of total serum magnesium should not be accepted as proof of adequate magnesium contents in the tissues.

Progressive impairments in renal function conduce to a progressively reduced renal excretion of magnesium [40]. Therefore, patients with CHF who present clinically significant renal insufficiency generally do not present magnesium deficit.

5. Prevention and treatment of magnesium depletion in CHF

Magnesium deficit must be prevented and treated in patients suffering from CHF, especially if they receive diuretics.

Maneuvers aimed at preventing magnesium deficit include a magnesium-rich diet and the control of coexisting factors that may favor the development of hypomagnesemia, such as high alcohol and caffeine intakes and diabetes mellitus; diuretics should be prescribed at the lowest possible doses, compatible with the attainment of clinical objectives (to which purpose salt intake must be optimized).

Patients presenting clinically covert or overt magnesium deficit should receive adequate magnesium therapy. Oral magnesium supplementation appears as the best option in patients with CHF and normal renal function. Potassium-retaining diuretics and fixed-dose combinations of these drugs with loop or thiazide-type diuretics should not be used to prevent routinely to prevent or treat magnesium deficit, for various reasons: the antimagnesiuretic action of potassium-retaining diuretics is less intense than their anti-kaliuretic action, and available fixed-dose combinations of kaliuretic and antikaliuretic diuretics are known to fail to prevent the development of changes indicating magnesium deficit in patients with CHF [49], [53]. Combinations of potassium-retaining and of common diuretics including doses of the former that would result in a lack of effect of the combination on magnesiuresis cause hyperkalemia and hyponatraemia.

6. Conclusions

Loop and thiazide-type diuretics increase the renal excretion of magnesium.

Loop and thiazide-type diuretics may conduce to magnesium deficit upon prolonged administration. The use of loop and thiazide-type diuretics is mandatory in chronic congestive heart failure.

Patients with chronic congestive heart failure and normal renal function under prolonged therapy with loop and/or thiazide-type diuretics are highly likely to develop magnesium deficit from such therapy. This possibility is enhanced by the coexistence of factors, other than diuretic therapy, which may concur to the development of magnesium deficit.

Intracellular magnesium correlates poorly with serum magnesium concentration. Ionized magnesium in serum and the intravenous magnesium loading test seem to be the most reliable procedures to evaluate the bodily magnesium state.

Magnesium deficit must be prevented with particular zeast in chronic CHF. Besides its usual general consequences, in the case of chronic CHF under prolonged therapy with loop or thiazide-type diuretics magnesium deficit predisposes to (digitalis-induced or not) cardiac arrhythmias, causes resistance to potassium replenishment in cases with hypokalemia, could impair cardiac pump function directly at the cardiac level, and could contribute to inordinate vasoconstriction at the level of the resistance vessels.

Oral magnesium suplementation appears as the best option to prevent and/or treat diuretic-induced deficit in patients with chronic CHF and normal renal function. Evaluating the prophylactic and therapeutic efficacies and the safety of such treatment is in order.

References

1. Schocken DD, Arrieta MI, Leaverton PE, Ross EA. (1992) Prevalence and mortality of congestive heart failure in the United States. *J Am Coll Cardiol*;**20**:301-306.
2. Cowie MR, Mosterd A, Wodd DA, Deckerst PA. Poole-Wilson PA, Sutton GC, Grobbee DE. (1997) The epidemiology of heart failure. *Eur Heart J*;**18**:208-225.
3. Ho KKL, Pinski Jl, Kannel WE. (1993) The epidemiology of heart failure: the Framingham Study. *J Am Coll Cardiol*; **22**(Suppl A):6A-13A.
4. The SOLVD Investigators. Effect of enalapril on survival in patients with reduced left ventricular ejection fraction and congestive heart failure. *N Engl J Med* (1991);**325**:293-302.
5. Reyes AJ. (1995) Pathophysiological basis of the use of diuretics in congestive heart failure. *Prog Pharmacol Clin Pharmacol*;**10**:75-85.
6. Achhammer I. (1990) Long-term efficacy and tolerance of torasemide in congestive heart failure. *Prog Pharmacol Clin Pharmacol*;**8**:127-136.
7. Dixon DW, Barwolf-Gohilke C, Gunnar RM. (1981) Comparative efficacy and safety of bumetanide and furosemde in long-term treatment of edema due to congestive heart failure. *J Clin Pharmacol*;**21**:680-687.
8. Düsing R, Piesche L. (1990) Second line therapy of congestive heart failure with torasemide. *Prog Pharmacol Clin Pharmacol*;**8**:381-391.
9. Oversohl K, Mehrabian G, Hanisch M, Schmitz H. (1985) Muzolimine as monotherapy in the treatment of patients with congestive heart failure. *Z Kardiol*;**74**(Suppl 2):27-31.
10. Podszus T, Piesche L. (1990) Effect of torasemide on pulmonary and cardiac hemodynamics after oral treatment of chronic heart failure. *Prog Pharmacol Clin Pharmacol*;**8/1**:157-166.
11. Sherman LG, Liang C, Baumgardner S, Charuzi Y, Chardo F, Kim CS. (1986) Piretanide, a potent diuretic with potassium-sparing properties, for the treatment of congestive heart failure. *Clin Pharmacol Ther*; **40**:587-594.
12. Stauch M, Stiehl L. (1990) Controlled, double-blind clinical trial of the efficacy and tolerance of torasemide in comparison with furosemide in patients with congestive heart failure-a multicenter study *Prog Pharmacol Clin Pharmacol*; **8/1**:121-126.
13. Verho M, Bückert C, Freude J, Laeger S, Pahnke K. (1985) The effects of piretanide, a potassium stable diuretic, on serum electrolytes in patients with congestive heart failure. *Pharmatherapeutica*;**4**:288-295.
14. Haerer W, Bauer U, Sultan N, Cernoch K, Fehske KJ, Hetzel M , Stauch M , Hombach V. (1990) Acute

and chronic effects of a diuretic monotherapy with piretanide in congestive heart failure-a placebo controlled trial. *Cardiovasc Drugs Ther:* **4**: 515-522.
15. Schäfer CE and Sievert H. (1987) Treatment of chronic congestive heart failure: effect of hydrochlorothiazide and trimaterene on cardiac performance. In: Andreucci VE and Dal Canton A, (eds.), *Diuretics: Basic, Pharmacologic and Clinical Aspects.* Boston: Martinus Nijhoff, pp 350-352.
16. Reyes AJ. (1987) Aspects of the mechanism of action of diuretics in heart failure. In: Andreucci VE, Dal Canton A, (Eds*), Diuretics: Basic, Pharmacologic and Clinical Aspects.* Boston, Martinus Nijhoff; pp 344-349.
17. Reyes AJ. (1987) Diuretics in heart failure. In: Puschett JB, Greenberg A, (eds.), *Diuretics II: Chemistry, Pharmacology, and Clinical Applications.* Elsevier Science, New York,; 332-344.
18. Reyes AJ. (1988) Therapy with diuretics in congestive heart failure. *Prog Pharmacol Clin Pharmacol*;**6/3**:167-192.
19. Reyes AJ: (1992) Actions of diuretics on the functional processes of congestive heart failure: from facts to heuristic hypotheses. *Prog Pharmacol Clin Pharmacol*; **9**:441-460.
20. Reyes AJ. (1993) Loop diuretics versus others in the treatment of congestive heart failure after myocardial infarction, *Cardiovasc Drugs Ther*;**7**:869-876.
21. Reyes AJ. (1993) The treatment of congestive heart failure with diuretics: turning points and unresolved issues. *J Heart Failure*; (Suppl), abstract 34.
22. Gillies A Morgan T, Myers J. (1980) Comparison of piretanide and chlorothiazide in the treatment of cardiac failure. *Med J Aust*;**1**:170-172.
23. Reyes AJ, Leary WP, Crippa G, Alcocer L. (1995) Untoward actions of diuretics in congestive heart failure. *Prog Pharmacol Clin Pharmacol*;**10/3**:132-158.
24. Mountokalakis T, Dourakis D, Mayopoulou-Symvoulidou D. (1989) Magnesium-sparing diuretics.1989, in Itokawa Y, Durlach J, (eds*.), Magnesium in Health and Disease.* John Libbey: London, 409-413.
25. Lasserre B, Spoerri M, Theubet MP, Mollet V. (1997) Magnesium dysbalance in clinical practice: 7-year experience from the Donmag study, *Adv Mag Res;* **1**:.281-293
26. Golf SW. (1994) Biochemistry of magnesium in man. In: Golf S, Dralle D, Vecciet TL.,(eds), *Magnesium 1993.*London: John Libbey, 31-41.
27. Lim P, Jacob E. (1972) Magnesium deficiency in patients on long-term diuretic therapy for heart failure. *Br Med J*; **3**: 620-622.
28. Abraham AS, Rosenman D, Meshulam Z, Balkin J, Zion M, Eylath U. (1986) Intracellular cations and diuretic therapy following acute myocardial infarction. *Arch Intern Med*;**146**:1301-1303.
29. Ryan MP, Ryan MF, Counihan TG. (1981) The effects of diuretics on lymphocyte magnesium and potassium. *Acta Med Scand* Suppl **647**:153-161.
30. Schwinger RG, Antoni DH. (1992) Triamterene may preserve lymphocyte magnesium and potassium in patients with congestive heart failure. *Magnesium Res*;**5**:29-32.
31. Ryzen E, Elkayan U, Rude Rk. (1986) Low blood mononuclear cell magnesium in intensive cardiac care unit patients. *Am Heart J*;**11**:475-480.
32. Sueta CA, Patterson JH, Adams KF. (1995) Antiarrhythmic action of pharmacological administration of magnesium in heart failure: a critical review of new data. *Magnesium Res*;**8**:389-401.
33. Gottlieb SS, Baruch L, Kulin ML, Bernstein JL, Fisher ML, Packer M. (1990) Prognostic importance of the serum magnesium concentration in patients with congestive heart failure. *J Am Coll Cardiol*;**16**:827-831.
34. Wester PO, Dyckner T. (1986) Intracellular electrolytes in cardiac failure. *Acta Med Scand*;**707**:33-36.
35. Gottlieb SS. (1989) Importance of magnesium in congestive heart failure. *Am J Cardiol*;**63**:39G-42G.
36. Spisak V. (1996) Serum and urinary concentrations of magnesium in patients with chronic heart failure. *Vnitr-Lek*;**42**:680-683.
37. Altura BM, Altura BT. (1977) Magnesium metabolism, atherogenesis and cardiovascular pathobiology. *Adv Magnes Res*;**1**:28-37.
38. Zoppi F, De-Gasperi A, Guagnellini E, Marocchi A, Mineo E, Pazzucconi F, Rossi C, Turrini D. (1996) Measurement of ionized magnesium with AVL 988/4 electrolyte analyzer: preliminary analytical and clinical results. *Scand J Clin Lab Invest;***224**:259-274.
39. Eichhorn EJ, Tandon PK, Di Bianco R, Timmis GC, Fenster Pe, Shannon J, Packer M, (1993) Clinical and prognostic significance of serum magnesium concentration in patients with severe chronic

congestive heart failure the PROMISE Study. *J Am Coll Cardiol*;**21**:634-640.

40. Popovtzer MM, Massry SG, Coburn JW, Kleeman CR. (1969) The interrelationship between sodium, calcium and magnesium excretion in advanced renal failure. *J Clin Med*; **731**: 763/771.
41. Wang RW. (1987) Magnesium deficiency. Pathogenesis, prevalence and clinical implications. *Am J Med*;**82**:24-29.
42. Reyes AJ. (1987) Interactions between magnesium and drugs in congestive heart failure. *Magnesium Bull*;**9**: 93-109.
43. Alcocer L, Reyes AJ. (1986) Negative effects of risk factor reduction in the heart. *Cardiol Clin*;**4**:153-161.
44. Reyes AJ. (1993) Renal excretory profile of loop diuretics: consequence for therapeutic application. *J Cardiovasc Pharmacol*; **41**(Suppl 3):S11-S23.
45. Reyes AJ. (1991) Effects of diuretics on outputs and flows of urine and urinary solutes in healthy subjects. *Drugs*; **41** (Suppl 3):35-59.
46. Leary WP, Reyes AJ, van der Byl K. Spironolactone is a facultative diuretic and antimagnesiuretic in man. In: Golf S, Dralle D, Vecciet L, (Eds*)*, (1984) *Magnesium 1993*. John Libbey, London 1994, pp 299-302.
47. Leary WP, Reyes AJ, Drug interactions with diuretics. *S Afr Med J*;**655**:455-461.
48. Leary WP, Reyes AJ, Van der Byl K. (1984) Effects of a combination of hydrochlorothiazide and amiloride on plasma magnesium in patients with essential hypertension. *Magnesium Bull*;**6**:127-132.
49. Reyes AJ, Leary WP, van der Byl K. Response of magnesiuresis to furosemide-amiloride and captopril.. In Golf S, Dralle D, Vecciet L, (eds.), (1994); *Magnesium 1993* London: John Libbey, 303/306.
50. Crawford RJ, Allman S, Kitchen SP, Richards HH. (1988) A long-term open study of a frusemide/amiloride combination ("Frumil") in elderly patients with congestive cardiac failure. *Curr Med Res Opin*;**10**:675-681.
51. Gullestad L, Midtvedt K, Dolva LO, Norselth . (1992) The magnesium loading test :reference values in healthy subjects. *Scand J Clin Lab Invest*;**52**:245-253.
52. Gullestad L,, Dolva LO, Waage A, Falch D, Fagerthun H, Kjecshus J. (1992);Magnesium deficiency diagnosed by an intravenous loading test. *Scand J Clin Lab Invest* **52**:245-253.
53. Webb EL, Godfrey JC, Rosenbaum R, Zisblatt M, Vukobich RA, Neiss ES.(1984) Chlorthalidone-triamterene: a potassium-sparing diuretic combination for the treatment of oedema. *J Int Med Res*;**12**:147-153.

MINDING MAGNESIUM WHILE TREATING ESSENTIAL HYPERTENSION WITH DIURETICS

A. J. REYES and L. ALCOCER*
Institute of Cardiovascular Theory, Sotelo 3908, Montevideo 11700, Uruguay; *National University and General Hospital, Tuxpan 16, Mexico City, 06760 DF, Mexico.

Oral formulations of diuretics classically used as once-daily antihypertensive pharmacotherapy, such as hydrochlorothiazide 25 mg, increase 24-hour natriuresis, kaliuresis and magnesiuresis (Table 1, [36]), and may thus conduce to magnesium deficit during prolonged treatment [1, 8, 14, 15, 31].

Bodily magnesium deficit has been incriminated as one of the unwanted actions of diuretics that contribute to the dose-dependent increase in cardiac sudden death caused by these agents [1, 11-13]. Fixed-dose combinations of a thiazide and a potassium-retaining diuretic may not raise 24-hour magnesiuresis [39], but they are highly potent natriuretics and may cause hyponatraemia (Fig. 1, Table 2, [36]). However, some recently developed effectively antihypertensive formulations of diuretics (e.g. torasemide 2.5 mg) do not change 24-hour natriuresis and magnesiuresis (Tables 3 and 4), as a result of a mild and biphasic action on renal excretions (Fig. 2), nor do these formulations increase plasma renin activity or aldosterone concentration (Table 5).

Unnecessarily high doses of diuretics have traditionally been used to treat essential hypertension. This practice recognizes three main sources.

First, when modern diuretics became available it was thought that their antihypertensive action is secondary to their natriuretic effect, and consequently no dose-response curves for the latter were evaluated. In actual fact, the dose-response curves for the natriuretic and for other renal excretory actions of diuretics and those for the antihypertensive effects of these drugs are not parallel; moreover, the natriuretic action of diuretics only accounts for 25-45 per cent of the decrease that occurs in systolic blood pressure (it has no bearing on the response of diastolic blood pressure) and for the rate at which blood pressure falls during the first two or three weeks of treatment (Figs. 3 and 4). Thus, some non-diuretic formulations of diuretics [28, 30], such as torasemide 2.5 mg (Tables 2 and 4, [28-30, 35]) and muzolimine 20 mg (this diuretic substance is no longer available, [20]), are efficacious as once-daily

T. Theophanides and J. Anastassopoulou (eds.), Magnesium: Current Status and New Developments, 189-198.

TABLE 1. Mean urinary electrolyte excretions (mmol) after single oral doses of placebo and of two formulations of diuretics in fourteen healthy subjects, on three conveniently separate days. An individually randomized, crossover and double-blind desing was followed. From Leary *et al.* [21], by courtesy of *Journal of International Medical Research*

	Sodium			Potassium			Magnesium		
	Hours after dosing			Hours after dosing			Hours after dosing		
	0 - 6	6 - 24	0 - 24	0 - 6	6 - 24	0 - 24	0 - 6	6 - 24	0 - 24
Placebo	36	111	147	18	33	50	1.2	4.1	5.3
Furosemide 80 mg	225***	49***	273***	34***	32	66**	3.0***	3.3*	6.3**
HCTZ 25 mg	116***	141*	257***	26	34	59*	1.9*	4.6	6.5*

HCTZ: hydrochlorothiazide.
Contrasts *vs.* placebo (p values): *<0.05; **<0.01; ***<0.001.

antihypertensives (Figs. 3 and 4, [2, 3, 10, 16, 34]).

The second source of the use of unnecessarily high doses of diuretics in antihypertensive treatment were some dose-finding studies on the response of high blood pressure to these drugs; the trials in question followed a design whereby a certain dose was used for a short period, usually between 1 and 4 weeks, at the end of which the dose was doubled if blood pressure had not achieved the value sought or had not undergone an arbitrarily defined decrease.

TABLE 2. Mean 24-hour diuresis (litre) and urinary electrolyte outputs (mmol) after single oral doses of placebo and of various formulations of diuretics in twelve healthy subjects, on four conveniently separate days. An individually randomized and crossover desing was followed. From Leary *et al.* [23], by courtesy of Elsevier Science Publishers B.V.

Treatment	Fluid	Sodium	Potassium	Magnesium
Placebo	3.19	185	54	6.5
HCTZ 25 mg	3.92***	304***	74**	7.6**
Spironolactone 100 mg	3.39	219	54	7.0
HCTZ 25 mg and spironolactone 100 mg	3.96***	349***	58	7.3

Contrasts *vs.* placebo (p values): **<0.01; ***<0.001.

This procedure was repeated at similar successive intervals. In actual fact, it may take as long as 12-14 weeks for a given oral formulation of a diuretic to exert its full antihypertensive action when used as once-daily monopharmacotherapy (Figs. 3 and 4). Early dose changes should not be carried out when dose-response curves for the antihypertensive effects of diuretics are evaluated, because the efficacy of lower doses may be underestimated and the conclusion that doses higher than necessary are required may be reached; higher doses will cause more frequent and intense undesired metabolic changes.

Third, unnecessarily high doses of diuretics are many times used because salt intake curtailment is disregarded as a therapeutic manoeuvre that enables to use lower doses to produce any given antihypertensive action.

The process whereby lower doses of diuretics are being used as to treat hypertension is progressing slowly. The lowest doses of traditional thiazide-type diuretics known to be effective are listed in Table 6. The effect of some of

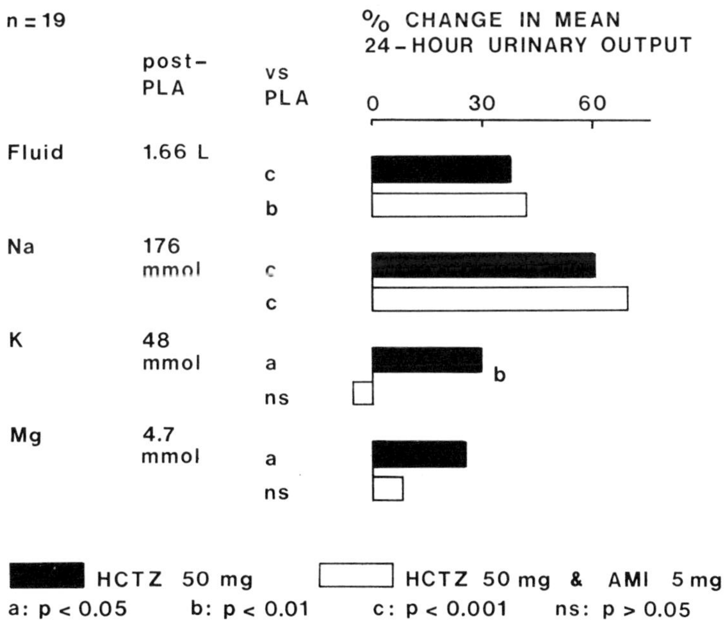

Figure 1. Nineteen healthy subjects took single oral doses of placebo (PLA), of 50 mg hydrochlorothiazide (HCTZ), and of a combination of 50 mg HCTZ and 5 mg amiloride (AMI), on three conveniently separate days. An individually randomized, crossover and double-blind desing was followed. Data from Leary *et al.* [19], by courtesy of *Current Therapeutic Research*; figure reproduced from Leary *et al.* [22], by courtesy of *Progress in Pharmacology and Clinical Pharmacology.*

TABLE 3. Mean 24-hour diuresis (litre) and urinary electrolyte outputs (mmol) after single oral doses of placebo and of two formulations of torasemide in fourteen healthy subjects, on three conveniently separate days. An individually randomized, crossover and double-blind design was followed. From Reyes *et al.* [33, 35], by courtesy of *Progress in Pharmacology and Clinical Pharmacology*

Treatment	Fluid	Chloride	Sodium	Potassium	Magnesium	Calcium
Placebo	2.95	168	180	54	5.0	4.8
Torasemide 2.5 mg	2.99	187	193	55	5.4	5.8
Torasemide 20 mg	4.15***	348***	321***	69**	6.2**	6.6***

Contrasts *vs.* placebo (p values): **<0.01; ***<0.001.

these doses on natriuresis and magnesiuresis have been investigated [4, 5], but not according to standardized criteria [28], except in the case of xipamide 5 mg which has been found to be a 24-hour natriuretic and magnesiuretic [32].

Present-day Antihypertensive Treatment with Diuretics

In conclusion, when the decision of treating essential hypertension with a diuretic is adopted, a low-dose (if possible non-natriuretic) formulation should be used, to avoid the unwanted increases in renal excretions caused by conventional doses. Fixed-dose combinations of a thiazide-type and a

TABLE 4. Mean 24-hour diuresis (litre) and urinary electrolyte outputs (mmol) after single oral doses of placebo and of three formulations of torasemide in sixteen healthy subjects, on four conveniently separate days. An individually randomized, crossover and double-blind design was followed. From A. J. Reyes, W. P. Leary and K. van der Byl, unpublished study

Treatment	Fluid	Chloride	Sodium	Potassium	Magnesium	Calcium
Placebo	3.08	187	180	56	6.0	5.7
Torasemide 2.5 mg	2.88	191	180	60	5.8	5.2
Torasemide 5 mg	3.24	240***	220**	61	6.3	6.2*
Torasemide 10 mg	3.47**	278***	240***	65*	6.5	6.5**

Contrasts *vs.* placebo (p values): *<0.05; **<0.01; ***<0.001.

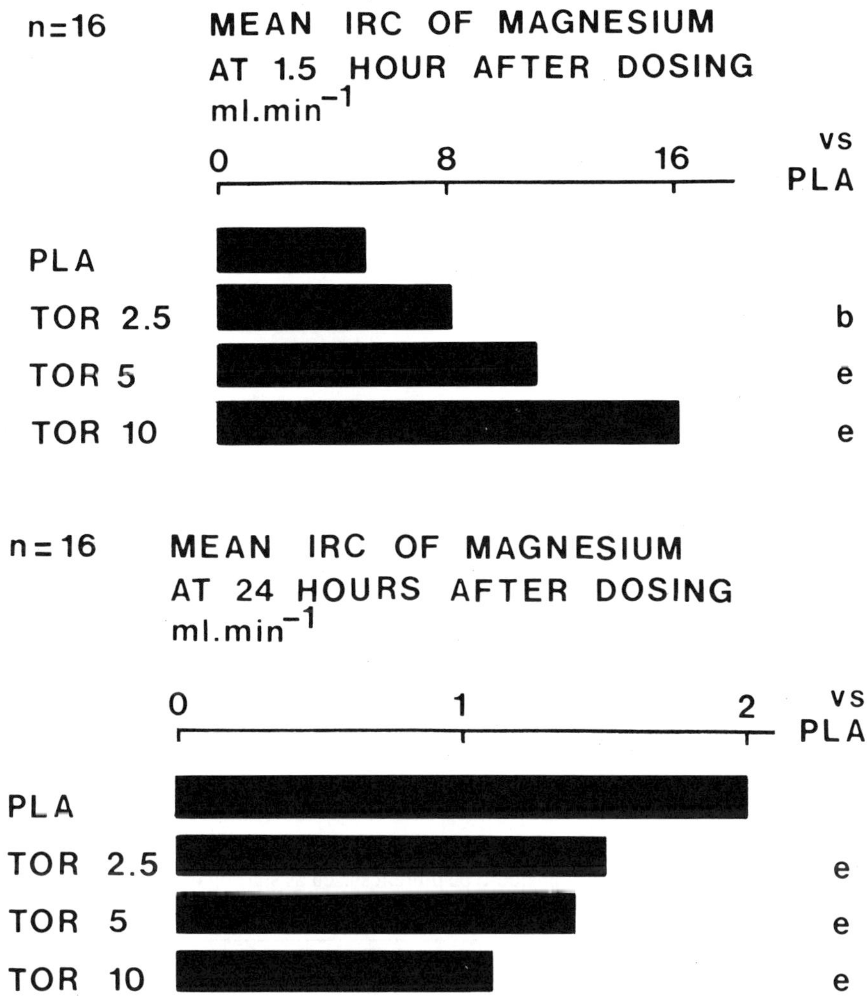

Figure 2. Mean instantaneous renal clearance of magnesium (IRC) in sixteen healthy subjects who took single oral doses of placebo (PLA) and of the 2.5, 5 and 10 mg formulations of torasemide (TOR), on four conveniently separate days. An individually randomized, crossover and double-blind design was followed. The IRC of magnesium was evaluated according to Reyes [28]. Contrasts *vs.* PLA (p values): $b < 0.01$; $e < 0.00001$. From A. J. Reyes, W. P. Leary and K. van der Byl, unpublished study.

TABLE 5. Mean plasma renin activity and aldosterone concentration after single oral doses of placebo and of three formulations of torasemide in sixteen healthy subjects, on three conveniently separate days. An individually randomized, crossover and double-blind design was followed. From A. J. Reyes, W. P. Leary and K. van der Byl, unpublished study.

Study variable	Treatment	Hours after dosing at hour 08.00			
		0	1.5	6	24
Plasma renin activity ($ng.ml^{-1}.h^{-1}$)	Placebo	2.1	3.5	2.5	2.4
	Torasemide 2.5 mg	2.8	2.4	2.4	3.0
	Torasemide 5 mg	1.9	3.2	3.1	2.1
	Torasemide 10 mg	2.2	4.3	4.1*	4.2*
Plasma aldosterone concentration ($nmol.L^{-1}$)	Placebo	0.47	0.18	0.23	0.53
	Torasemide 2.5 mg	0.47	0.23	0.29	0.52
	Torasemide 5 mg	0.46	0.23	0.25	0.57
	Torasemide 10 mg	0.50	0.41	0.43	0.99

Contrasts *vs.* placebo (p values): *<0.05; **<0.01.

TABLE 6. Once-daily doses of thiazide-type diuretics used as starting or maintenance monopharmacotherapy in uncomplicated essential hypertension

Diuretic substance	Dose used after drug availability (mg)	Dose currently used in 1997 (mg)	Dose with proven efficacy (mg)	Reference number for previous column
Bendrofluazide	5-10	2.5	1.25	7
Chlorthalidone	50-100	25	12.5	17
Cyclopenthiazide	0.25-0.5	0.25	0.125	24
Hydrochlorothiazide	50-100	25	12.5	5
			12.5	6
Metolazone	5-10	2.5	0.5-1	9
Xipamide	20-40	10-20	5	26
			5	18

potassium-retaining diuretic should not be used, since they may provoke hyponatraemia in some patients, particulary in elderly women during summer [36].

If a non-diuretic and non-magnesiuretic dose of a diuretic is not effective after 14 weeks of monotheraphy, a low-potency natriuretic but still non-

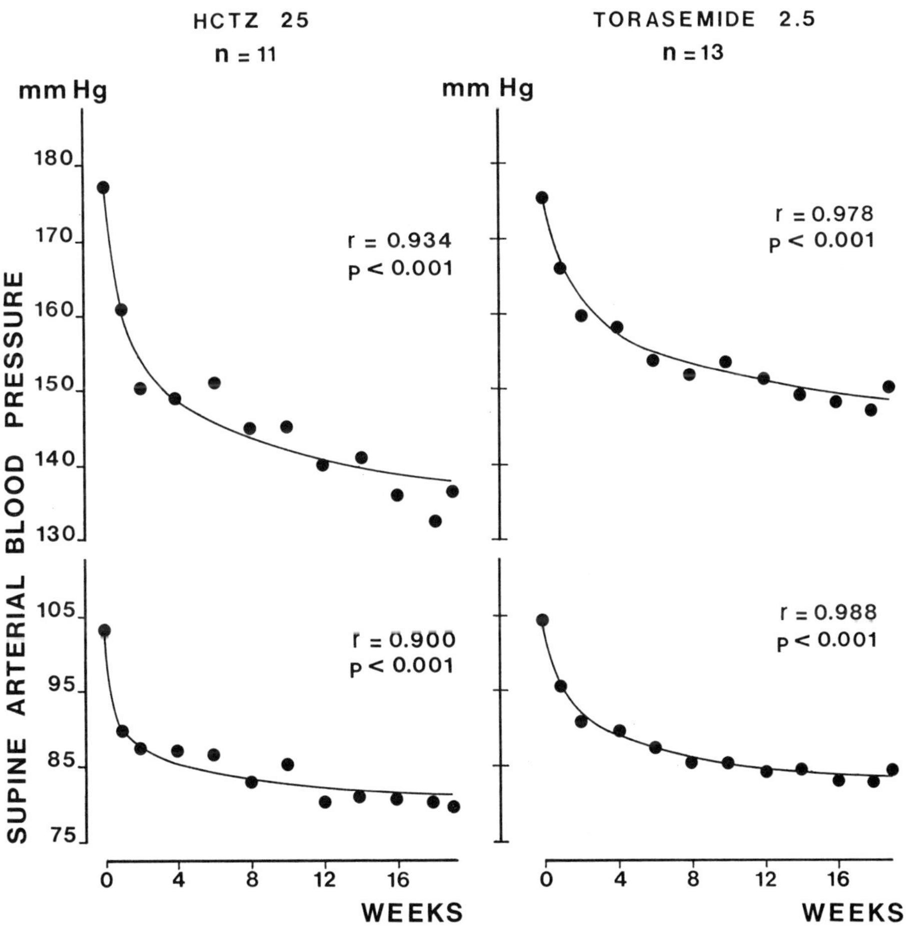

Figure 3. Mean values of blood pressure at the end of a 4-week placebo run-in (week 0) and during antihypertensive monopharmacotherapy with 25 mg hydrochlorothiazide (HCTZ) and with 2.5 mg torasemide, once daily, in eleven and thirteen elderly patients respectively. The study was individually randomized and double-blind. The Montevideo Mathematical Model [27] (curves) was fitted to mean values of blood pressure as functions of time; the correlation coefficients (r) stand for the relationship between linearized mean values of blood pressure and values of time. Data from Reyes *et al.* [34]; figure reproduced from Metz *et al.* [25], by courtesy of *Progress in Pharmacology and Clinical Pharmacology*.

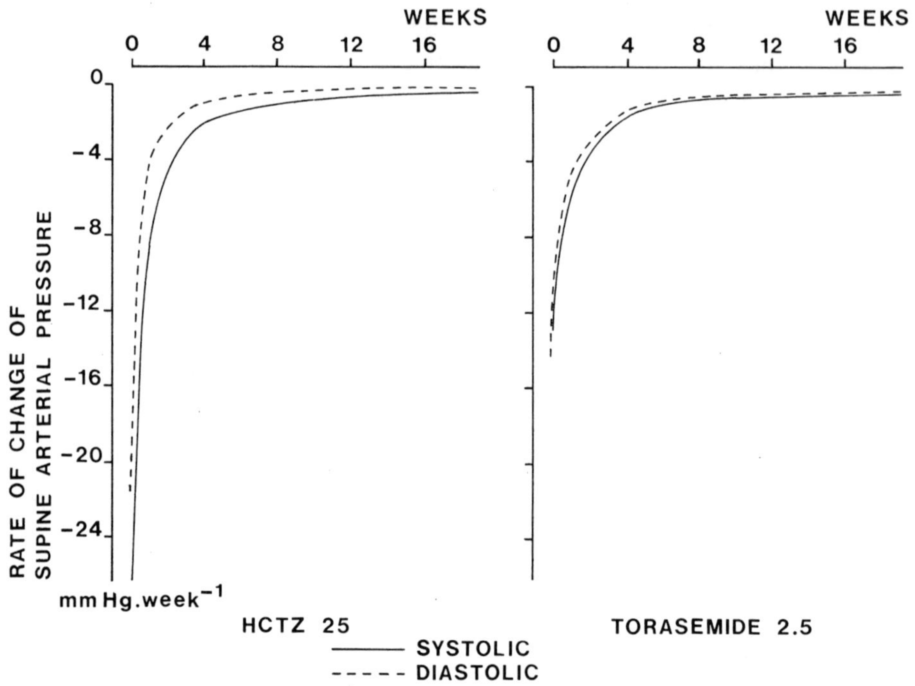

Figure 4. Rate of change of blood pressure as a function of time during antihypertensive monopharmacotherapy with 25 mg hydrochlorothiazide (HCTZ) and with 2.5 mg torasemide, once daily, in eleven and thirteen elderly patients respectively. The functions depicted were evaluated as the first derivatives of corresponding functions in Fig. 3. From Metz *et al.* [25], by courtesy of *Progress in Pharmacology and Clinical Pharmacology.*

magnesiuretic [37] formulation, such as torasemide 5 mg or cicletanine 50 mg [38], may be used. If blood pressure is not controlled after 12-14 weeks, another drug should be added or substitute for diuretics.

References

1.Alcocer L, Reyes AJ: Negative effects of risk factor reduction on the heart. *Cardiol Clinics* 1986;**4**,153-161.

2.Alcocer L, Reyes AJ: Antihypertensive treatment with diuretics in elderly patients. *Prog Pharmacol Clin Pharmacol* 1995;**10/3**,61-74.

3. Alcocer L, Parra JZ, Campos E, Calvo C, Trujillo C: Non-diuretic doses of torasemide are as effective as regular diuretic doses of chlorthalidone for the treatment of arterial hypertension. *Am J Hypertens* 1995;**8**,77A.
4. Beermann B, Groschinsky-Grind M: Pharmacokinetics of hydrochlorothiazide in man. *Eur J Clin Pharmacol* 1977;**12**,297-303.
5. Beermann B, Groschinsky-Grind M: Antihypertensive effects of various doses of hydrochlorothiazide and its relation to the plasma level of the drug. *Eur J Clin Phamacol* 1978;**13**,195-201
6. Berglund G, Andersson O: Low doses of hydrochlorothiazide in hypertension. Antihypertensive and metabolic effects. *Eur J Clin Pharmacol* 1976;**10**,177-182.
7. Carlsen JE, Kober L, Torp-Pedersen C, Johansen C: Relations between dose of bendrofluazide, antihypertensive effects, and adverse biochemical effects. *Br Med J* 1990;**300**,975-978.
8. Cocco G, Iselin HU, Strozzi C, Cesana B, Baumeler HR: Magnesium depletion in patients on long-term chlorthalidone therapy for essential hypertension. *Eur J Clin Phamacol* 1987;**32**,335-338.
9. Curry CL, Janda SM, Harris S, MacKay JH, Nugent CA; Ryan J, Schnaper H, Schoenberger J: Clinical studies of a new, low-dose formulation of metolazone for the treatment of hypertension. *Clin Ther* 1986;**9**,47-62.
10. Dupont AG, Schoors D, Sin RO, Vanhaelst L: Antihypertensive efficacy of low dose torasemide in essential hypertension: a placebo controlled study. *J Human Hypertens* 1988;**2**,265-268.
11. Frishman WH: Results of hypertension treatment with low-dose combinations of a β-adrenergic blocker and a diuretic. *Am J Hypertens* 1997;**10**,17S-23S.
12. Grobbee DE, Hoes AW: Non potassium-sparing diuretics and risk of sudden cardiac death. *J Hypertens* 1995;**13**,1539-1545.
13. Hoes AW, Grobbee DE, Lubsen J, Man in´t Veld AJ, van der Does E, Hofman A: Diuretics, beta-blockers and the risk for sudden cardiac death in hypertensive patients. *Ann Intern Med* 1995;**123**,481-487.
14. Hollifield JW: Electrolyte disarray and cardiovascular disease. *Am J Cardiol* 1989;**63**,21B-26B.
15. Hollifield JW: Thiazide treatment of systemic hypertension: effects on serum magnesium and ventricular ectopic activity. *Am J Cardiol* 1989;**63**,22G-25G.
16. Kirsten R, Molz KH, Tzonev I, Nelson K: Clinical evaluation of muzolimine and indapamide during treatment for essential hypertension. *Z Kardiol* 1985;**74**(Suppl 2),66-72.
17. Korduner I, Käbin I, Hagbarth G: Low-dose chlorthalidone treatment in previously untreated hypertension. *Curr Ther Res* 1981;**29**:208-215.
18. Leary WP, Reyes AJ: Effects of low doses of xipamide given as monotherapy in essential hypertension. *Curr Ther Res* 1983;**34**,888-899.
19. Leary WP, Reyes AJ, van der Byl K: Effects of a combination of hydrochlorothiazide and amiloride on urinary magnesium excretion in healthy adults. *Curr Ther Res* 1984;**35**,293-300.
20. Leary WP, Reyes AJ, van der Byl K: The effects of single doses of muzolimine upon urinary solute and fluid excretion. *Z Kardiol* 1985;**74**(Suppl 2),135-140.
21. Leary WP, Reyes AJ, van der Byl K: Renal excretory actions of furosemide, of hydrochlorothiazide and of the vasodilator flosequinan in healthy subjects. *J Int Med Res* 1990;**18**,120-141.
22. Leary WP, Reyes AJ, van der Byl K: Interactions between different diuretics and between diuretics and other drugs on renal excretions in man: mechanisms and clinical implications. *Prog Pharmacol Clin Pharmacol* 1992;**9**,317-360.

23. Leary WP, Reyes AJ, van der Byl K: The antildosterone substance spironolactone is a facultative natriuretic, antikaliuretic and antimagnesiuretic in man. In: Puschett JB, Greenberg A, (eds.), *Diuretics IV: Chemistry, Pharmacology and Clinical Applications.* Amsterdam: Elsevier Science, 1993; 391-394.
24. McVeigh G, Galloway D, Johnston D: The case for low-dose diuretics in hypertension: comparison of low and conventional doses of cyclopenthiazide. *Br Med J* 1988;**297**,95-98.
25. Metz P, Alcocer L, Leary WP, Ferreira MI, Reyes AJ: Antihypertensive therapy with diuretics in the elderly: clinical approach. *Prog Pharmacol Clin Pharmacol* 1992;**9**,473-495.
26. Pasquel R, Tribble PW, Simon A: Hypotensive effects of xipamide in essential hypertension. Crossover comparison with hydrochlorothiazide. *J Clin Pharmacol* 1981;**21**,316-322.
27. Reyes AJ: Formal comparison of the antihypertensive effects of torasemide and other diuretics by the Montevideo Mathematical Model. *Arzneimittelforschung* 1988;**38/I**,194-199.
28. Reyes AJ: Effects of diuretics on outputs and flows of urine and urinary solutes in healthy subjects. *Drugs* 1991;*41*(Suppl 3),33-59.
29. Reyes AJ: Effects of diuretics on renal excretory function. *Eur Heart J* 1992;**13**(Suppl G),15-21.
30. Reyes AJ: Renal excretory profile of loop diuretics – consequences for clinical applications. *J Cardiovasc Pharmacol* 1993;**22**(Suppl 3),S11-S23.
31. Reyes AJ, Leary WP: Cardiovascular toxicity of diuretics related to magnesium depletion. *Human Toxicol* 1984;**3**,351-372.
32. Reyes AJ, Leary WP: The magnesiuric effects of several single doses of xipamide in healthy adults. *Braz J Med Biol Res* 1984;**17**,285-291.
33. Reyes AJ, Leary WP, van der Byl K: Excretion of urinary fluid and solutes after single doses of furosemide and hydrochlorothiazide and of four single doses of the diuretic torasemide in healthy man. *Prog Pharmacol Clin Pharmacol* 1990;**8/1**,47-71.
34. Reyes AJ, Chiesa PD, Santucci MR, Batista LB, Olhaberry JV, Mosler Al, Mignone AM, Leary WP, Queiruga G: Hydrochlorothiazide versus a non-diuretic dose of torasemide as once-daily antihypertensive monopharmacotherapy in elderly patients. *Prog Pharmacol Clin Pharmacol* 1990;**8/1**,183-209.
35. Reyes AJ, Leary WP, van der Byl K: The urinary yield (UY) mathematical model of accumulated renal outputs of urine and of urinary solutes after administration of placebo, diuretics or any other drug. Continuous flows of urine and urinary solutes and variables derived thereof: maximal urine and urinary solute flows, times from dosing to them, and instantaneous renal clearance and instantaneous fractional excretions of urinary solutes. *Prog Pharmacol Clin Pharmacol* 1992;**9**,173-218.
36. Reyes AJ, Leary WP, Crippa G, Alcocer L: Untoward actions of diuretics in congestive heart failure. *Prog Pharmacol Clin Pharmacol* 1995;**10/3**,131-158.
37. Reyes AJ, Leary WP, van der Byl K: Responses of 24-hour magnesiuresis to eighteen formulations of thiazide-type or loop diuretics. *Cardiovasc Drugs Ther* 1997;**11**,361.
38. Tarrade T, Berthet P, Paillasseur JL, Bosquet D, Allard M: Efficacité antihypertensive et tolérance du ciclétanine. *Arch Mal Coeur* 1989;**82**:91-97.
39. van del Byl K, Reyes AJ, Leary WP: Responses of 24-hour magnesiuresis to potassium-retaining diuretics, to combinations of diuretics and to combinations of diuretics and other drugs. *Cardiovasc Drugs Ther* 1997;**11**,362.

DIURETICS AND MAGNESIUM: CLINICAL PHARMACOLOGY IN HEALTHY SUBJECTS

A.J. REYES
Institute of Cardiovascular Theory,
Sotelo 3908, Montevideo 11700, Uruguay

Single doses of the majority of common oral formulations of loop and of early distal tubular diuretics raise 24-hour magnesiuresis in healthy subjects; the actions of 40 mg furosemide (20), of 50 mg hydrochlorothiazide (1), and of 100 mg chlorthalidone (21) constitute archetypal examples.

The responses of 24-hour urinary magnesium output to single oral doses of diuretics (Tables 1-4) are being evaluated for the first time or re-investigated as part of an investigative series in progress, in which a strictly formal methodology is observed (2, 13-19, 22, 23, 25, 26). In each study, healthy subjects receive placebo and one or more active formulations, on conveniently separate days, according to an individually randomized, crossover and double-blind design (16).

Some formulations of loop diuretics (e.g. torasemide 2.5 mg) are non-diuretic, non-natriuretic and non-magnesiuretic (Table 1). Some formulations of diuretics that elevate 24-hour natriuresis by less than 45 per cent, and are consequently regarded as low-potency formulations (17, 18), do not increase 24-hour magnesiuresis (Tables 1 and 2). Nearly all high-potency formulations of common diuretics (i.e. formulations that increase 24-hour natriuresis by more than 45 per cent) elevate 24-hour magnesiuresis (Tables 1 and 2). Magnesiuretic formulations of loop diuretics increase renal magnesium output markedly during the first few hours after the first dose, but this rise is followed by an undershoot (Table 3). Thiazide-type diuretics increase magnesiuresis less than loop diuretics do early after dosing, but the thiazide-induced elevation is more prolonged than its loop-diuretic counterpart and it is not followed by a rebound (Table 3). Thus, furosemide 80 mg and hydrochlorothiazide 25 mg exert similar 24-hour magnesiuretic actions, despite the marked early-after-dosing rise in urinary magnesium output provoked by the loop diuretic formulation (Table 3).

Potassium-retaining diuretics tend to retain magnesium

T. Theophanides and J. Anastassopoulou (eds.), Magnesium: Current Status and New Developments, 199-208.

TABLE 1. Responses of 24-hour urinary sodium (Na) and magnesium (Mg) outputs to single oral doses of loop (furosemide-type) diuretics in healthy subjects

Study	Number of subjects	Diuretic substance	Dose (mg)	Mean 24-hour excretion after placebo (mmol)		% change in mean 24-hour excretion after diuretic		Reference number
				Na	Mg	Na	Mg	
D	14	Furosemide	40	147	3.8	52***	45*	28
F	14	Furosemide	80	147	5.3	86***	20**	10
M1	10	Muzolimine	20	186	4.7	7	-6	7
M1	10	Muzolimine	30	186	4.7	33**	9	7
M2	10	Muzolimine	40	225	4.8	17*	11	7
N	14	Torasemide	2.5	180	5.0	7	8	28
O	16	Torasemide	2.5	180	6.0	0	-2	31
D	14	Torasemide	5	147	3.8	28	12	28
O	16	Torasemide	5	180	6.0	22**	5	31
D	14	Torasemide	10	147	3.8	26	3	28
O	16	Torasemide	10	180	6.0	34***	9	31
N	14	Torasemide	20	180	5.0	78***	24**	28

Contrasts vs. placebo means (p values): * < 0.05; ** < 0.01; *** < 0.001.

TABLE 2. Responses of 24-hour urinary sodium (Na) and magnesium (Mg) outputs to single oral doses of early distal tubular (thiazide-type) diuretics in healthy subjects

Study	Number of subjects	Diuretic substance	Dose (mg)	Mean 24-hour excretion after placebo (mmol)		%change in mean 24-hour excretion after diuretic		Reference number
				Na	Mg	Na	Mg	
B	14	Cicletanine	50	156	5.4	26*	0	29
B	14	Cicletanine	100	156	5.4	38***	15*	29
B	14	Cicletanine	150	156	5.4	49***	8	29
C	9	Clopamide	5	154	5.0	85***	27**	27
D	14	Hydrochlorothiazide	25	147	3.8	86***	84***	28
G	13	Hydrochlorothiazide	25	151	4.1	33***	73***	6
F	14	Hydrochlorothiazide	25	147	5.3	75***	23*	10
I	12	Hydrochlorothiazide	25	185	6.5	64***	17**	12
J	19	Hydrochlorothiazide	50	177	4.7	61***	24*	4
Z	12	Hydrochlorothiazide	50	146	5.2	110***	24*	5
P	13	Xipamide	5	151	4.0	120***	28**	24
P	13	Xipamide	10	151	4.0	111***	50***	24
P	13#	Xipamide	20	151	4.0	116***	40***	24

#Active-medication data from twelve subjects.
Contrasts vs. placebo means (p values): * 0.05; ** < 0.01; *** < 0.001.

TABLE 3. Mean urinary excretions (mmol) of chloride (Cl), sodium (Na), potassium (K), and magnesium (Mg) after administration of single oral doses of placebo, of furosemide 80 mg and of hydrochlorothiazide (HCTZ) 25 mg to fourteen healthy subjects, on three conveniently separate days. Dosing took place at hour 08.00. An individually randomized, crossover and double-blind design was followed. From reference number 10, by courtesy of Journal of International Medical Research

Hours after dosing	Substance	Dose (mg)	Cl	Na	K	Mg
0 to 3	Placebo		19	16	8	0.7
	Furosemide	80	195***	182***	22***	2.2***
	HCTZ	25	53***	48***	10	0.8
3 to 6	Placebo		21	19	9	0.5
	Furosemide	80	48**	43**	12	0.8
	HCTZ	25	79***	68***	16	1.1**
6 to 9	Placebo		36	34	10	0.9
	Furosemide	80	19***	17***	10	0.5**
	HCTZ	25	55**	52**	10	1.1
9 to 12	Placebo		26	25	7	0.6
	Furosemide	80	10**	12**	6	0.4
	HCTZ	25	44**	41**	11	0.9*
12 to 24	Placebo		54	52	16	2.6
	Furosemide	80	15***	20***	15	2.3
	HCTZ	25	53	49	12	2.7
0 to 24	Placebo		155	147	50	5.3
	Furosemide	80	288***	273***	66**	6.3**
	HCTZ	25	283***	257***	59*	6.5*

Contrasts vs. placebo (p values): * < 0.05; ** < 0.01; *** < 0.001. Contrasts between 24-hour urinary outputs of chloride, sodium, potassium and magnesium after the two diuretics: $p > 0.05$.

TABLE 4. Responses of 24-hour urinary sodium (Na) and magnesium (Mg) outputs to single oral doses of potassium-retaining diuretics and of combinations of a kaliuretic and a potassium-retaining diuretic in healthy subjects

Study	Number of subjects	Diuretic substance(s)	Dose (mg)	Mean 24-hour excretion after placebo (mmol)		% change in mean 24-hour excretion after diuretic		Reference number
				Na	Mg	Na	Mg	
A	13	Amiloride	5	121	4.0	38*	-5	3
A	13	Amiloride	10	121	4.0	66***	-7	3
I	12	Spironolactone	100	185	6.5	18	8	12
Q	14	Furosemide and amiloride	40 5	146	4.9	75***	-1	30
J	19	Hydrochlorothiazide and amiloride	50 5	177	4.7	70***	8	4
I	12	Hydrochlorothiazide and spironolactone	25 100	185	6.5	88***	13	12

Contrasts vs. placebo means (p values): * < 0.05; *** < 0.001.

TABLE 5. Mean urinary excretions (mmol) of sodium (Na), potassium (K), calcium (Ca), and magnesium (Mg) on the first and seventh experimental days of once-daily oral administration of furosemide 40 mg to fourteen healthy subjects. Dosing took place at hour 08.00. From WP Leary, AJ Reyes, K van der Byl: unpublished results

Hours after dosing	Treatment day	Na	K	Ca	Mg
0 to 3	1	146	24	2.6	1.5
	7	100*	19	2.5	1.6
3 to 6	1	34	9	1.1	0.7
	7	20	7	1.1	0.6
6 to 9	1	25	9	0.9	0.6
	7	23	6	1.0	0.5
9 to 12	1	13	6	0.6	0.5
	7	16	5	0.9	0.5
12 to 24	1	29	14	2.1	2.3
	7	23	11*	2.0	2.2
0 to 24	1	247	62	7.4	5.6
	7	181***	50*	7.5	5.5

Contrasts vs. day 1 (p values: * < 0.05; *** < 0.001.

(Table 4), but to a less extent than they retain potassium (15).

During repeated once-daily administration, the rises in 24-hour natriuresis and kaliuresis produced by the first dose of natriuretic formulations of loop diuretics wane (11); the elevation in 24-hour magnesiuresis persists, unless bodily magnesium undergoes a reduction (Table 5).

Available fixed-dose combinations of a kaliuretic and a potassium-retaining diuretic generally do not elevate, or tend not to elevate, magnesiuresis (Table 4).

The responses of 24-hour magnesiuresis to combinations of diuretics and various cardiovascular drugs are shown in Table 6.

TABLE 6. Responses of 24-hour urinary sodium (Na) and magnesium (Mg) outputs to single oral doses of combinations of a diuretic and another or other drug(s) in healthy subjects

Study	Number of subjects	Substances	Dose (mg)	Mean 24-hour excretion after placebo (mmol)		% change in mean 24-hour excretion after active treatment		Reference number
				Na	Mg	Na	Mg	
C	9	Clopamide and pindolol	5 10	154	5.0	99***	-1	27
U	14	Clopamide, reserpine and dihydro-ergocristine	5 0.1 0.5	144	3.8	88***	40**	9
Q	14	Furosemide, amiloride and captopril	40 5 25	146	4.9	99***	8	30
W	10	Hydrochlorothiazide and acebutolol	12.5 200	133	5.5	27*	4	8
G	13	Hydrochlorothiazide and captopril	25 100	151	4.1	75***	45***	6
Z	12	Hydrochlorothiazide and sotalol	50 320	146	5.2	66**	20*	5
U	14	Hydroflumethiazide, *Rauwolfia serpentina* and potassium chloride	50 50 625	144	3.8	85***	66**	9

Contrasts vs. placebo means (p values): * <0.05; ** <0.01; *** <0.001.

REFERENCES

1. Leary WP, Reyes AJ: The magnesiuric effect of a single dose of hydrochlorothiazide in healthy adults. Curr Ther Res 1982;32,425-431.
2. Leary WP, Reyes AJ: Diuretic-induced magnesium losses. Drugs 1984;28 (Suppl 1),182-187.
3. Leary WP, Reyes AJ, van der Byl K: Urinary magnesium and zinc excretions after two different single doses of amiloride in healthy adults. Curr Ther Res 1983;34, 205-216.
4. Leary WP, Reyes AJ, van der Byl K: Effects of a combination of hydrochlorothiazide and amiloride on urinary magnesium excretion in healthy adults. Curr Ther Res 1984;35,293-300.
5. Leary WP, Reyes AJ, van der Byl K: Effects of hydrochlorothiazide plus sotalol on acute urinary electrolyte excretion in normal subjects. S Afr Med J 1984; 66,680-681.
6. Leary WP, Reyes AJ, van der Byl K, Acosta Barrios TN: Effects of captopril, hydrochlorothiazide, and their combination on timed urinary excretions of water and solutes. J Cardiovasc Pharmacol 1985;7 (Suppl 1), S56-S62.
7. Leary WP, Reyes AJ, van der Byl K: The effects of single doses of muzolimine upon urinary solute and fluid excretion. Z Kardiol 1985;74 (Suppl 2),135-140.
8. Leary WP, Reyes AJ, van der Byl K: Effects of an acebutolol-hydrochlorothiazide combination on urinary solute excretion in healthy adults. S Afr Med J 1986; 69,98-100.
9. Leary WP, Reyes AJ, van der Byl K: Effects of single doses of two antihypertensive rauwolfia-diuretic combinations on urinary excretion. S Afr Med J 1986;70, 95-98.
10. Leary WP, Reyes AJ, Wynne RD, van der Byl K: Renal excretory actions of furosemide, of hydrochlorothiazide and of the vasodilator flosequinan in healthy subjects. J Int Med Res 1990;18,120-141.
11. Leary WP, Reyes AJ, van der Byl K: Response of urinary excretions to a combination of furosemide and perindopril upon once-daily administration in healthy subjects. In: JB Puschett, A Greenberg, (eds.), Diuretics IV: Chemistry, Pharmacology and Clinical Applications. Amsterdam: Elsevier Science, 1993;591-594.
12. Leary WP, Reyes AJ, van der Byl K: The antialdosterone substance spironolactone is a facultative natriuretic, antikaliuretic and antimagnesiuretic in man. In: JB Puschett, A Greenberg, (eds.), Diuretics IV: Chemistry, Pharmacology and Clinical Applications. Amsterdam: Elsevier Science, 1993;391-394.

13. Reyes AJ: Deleterious effects of antihypertensive treatment on magnesium turnover. Prog Pharmacol 1985; 6/1,51-87.
14. Reyes AJ: Interactions between magnesium and drugs in congestive heart failure. Magnes Bull 1987;9,93-109.
15. Reyes AJ: Mechanisms and extent of the decrease in magnesiuresis induced by antikaliuretic diuretics in man. In: Y Itokawa, J Durlach, (eds.), Magnesium in Health and Disease. London: John Libbey, 1989;415-422.
16. Reyes AJ: Effects of diuretics on outputs and flows of urine and urinary solutes in healthy subjects. Drugs 1991;41 (Suppl 3),35-59.
17. Reyes AJ: Effects of diuretics on renal excretory function. Eur Heart J 1992;13 (Suppl G),15-21.
18. Reyes AJ: Renal excretory profile of loop diuretics — consequences for therapeutic application. J Cardiovasc Pharmacol 1993;22 (Suppl 3),S11-S23.
19. Reyes AJ: Momentous clinicopharmacological variables concerning diuretics. Prog Pharmacol Clin Pharmacol 1995;10/3,1-22.
20. Reyes AJ, Leary WP: The magnesiuric effect of a single dose of furosemide in healthy adults. Curr Ther Res 1982;32,406-416.
21. Reyes AJ, Leary WP: Urinary magnesium and zinc excretions after monodosing healthy volunteers with chlorthalidone. Curr Ther Res 1982;32,128-137.
22. Reyes AJ, Leary WP: Cardiovascular toxicity of diuretics related to magnesium depletion. Human Toxicol 1984;3,351-372.
23. Reyes AJ, Leary WP: Diuretics and magnesium. Magnes Bull 1984;6,87-99.
24. Reyes AJ, Leary WP: The magnesiuretic effect of several single doses of xipamide in healthy subjects. Bras J Med Biol Res 1984;17,285-291.
25. Reyes AJ, Leary WP: Clinicopharmacological reappraisal of the potency of diuretics. Cardiovasc Drugs Ther 1993;7,23-28.
26. Reyes AJ, Leary WP: Renal excretory responses to single and repeated administration of diuretics in healthy subjects. Cardiovasc Drugs Ther 1993;7,29-44.
27. Reyes AJ, Leary WP, van der Byl K: Blunting of diuretic-induced increases in urinary magnesium and potassium outputs by beta-adrenergic blockade in healthy subjects. Magnes Bull 1985;7,121-139.
28. Reyes AJ, Leary WP, van der Byl K: Excretion of urinary fluid and solutes after single doses of furosemide and hydrochlorothiazide and of four different single doses of the diuretic torasemide in healthy subjects. Prog Pharmacol Clin Pharmacol 1990;8/1,47-71.
29. Reyes AJ, Leary WP, van der Byl K: Urinary excretory

actions of various doses of cicletanine during once-daily administration. In: JB Puschett, A Greenberg, (eds.), Diuretics IV: Chemistry, Pharmacology and Clinical Applications. Amsterdam: Elsevier Science, 1993; 383-386.

30. Reyes AJ, Leary WP, van der Byl K: Responses of urinary sodium, potassium and magnesium excretions to the addition of captopril to a combination of furosemide and amiloride. In: JB Puschett, A Greenberg, (eds.), Diuretics IV: Chemistry, Pharmacology and Clinical Applications. Amsterdam: Elsevier Science, 1993; 595-598.

31. Reyes AJ, Leary WP, van der Byl K: Responses of magnesiuresis to the diuretic torasemide. In: S Golf, D Dralle, L Vecchiet, (eds.), Magnesium 1993. London: John Libbey, 1994;291-294.

ALTERATIONS OF SERUM MAGNESIUM UNDER CHRONIC THERAPY WITH DIURETICS AND/OR ANGIOTENSIN - CONVERTING ENZYME INHIBITORS IN HYPERTENSIVE PATIENTS.

A. D. EFSTRATOPOULOS, S. M. VOYAKI, M.A. MEIKOPOULOS.
Hypertension Unit & 3rd Medical Dept. of General Hospital of Athens, 154, Mesogion Ave, Athens-11527- Greece

1. INTRODUCTION

Diuretic drugs (both thiazide and loop diuretics) have been the mainstay of treatment for arterial hypertension and congestive heart failure [1,3]; however, the prolonged use of thiazide diuretics and (to a lesser extent) of loop-acting diuretics is accompanied by a variable depression of serum and tissue levels of potassium and magnesium and reduced concentrations of sodium-potassium pumps in muscle [4-6]. These diuretic - induced electrolytic disturbances have been associated with cardiac arrhythmias (ventricular ectopic activity [7,8], torsade de pointes [9] and sudden death), especially in elderly patients with hypertension, left ventricular hypertrophy, and/or congestive heart failure [9,10], although in a Editorial Review the cardiotoxicity of diuretics is thought to be the result of "much speculation and little substance" [11].
The negative potassium and magnesium balances resulting from diuretic therapy are due primarily to stimulation of the renin - angiotensin - aldosterone system (or secondary hyperaldosteronism [12], and attempts have been made to compat these electrolyte disturbances by giving potassium supplementation, or with the use of anti-kaliuretics (spironolactone, triamterene, amiloride) [13-15], or ACE-inhibitors [13-16].
Aim of the present study was the evaluation (on a long term basis) of serum potassium and magnesium levels of patients with mild or moderate essential hypertension, who were untreated or treated with diuretic monotherapy or with the combination of diuretic and ACE-inhibitor.

2. METHODS

2.1 SUBJECTS

A total of 103 hypertensive patients (39M/64F, 25-82 year-old), suffering from essential hypertension of mild-moderate dergee (WHO stage I-II) were followed for a period of 14-37 months; 31 patients (13M/18F) formed the control group (treatment was given), 22 patients (12M/10F) comprised the group-A (treatment with diuretics

T. Theophanides and J. Anastassopoulou (eds.), Magnesium: Current Status and New Developments, 209-214.

monotherapy), and 50 patients (13M/37F) formed group-B, in **which** antihypertensive treatment consisted from a combination (free combination) of a diuretic and an ACE-inhibitor.

The diuretics given were : Lasix (40mg/day), or Chlorothalidone (25mg every 2nd day) or Polythiazide (0.5 mg every 2nd day). The ACE-inhibitors accompanied the diuretics in Group-B, were: Captopril (50-75mg/d), or Enalapril (5-10 mg/d).

Serum levels of magnesium, potassium and sodium were measured twice (with an Absorption Spectrophotometer) before the start of treatment and every 1-2 months there after; the mean values of the last 3 measurements were compared with the mean values at the baseline (all electrolytes are given as mmol/L, mean+SD).

2.2 *Statistical Analysis*

Values are expressed as means+SD student's and two-tailed t-test for paired data was used to assess the significance of values obtained before and after treatment. One way analysis of variance followed by Bonferroni's method was used to compare values between the three groups (18). As statistically significant was considered p values <0.005.

3. RESULTS

Blood pressure was controlled adequately in all three groups of patients (Table1)..Alterations of serum electrolytes (Mg++, K+ and Na+) in the 3 groups of patients are shown in Table 2; there was no significant change in Table 2; there was no significant change for any of the serum electrolytes in the patients of control group. Monotherapy with diuretics was accompanied by a significant lowering of serum magnesium (from 0.872+0.05 to 0.814+0.06 mmol/L, p=0.001), with a concomitant decrease of serum potassium (from 4.35+0.4 to 4.1+0.3 mmol/L, P=0.024).

TABLE 1 : Alterations of BP (mmHg, M+SEM) in all groups of hypertensives studied.

Groups	N	SBP/DBP (mmHg) Pre - treatment	SBP/DBP (mmHg) During treatment	P
- Control	31	153+2.9/93+1.9	136+2.1/84+1.4	<0.001
- Group - A	22	164+3.7/90+3.2	147+3.2/83+2.2	<0.001
- Group - B	58	175+3/94+2	145+1.8/81+1.3	<0.001
Total	103	164+3.2/93+2.4	143+2.4/83+1.6	<0.001

The combined treatment by ACE-inhibitors and diuretics wasn't accompanied by alterations of serum electrolytes during chronic therapy (Table 2).

TABLE 2: Chronic electrolytic alterations of Na+, K+ and Mg++ (mmol/L, Mean+SD) in the three groups of hypertensive patients.

Groups	N	Mg++ Pre- On Treatment	P	K+ Pre - On Treatment	P	Na+ Pre - On Treatment	P
1. Control	31	0.87+0.09	NS	4.57+0.4 4.46+0.4	NS	143+4 144+4.2	NS
2. Diuretics	22	0.87+0.06	0.001	4.35+0.4 4.1+0.3	0.024	141+4 142+6.7	NS
3. ACEIs + Diuretics	50	0.87+0.05	NS	4.35+0.4 4.35+0.4	NS	144+5 143+3	NS
		0.814+0.06					
		0.851+0.1					
		0.863+0.06					
Total	103						

In comparison to control group, treatment with diuretics resulted in significant hypokalemia and hypomagnesemia, while the co-administration of ACE-inhibitors protected from these electrolytic alterations (Table 2). Although diuretics induced both hypokalemia and hypomagnesemia during chronic therapy, hypomagnesemia was more pronounced than hypokalemia, as was shown by a negative significant correlation of the differences from the values at baseline between these two electrolytic changes in serum (r=-0.438, P=0.041) (Figure 1).

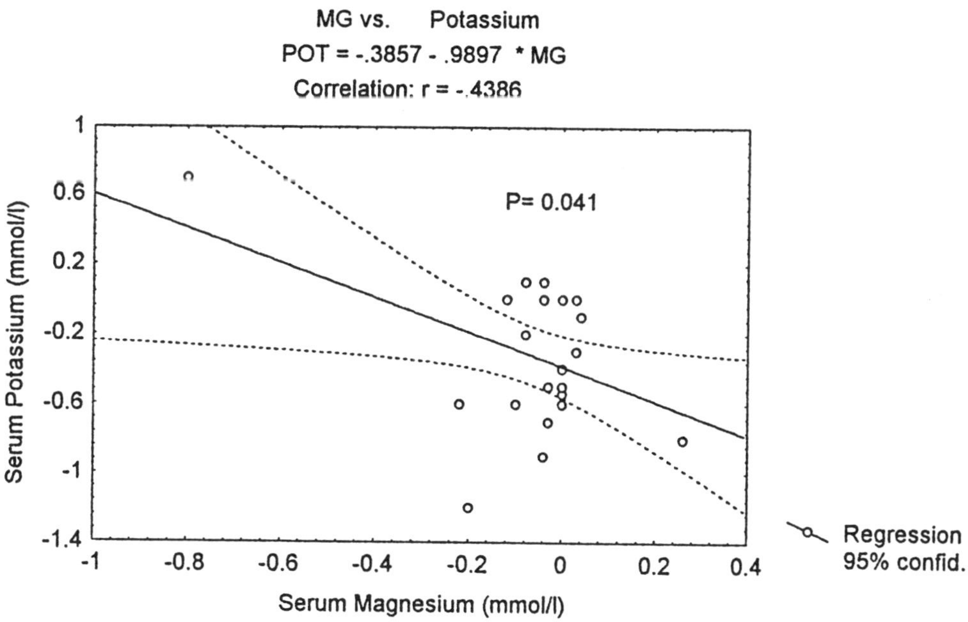

Figure 1 : Correlation of alterations of serum magnesium and potassium in group of hypertensives treated with diuretics.

4. DISCUSSION

Although diuretics (especially thiazide diuretics) have a cornestone place in the management of hypertension and heart failure, there is a controversy as far as their safety is concern, during chronic therapy [9,11,19,20]. The most persistently advanced hypothesis is that the benefit of treatment with diuretics is offset by the diuretic-induced metabolic abnormalities; disturbances of potassium and magnesium metabolism are proposed as possible causes of serious cardiac arrhythmias, whilst hyperlipidemia, impaired carbohydrate tolerance and hyperuricemia are assumed to be atherogenic and to predispose to myocardial infarction. The symptomatic cardiac arrhythmias related to electrolytic disturbances induced by the diuretic treatment, is more common in the elderly patients [21].

Although the major electrolytic disturbance implicated in cardiac arrhythmogenesis is diuretic - induced severe hypokalemia (<2.5 mmol/L), magnesium deficiency is also regarded to be as important as hypokalemia [22].

Magnesium is an essential cofactor for many enzymatic reactions, especially those involved in energy metabolism (Protein synthesis, DNA synthesis, DNA and m-RNA transcription). In tissues, the intracellular Mg++ is about 10mmol/L, while in plasma about 0.8 -1.1mmol/L.

Deficiency of magnesium is a major risk factor for the occurence of severe cardiac arrhythmias [14], especially in patients with congestive heart failure; furthermore, protracted deficits of magnesium can result in severe neurological disturbances [22].

Magnesium is a known activator of Na-K-ATPase, which provides the energy for the sodium - potassium pump, while it is a natural calcium channel blocking agent; thus lack of Mg++ may impair the pumping of Sodium out of the cell and potassium into the cell. The correlation between serum potassium and muscle potassium it seems to be dependent on the content of muscle magnesium, while muscle potassium cannot be corrected by potassium supplementation if there is a concomitant magnesium deficiency [23], which often seems to be the case following long-term (more than 3 years) treatment with most kinds of diuretics. A magnesium deficit may exist while plasma Mg concentration is normal; therefore other techniques for Mg assessment, such as the repletion test, as well as red blood cell and lymphocyte concentrations have been used [16, 24, 25].

The anticipation of diuretic-induced deficiency of potassium and magnesium, includes the supplementation of diuretic therapy with potassium and magnesium, or the administration of potassium (and magnesium) sparing diuretics, such as triamterene and amiloride [16,23,26]. Because the diuretics - induced electrolytic disturbances are thought to be the result of the RAAS stimulation [12,21] and sympathoadrenergic activation, spironolactone and various ACE inhibitors have been used to preserve potassium and magnessium, homoeostasis.

Enalapril in a low dose (10mg/day) is capable to reverse the magnesium and potassium wastage induced by the daily use of 50mg of hydrochlorothiazide [12]; the same protective effect of ACEIs co-administration with diuretics was shown for perindopril [27], for Ramipril [28], and other ACE-inhibitors.

Moreover, ACE-inhibitors appear to exhibit a specific spectrum of effects in patients with congestive heart failure that may contribute to prolongation of survival of these patients; in particular, correction of hyponatremia, positive potassium and magnesium balance, reduction in the activity of vasopressor and volume retaining systems, as well as their potentional anti-ischemic and antiarrhythmic effects are the factors for the beneficial effect of ACEIs in these patients [29, 30].

The results of our study are confirming the observations of previous investigators as far as the preservation of potassium and magnesium homeostasis is concern, by the co-administration of ACE-inhibitors (in this study Captopril and Enalapril) with diuretics. Interesting was the finding that hypomagnesemia was relatively more pronounced than hypokalemia during chronic treatment with diuretic monotherapy in our hypertensive patients.

In conclusion, from this study it is evident that chronic monotherapy with diuretics results in significant hypomagnesemia even with normal serum levels of potassium, while ACE-inhibitors Captopril, Enalapril) when were given with diuretics they reversed completely the hypomagnesemic action of diuretics and sustained the normal levels of serum magnesium.

References

1. Joint National Committee on the Detection, Evaluation and Treatment of High Blood Pressure (1993) The Fifth Report of the Joint National Committee on Detection, Evaluation, and Treatment of High Blood Pressure (JNC-V), *Arch. Intern. Med.* **153**, 154-183.
2. Chalmers J and Zanchetti A. (1996) The 1996 report of a World Health Organization expert committee on hypertension control, *J Hypertens* **14**, 929-33.
3. Cohn, J.N.(1996) Management of chronic heart failure, *N. Engl. J. Med.* **335**,490-498.
4. Dycker, T. and Wester, P.O. (1987) Potassium/ Magnessium depletion in patients with cardiovascular disease, *Amer. J. Med.* **82**(Suppl. 3A),11-17.
5. Dorup, I., Skajaa, K., Clausen, T. and Kjelden, K. (1988) Reduced concentrations of potassium, magnesium and sodium - potassium pumps in human skeletal during treatment with diuretics, *Brit. Med. J.* **296**, 455-58.
6. Nicholls, M.G. (1990) Interaction of diuretics and electrolytes in congestive heart failure, *Am. J. Cardiol.* **65**,17E-21.
7. Holland, O.B, Kuhnert, L., Pollard, J., Padia, M., Anderson, RJ and Blomqvist, G. (1988) Ventricular ectopic activety with diuretic therapy, *Am. J. Hypertens.* **1**, 380-85.
8. Hollifield J.W. (1989) Thiazide treatment of systemic hypertension : effects on serum magnesium and ventricular ectopic activity, *Am J Cardiology* **63**, 22G-25.
9. Singh ,B.N., Hollenberg ,N.K., Poole - Wilson, P.A. and Robertson, J.S. (1992) Diuretic induced potassium and magnesium deficiency : relation to drug - induced QT prolongation, cardiac arrhythmias and sudden death, *J Hypertens.* **10**,301-16.
10. Nicholls, M.G. (1988) Age related effects of diuretics in hypertensive subjects, *J Cardiovas Pharmacol.* **12** (Supp.8), 51-59.
11. McInnes, G.J., Yeo, W.W., Ramsay, L.E. and Moser M. (1992) Cardiotoxicity and diuretics: much speculation - little substance, *J Hypertens.* **10**,317-335.
12. Melby J.C. (1986) The renin - angiotensin - aldosterone complex, *Am. J. Med.* **81** (Suppl.4C),8-12.
13. Prichard, B.N., Owens, C.W., and Woolf, A.S (1992) Adverse reactions to diuretics, *Eur. Heart J.* **13** (Suppl.G),96-103.
14. Wester, P.O. (1992) Electrolyte balance in heart failure and the role for magnesium ions, *Am J Cardiol.* **70**, 4C-49C.
15. Priewer H, Ullrich F, Mutschler E. (1996) Potassium and magnesium retaining potency and efficacy of triamterene and analogues in concious saline - loaded rats, *Magnes Res.* **9** , 33-40.

16. Malini P.L., Strocchi E., Valtancoli G., Ambrosioni E. (1990) Angiotensin converting enzyme inhibitors, thiazide diuretics and magnesium balance. A preliminary study *Magnes Res.* **3**, 193-96.
17. American Society of Hypertension (1992) : Recommendations for routine blood pressure measurement by indirect cuff sphygmomanometry, *Am J Hypertens.* **5**, 207-9.
18. Wallenstein S., Zucker C.L., and Fleiss J.L. (1980) Some statistical methods useful in circulation research, *Circ Res.* **47**,1-9.
19. Siegel D., Hulley B., Black D.M., Oheitlin M.D., Sebastian A, Seeley D.G. (1992) Diuretics, serum and intracellular electrolyte levels, and ventricular arrhythmias in hypertensive men, *JAMA* **67**, 1083-89.
20. Freis E.D. (1990) The cardiotoxicity of thiazide diuretics : Review of the evidence, *J Hypertens.***8** (Suppl.),S23-32.
21. Ikram H. (1987) Arrhythmias, electrolytes, and ACE inhibitor therapy in the Elderly, *Gerontology* **33** (Suppl.1),42-47.
22. Morris M.E. (1992) Brain and CSF magnesium concentrations during magnesium deficit in animals and humans : neurologic symptoms, *Magnes Res.*S: 303-313.
23. Wester P.O., Dyckner T, (1981) Diuretic treatment and Magnesium losses, *Acta Med. Scand.* **209** (Suppl.), 145-52.
24. Delhnmean A., Granry J.C., Monrigal J.P,, Costerousse F. (1995) Indication for the use of magnesium in anesthesia and intensive care, *Ann. Fr. Anesth. Reanim.* **14**, 406-16.
25. O' Keeffe, Grimes H., Finn J., McMurrough P., Daly K. (1992) Effect of captopril therapy on lymphocyte potassium and magnesium concentrations in patients with congestive heart failure, *Cardiology* **80**, 100-105.
26. Schminger R.H.and Erdmann E. (1992) Heart failure and electrolyte disturbances, *Methods, Find Exp Clin Pharmacol.* **14**, 315-25.
27. Leary W.P., Reyes A.J., van der Byl K. (1992) Effects of angiotensin-converting enzyme inhibitors on urinary excretions : Interactions with diuretics, *Am J Med,* **.92** (Suppl 4B),4B-68S.
28. .Simunic M., Rumboldt Z., Ljutic D., Sardelic S. (1995) Ramipril decreases chlortha-lidone - induced loss of magnessium and potassium in hypertensive patients. *J Clin Pharmacol* **35**, 1150-55.
29. Riegger A.J. (1989) ACE inhibitors in congestive heart failure, *Cardiology,* **76** (Suppl.2), 42-49.
30. Dietz R., Osterziel K.J. (1988) Effects of angiotensin-converting enzyme inhibitors in heart failure, *Z Kardiol.* 77(Suppl.3),89-93.

CARDIAC ARRHYTHMIAS IN MAGNESIUM DEFICIENCY

Studies on rat model

Motoomi NAKAMURA, M.D., Ph. D.
The Graduate School of Health and Nutrition Sciences, Nakamura-GAKUEN University, 5-7-1, Befu, Jonan-ku, Fukuoka, Japan

1. ABSTRACT To clarify malignant arrhythmias in magnesium (Mg) deficiency, we used weanling rats in study 1 and an adult rats in study 2. In study 1, weanling rats were fed either an Mg-deficient diet or a control diet for approximately 2 weeks and then electroencephalogram (EEG) and electrocardiogram (ECG) were recorded during auditory stimulation until death. EEG changes were abrupt attenuation of EEG background activity at the onset of tonic convulsion and then spike-wave complex at the falling down stage in association with marked bradyarrhythmias. When flattening of EEG became after generalized seizures, death occurred within a few minutes but consistent ventricular tachyarrhythmias were not observed. NMDA receptor blockers given intraperitoneally prevented significantly seizure occurrence and sudden death. These results may suggest that sudden death in weanling Mg-deficient rats results from a neurologic trigger.

In study 2, ventricular tachyarrhythmias were induced in adult rats fed either an Mg-deficient or a control diet for 20 days, by intravenous infusion of epinephrine continuously and increasing dose in a stepwise manner to a maximum dose of approximately 200 μg/kg. The incidence of ventricular tachycardia was significantly higher in the Mg deficient rats than that in the control rats. These results suggest that Mg deficiency enhances arrhythmogenic propensity of epinephrine.

Key Words: magnesium deficiency, sudden death, arrhythmia, rat, neurologic trigger, epinephrine

2. BACK GROUND:

Attention has focused on the relation between magnesium (Mg) deficiency and ventricular tachyarrhythmia or sudden cardiac death (1-3). Over the past few decades a consensus emerged that low Mg content in drinking water is associated with high rates of sudden death (3). Some cardiologists believe that pharmacological dose of Mg shows preventive effects empirically as well as experimentally in the treatment of various tachyarrhythmias, particularly of ventricular tachycardia and torsade de points

T. Theophanides and J. Anastassopoulou (eds.), Magnesium: Current Status and New Developments, 215-221.

in the patients with or without hypomagnesemia in various cardiovascular diseases including acute myocardial infarction, congestive heart failure and vasospastic angina (4,5). However, an association between Mg deficiency and ventricular tachyarrhythmia or sudden death is still not clear and remains to be clarified. Thus, the aims of our study are to clarify an association between either sudden death or malignant arrhythmias and Mg-deficiency in the rats.

3. Study 1: WEANLING RATS MODEL

3.1 *Abnormal behaviors and arrhythmias in Mg-deficient weanling rats*

To elucidate the mechanism of behavioral alterations in severe Mg-deficiency in weanling rats, changes in electroencephalogram (EEG) and electrocardiogram (ECG) were recorded during auditory stimulation. White noise (1 sec on and 1 sec off) was given to induce behavioral changes. The Mg-deficient rats developed graded behavioral changes in response to white noise, showing running-jumping (stage 1), followed by tonic limb convulsion (stage 2) and finally by falling down on the floor (stage 3). The EEG demonstrated an abrupt attenuation of the EEG background activity at the onset of tonic limb convulsion in association with prolonged bradyarrhythmias as shown in Fig 1. At the falling down stage, spike activity initiated in the deeper brain and propagated spread to the neocortices (stage 3), suggesting that the behavioral changes induced by auditory stimulation in the Mg-deficient weanling rats are due to seizures arising in deeper brain structures and projecting secondary to the neocortices. The ECG changes occurred at the appearance of the EEG changes and were mainly bradyarrhythmias, such as sinus arrest with idioventricular rhythm, sinus bradycardia and complete atrio-ventricular block related to seizures as reported previously (6,7).

3.2 *Prevention of seizures in Mg-deficient weanling rats by N-methyl-D-aspartate (NMDA) and α-amino-3-hydroxy-5-methyl-4-isoazole propionic acid (AMPA) receptor blockers*

Since Mg ion present in the extracellular space produces a well characterized voltage-dependent block of the open NMDA receptor-gated channel and reduces channel opening (8,9), a marked reduction in Mg concentration of the extracellular space will disinhibit the NMDA receptor receptor, resulting Ca influx and the increased intracellular Ca leads to increased neuronal excitability and susceptibility to seizure.

The mean concentration of serum-Mg in the Mg-deficient weanling rats was approximately 0.2~0.3mM/L which was a concentration close to that found in the cerebrospinal fluid of Mg-deficient sheep developing seizure (10). In order to study effects of NMDA-and AMPA-receptor blockers on the induction of seizure in the Mg-deficient rats, we had to develop a procedure in which seizures can be repeatedly induced during 1-2 hours. Caddell reported that Mg-deficient rats surviving after generalized seizures episode usually resisted further audiogenic seizure for at least 24 hours (11). By stopping an auditory stimulation as quickly as possible after occurrence of stage 1 of seizures (running), we could induce stage 1 or 2 seizures repeatedly for 1-

2 hours and thus provide an opportunity to examine effects of various receptor blockers on changes in behabiors, ECG and/or EEG associated with seizures in vivo as reported previously (7).

As the NMDA-receptor blocker, DL-2-amino-7-phosphonoheptanoic acid (APH) and MK 801 maleate, and as the AMPA receptor blocker, 2, 3-dihydroxy-6-nitro-7-sulfamoyl-benzo (F)-quinoxaline (NBQX) which were proved to permeate the blood brain barrier, were given intraperitoneally at the comparable dosage used previously in other seizure models. APH is a competitive NMDA receptor blocker which binds the glutamate binding sits of the receptor while MK 801 is a non-competitive NMDA receptor blocker which binds the phencyclidine site of the NMDA receptor. Both NMDA-receptor blockers and AMPA-receptor blocker (NBQX) successfully prevented seizure occurrence. The amount of APH required to suppress seizure was 36-72 mg/kg body weight and the duration of preventive effects was 1-2 hours, a greater dosage and shorter duration than for MK 801 which required only 1.35 mg/kg body weight and was effective for 3-6 hours. The amount of NBQX was 15 or 30 mg/kg B.W. for 3 times separated by 10 or 15 min interval respectively as reported previously (12). Both dosages of NBQX prevented seizure induction for 2 hours, probably due to its sedative effects.

3.3 *Sudden death in Mg-deficient weanling rats results from a neurologic trigger and brain dysfunction*

In 9 of 25 Mg-deficient weanling rats, both EEG and ECG were recorded continuously from the beginning of seizure induction until death. When flattening of EEG became after generalized seizures occurrence, all 9 Mg-deficient rats did not recover and death occurred within a few minutes as shown in Fig.1. On the other hand, malignant ventricular tachyarrhysmias, such as ventricular fibrillation and/or sustained ventricular tachycardia were not observed in our study. In 15 Mg-deficient weanling rats, we recorded ECG continuously by using Holter type ECG recorder (Marquet Electronic Inc., Wisconsin, U. S. A, model 8500). We made the motor speed of the Holter-recorder twice the usual speed in order to record ECG with high heart rate, such as 500 per min in the weanling rats. Six of 15 rats died spontaneously during Holter ECG recording and showed consistently bradyarrhythmias until death as shown in Fig 2. Therefore, we did not find any ventricular tachyarrhythmias in the weanling Mg-deficient rats during seizures and sudden death. Thus, these results may support idea that sudden death in Mg-deficient weanling animals did not result from ventricular tachyarrhythmia but resulted from brain dysfunction after episodes of seizures as reported previously (13).

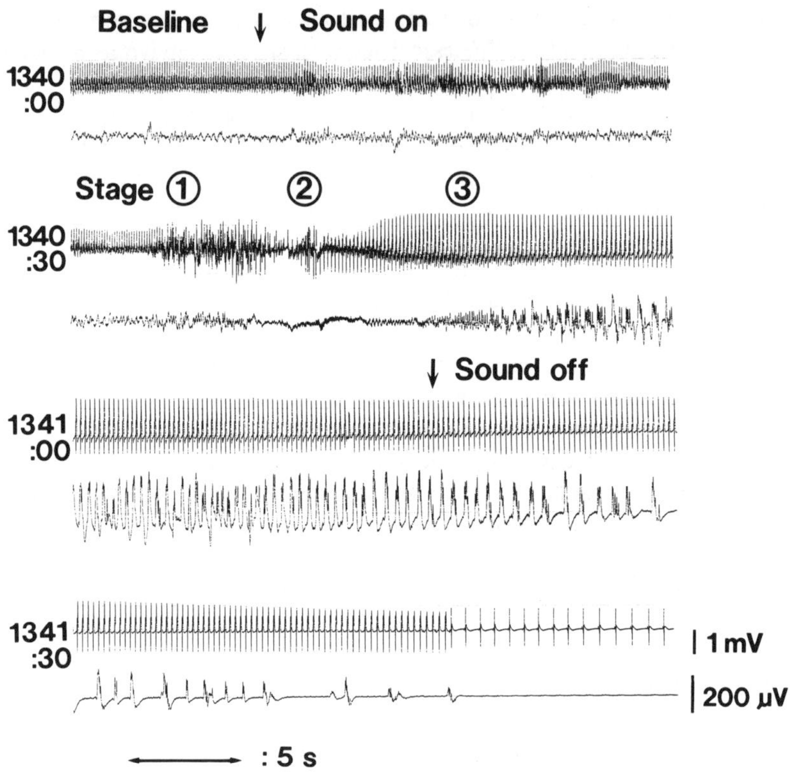

Fig 1. Changes in ECG and EEG during auditory stimulation in an Magnesium-deficient rat. ECG (top) and EEG (bottom) from the left sensorimotor cortex. The first arrow indicates the start of the white-noise stimulation and the second one the cessation. Stage ①,②and ③ indicate the occurrence of stage 1 (running), 2 (tonic conclusion) and 3 (falling down) seizures respectively. An abrupt attenuation of EEG and bradyarrhythmia occurred in stage 2 and a high voltage rhythmic spike-wave complex in EEG with bradyarrhythmia in stage 3.
Note the flattening of EEG became approximately one min after seizure occurrence and the rat died soon after with slowing idioventricular rhythm. Number in ordinate indicates the time of the experimental day.

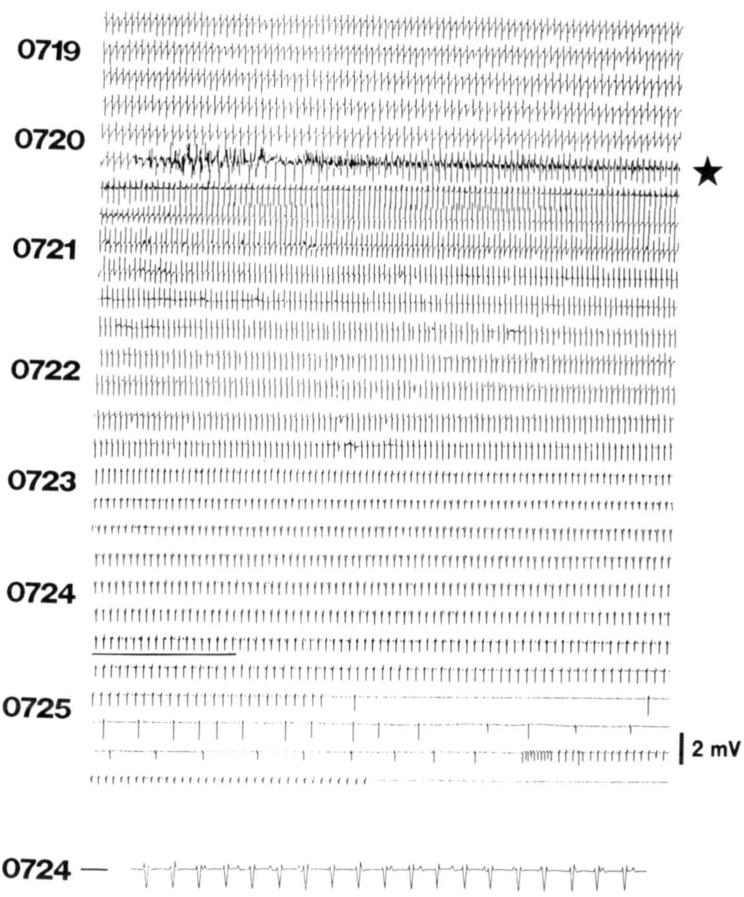

Fig 2. Changes in ECG Holter recording in an Mg-deficient weanling rat
At 0720 (A.M.), ECG was contaminated markedly by electromyogram suggesting a seizure occurrence with bradyarrhythmias, indicated by a star. At 0724 (A.M.), ECG underlined was enlarged in the bottom line and showed atrio-ventricular dissociation. And then a slow idioventricular rhythm was followed and the rat died soon after. Number in ordinate indicates the time of the experimental day.

4. STUDY 2: ADULT RATS MODEL

4.1 *Hypomagnesemia in adult rats promotes induction of ventricular tachycardia by intravenous administration of epinephrine*

Ventricular tachycardia was induced in the adult rats fed either a control or an Mg-deficient diet for 20 days, by an intravenous and continuous administration of epinephrine starting at 5 μg/kg/min and then increasing the dose in stepwise manner by 5 μg/kg/min at 5 minutes intervals to total dose of approximately 200 μg/kg with continuous monitorings of ECG and blood pressure under the conscious and the unrestrained condition. QT interval in the Mg-deficient rats was significantly longer than that in the normal control. Sustained ventricular tachycardia, ventricular fibrillation, asystole and sudden death were not found in both the control and Mg-deficient adult rats before epinephrire infusion. The incidence of sustained ventricular tachycardia was significantly higher in the Mg-deficient rats (36 of 42 rats) than that in the control rats (13 of 30 rats) after continuous infusion of epinephrine according to Fisher's exact test ($P<0.01$). The cumulative frequency of both non-sustained and sustained ventricular tachycardia was significantly higher in the Mg-deficient rats than that in the control rats according to Mantel-Cox test ($P<0.01$). These results suggest that Mg deficiency enhances the arrhythmogenic propensity of epinephrine.

5. CONCLUSION

Mg-deficiency enhances the susceptibility of the adult rats to epinephrine-induced ventricular tachyarrhythmias. On the other hand, sudden death in Mg-deficient weanling rats results from brain dysfunction after seizures occurrence, but not from ventricular tachyarrhythmias.

References

1. Surawicz B. (1989) Is hypomagnesemia or magnesium deficiency arrhythmogenic? J. Am. Coll. Cardiol. **14**, 1093-1096.
2. Millane T. A., Ward D. E., Camm A. J. (1992) Is hypomagnesemia arrhythmogenic? Clin. Cardiol. **15**, 103-108.
3. Eisenberg M. J., Lerman B. N. (1992) Magnesium deficiency and sudden death. Am. Heart J. **124**, 544-549.
4. Keller P. K., Aroson R. S (1990) The role of magnesium in cardiac arrhythmias. Prog. Cardiovasc. Dis. **32**, 433-448.
5. Keren A, Tzivoni D. (1990) Magnesium therapy in ventricular arrhythmias. PACE. **13**, 937-945.
6. Goto Y., Nakamura M., Abe S., Kato M. and Fukui M (1993) Physiological correlates of abnormal behaviors in magnesium-deficient rats. Epilep. Res., **15**, 81-89.
7. Nakamura M., Abe S., Goto Y., Chishaki A., Akazawa K., Kato M. (1994) In vivo assessment of white-noise-induced seizure in magnesium-deficient rats by N-methyl-D-aspartate receptor blockers. Epilepsy Res. **17**: 249-256.
8. Nowak, L, Bregestovski P. Ascher P., (1984) Magnesium gates glutamate-activated channels in mouse central neurones. Nature **307**, 462-465.
9. Mayer, ML, Westbrook, GL, Gluthrie PB, (1984) Voltage-dependent block of Mg^{2+} of NMDA responses in spinal cord neurons. Nature **309**, 261-263.
10. Leaver, DD, Parkinson GB, Schneider KM, (1987) Neurological consequences of magnesium deficiency : Correlations with epilepsy. Clin. Exp. Pharmacol. Physiol. **14**, 361-370.
11. Caddell JL., (1978) Exploring the magnesium-deficient weanling rat as an animal model for the sudden infant death syndrome: physical, biochemical, electrocardiographic and gross pathologic changes. Pediat. Res., **12**, 1157-1166.
12. Nakamura M., Abe S., Akazawa K. (1995) AMPA-, but not NMDA-, receptor blocker, NBQX, prevents seizure induction in magnesium deficient rats. Magenesium Res., **8**, 55-56.
13. Nakamura M., Abe S., Goto Y., Chishaki A. (1995) Sudden sound-induced death in magnesium-deficient rats after repetitive episodes of seizure result from brain dysfunction. Magnesium Res. **8**, 47-53.
14. Tomiyasu T, Chishaki A, Nakamura M.
Hypomagnesemia in adult rats promotes the induction of ventricular tachycardia by the administration of epinephrine (unpublished)

HYPOMAGNESEMIA IN LONG QT SYNDROME

K Hoshino MD, K Ogawa MD, R Kitazawa MD, Y Nakamura MD and R Uehara MD.
Department of Pediatric Cardiology, Saitama Children's Medical Center
2100 Magome Iwatsuki Saitama Japan 339

1. Object

The successful use of magnesium sulfate (MgSO4) for torsade de pointes (TdP) in patients with long QT syndrome (LQTS) was first reported in 1984 by Tzivoni et al. In our institute, four patients who developed TdP with LQTS were treated with intravenous injection of MgSO4 and it was effective in all four. We studied the dynamic and clinical application of magnesium (Mg) in patients with LQTS.

2. Patients

The diagnosis of LQTS was based on Schwartz's criteria and the patients were given oral β-blocker (LQTS group). Patients having QT prolongation alone were diagnosed as long QT (LQT group). The control subjects included patients with history of Kawasaki disease (MCLS) who did not have coronary association and patients after surgery of atrial septal defect (ASD) who did not show congestive heart failure (Control group).

3. Methods

Serum magnesium (SMg) level was detected by enzyme assay. Total body Mg deficiency was detected by Mg retention (MgR) of low dose Mg (0.2mEq/kg lean body weight) administered intravenously over 4 hours. We assessed the SMg level and Mg retention in each three groups and the correlation with QTc interval. In addition, we assessed the correlation of clinical factors (syncope history, age at initial presentation, heart rate at initial presentation, family

T. Theophanides and J. Anastassopoulou (eds.), Magnesium: Current Status and New Developments, 223-225.

history, presence of premature ventricular contraction(PVC), presence of Twave abnormality and therapy) with SMg level and Mg retention in each three groups.

4. Results

4.1. SMG LEVEL

4.1.1. *SMg level and QTc interval in three groups :*

SMg level was 1.93±0.15mg/dl (n=37) in LQTS, 2.03±0.16mg/dl (n=19) in LQT and 2.06±0.17mg/dl (n=60) in controls. The value of LQT group did not differ from those of control group. But the value of LQTS group was significantly lower than those of LQT group ($P<0.05$) and control group ($P<0.001$).

There was a significant negative correlation between SMg level and QTc interval in LQTS group(γ=-0.56, $P<0.01$).

4.1.2. *Factors affecting the SMg level in the LQTS group*

SMg level was significantly lower in patients with Twave abnormality than those without it ($P<0.05$). It was significantly lower in patients without family history than those with family history ($P<0.05$). SMg level showed no correlation with syncope history, age at initial presentation, heart rate at initial presentation nor presence of PVC.

SMg level did not change after oral β-blocker in 10 patients with LQTS.

4.2. MG RETENTION TEST

4.2.1. *Mg retention and QTc interval in three groups*

MgR was 57.5±21.2% (n=26) in LQTS, 33.2±14.6% (n=8) in LQT and 32.8±14.2% (n=8) in controls. The value of LQT group did not differ from those of control group. But the value of LQTS group was significantly higher than those of LQT group ($P<0.01$) and control group ($P<0.01$).

There was no significant correlation between MgR and QTc interval in three groups. But there was a significant positive correlation between MgR and QTc interval in LQTS patients without family history (γ=0.62, $P<0.05$).

4.2.2. *Factors affecting the MgR in the LQTS group*

MgR was significantly higher in patients with syncope history than those without it ($P<0.05$). But MgR showed no correlation with age at initial presentation, heart rate at initial presentation, presence of PVC, presence of

Twave abnormality nor family history.
MgR did not change after oral β-blocker in 10 patients with LQTS.

4.3. ORAL ADMINISTRATION OF MG

Three LQTS patients with repetitive syncope episodes in spite of oral β-blocker were given oral MgO in a dose of 10-20mg/kg/day, which resulted in disappearance of episodes. MgR gradually decreased after the administration of MgO and this suggests an improvement of Mg deficiency after oral MgO administration. In one patient, ionized Mg level was raised after oral administration and it was suspected that Mg deficiency might be improved.

5. Conclusion

5.1. SMG LEVEL

SMg level was significantly lower in LQTS group (especially in patients without family history or with Twave abnormality). There was a significant negative correlation between SMg level and QTc interval in LQTS group. But SMg level does not always reflect physiological activity of Mg. Thus, it is difficult to assess the role of SMg in patients with LQTS.

5.2. MG RETENTION TEST

MgR was significantly higher in LQTS group (especially in patients with syncope history). Total body Mg deficiency in these patients was suggested. There was a significant positive correlation between MgR and QTc interval in patients with LQTS without family history, and it was suspected that Mg deficiency may become a cause of QT prolongation in these patients. Thus, we suspect that Mg may be deficient in some patients with LQTS and it may become a exciting cause of TdP.

5.3. ORAL ADMINISTRATION OF MG

In some patients, oral administration of MgO may prevent TdP.

EFFECT OF METABOLIC INHIBITORS AND ISCHAEMIA ON CONTRACTILE FORCE AND MAGNESIUM EFFLUX IN THE ISOLATED PERFUSED RAT HEART

B.I. HUSTLER, J.J. WARING AND J. SINGH
Department of Applied Biology, University of Central Lancashire, Preston, Lancashire, U.K.

1. SUMMARY

This study investigates the effects of ischaemia using reduced flow rate and metabolic inhibition by rotenone, KCN, oligomycin and DNP on contractile force and net Mg^{2+} efflux from the isolated perfused rat-heart. There was a significant (Student's t-test; $P < 0.01$) decrease in force of contraction and coronary pressure during 10 min of ischaemic conditions and this was accompanied by a significant ($P < 0.001$) increase in Mg^{2+} efflux. Perfusion of the heart for 10 min with Mg^{2+}-free KH containing metabolic inhibitors either 10^{-5}M rotenone, 10^{-7}M KCN, or 10^{-7} Oligomycin B resulted in a decrease in contractile force and a small but not significant ($P > 0.01$)increase in net Mg^{2+} efflux. In contrast, perfusion with DNP, an oxidative phosphorylation uncoupler, resulted in a significant decrease in contractile force accompanied by a highly significant (increase in Mg^{2+} efflux. The results indicate that both ischaemia and loss of the mitochondrial proton gradient are associated with Mg^{2+} efflux, and a concomitant decrease in contractile force in the isolated heart.

2. INTRODUCTION

Magnesium (Mg^{2+}) is present in all living cells where it is the second most abundant divalent cation. Its physiological role is primarily intracellular and preservation of intracellular Mg^{2+} concentration ($[Mg^{2+}]_i$) is essential for normal functioning of the heart[1]. Mg^{2+} is an essential cofactor for over 300 enzymes and as such it plays a regulatory role in many cellular process including many of those involved in the glycolytic process, those catalysing the transfer of phosphate and most of those utilizing ATP[2]. In essentially all

T. Theophanides and J. Anastassopoulou (eds.), Magnesium: Current Status and New Developments, 227-230.

reactions where ATP is a substrate, the true substrate is Mg^{2+}-ATP. Several cardiac disorders are associated with hypomagnesemia but the molecular function of Mg^{2+} in health and disease has yet to be determined. The aim of this study was to investigate the effects of (a) experimental ischaemia induced by a reduced perfusion flow rate of 1 ml min^{-1} and (b) inhibition of the electron transport chain on magnesium efflux and contractile force in the isolated paced (20V, 4 - 6 Hz, 1 msec) perfused (5 ml min^{-1}, 37°C) rat-heart in an attempt to understand the role of ATP in magnesium homeostasis in the heart.

3. METHODS

Adult Sprague-Dawley rats of either sex weighing 250 to 300 g were used throughout this study. The procedure for perfusion of the isolated rat-heart, collection of effluent samples following stimulation and the analysis of effluent samples for Mg^{2+} content by a spectroscopic method have been described in detail elsewhere[3]. Lactate dehydrogenase (LDH) activity in effluent samples from ischaemic hearts was measured using a standard LDH assay (Sigma LD-L, 228-10).

4. RESULTS AND DISCUSSION

Figure 1 shows original force traces when the heart was perfused under control conditions (flow rate 5 ml min^{-1}), under ischaemic conditions (flow rate 1 ml min^{-1}), and during stimulation with either 10^{-5}M rotenone, 10^{-7}M potassium cyanide (KCN) 10^{-7}M oligomycin B or 5 x 10^{-5} M dinitrophenol (DNP). Perfusion with 10^{-4} M DNP produced a force trace similar to that of 5 x 10^{-5} M DNP and is therefore not shown. Figure 2 shows the net total Mg^{2+} efflux during 10 min of stimulation with Mg^{2+}-free KH under control and ischaemic conditions and in the presence of 10^{-5}M rotenone, 10^{-7}M potassium cyanide (KCN) 10^{-7} M oligomycin B, 10^{-4} M DNP and 5 x 10^{-5} M DNP. Under ischaemic conditions contractile force decreased to 39.9 ± 2.5% of the original force (n=6) and coronary pressure decreased from 30 mmHg to 12 mm Hg. Total net Mg^{2+} efflux during this period was 523.56 ± 27.93 ng Mg^{2+} (100 mg $tissue)^{-1}$ compared to 376.11 ± 17.92 ng Mg^{2+} (100 mg $tissue)^{-1}$ for control (n=7). There was no significant change in LDH activity in effluent samples from ischaemic hearts, thus indicating that Mg^{2+} efflux was not due to damage of the heart. Contractile forces during perfusion with 10^{-5}M rotenone, 10^{-7}M potassium cyanide (KCN) 10^{-7} M oligomycin B were 72.89 ± 7.11%, 64.7 ± 6.8% and 81.80 ± 3.0% of the original force, respectively, whilst net total Mg^{2+} efflux was 457.59 ± 20.54, 458.46 ± 26.54 and 447.84 ± 21.92 ng Mg^{2+} (100

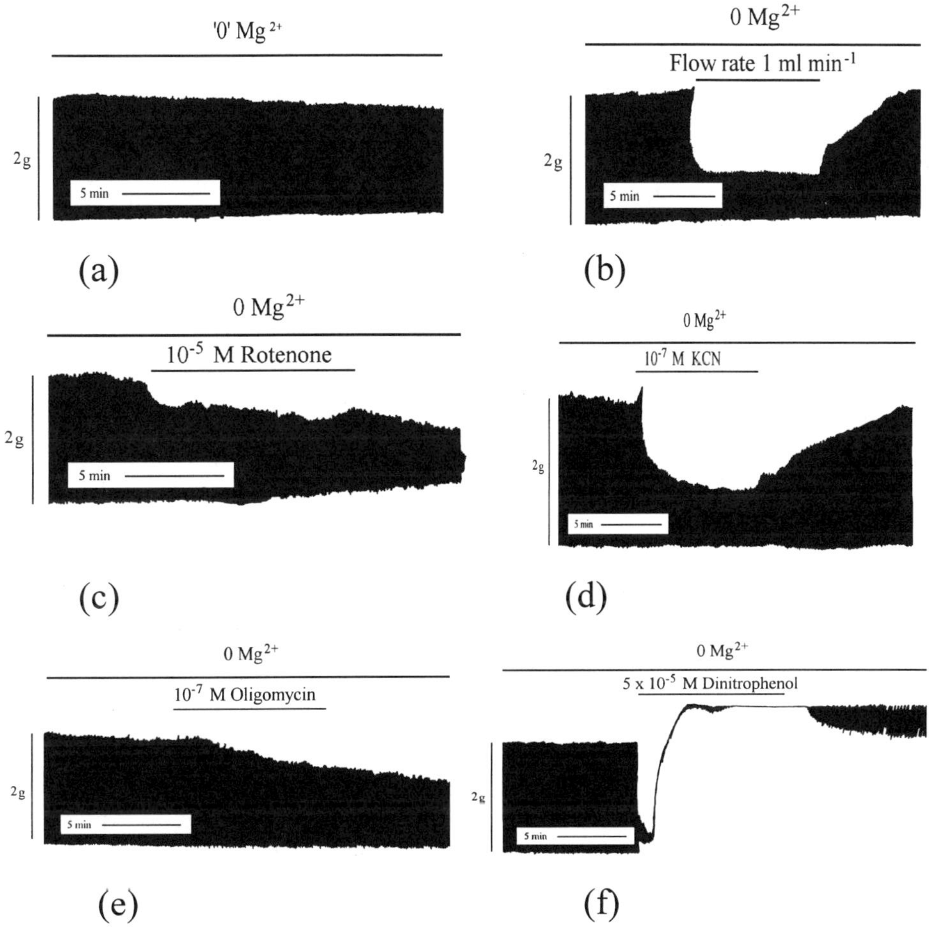

Figure 1. Original chart recording of contractile force during perfusion of the isolated rat heart for 10 min with (a) Mg^{2+} free KH (control) (b) Mg^{2+} free KH at a flow rate of 1 ml min^{-1} (ischaemic) (c) Mg^{2+}-free KH containing 10^{-5} M rotenone, (d) Mg^{2+}-free KH containing 10^{-7} M KCN, (e) Mg^{2+}-free KH containing 10^{-7}M oligomycin B and (f) Mg^{2+}-free KH containing 5 x 10^{-5} M DNP. Time and force calibrations are shown in the figure. These traces are typical of 3 to 7 such experiments for each parameter. Hearts were perfused at 5 ml min^{-1} and paced electrically (4-6 Hz frequency, 20 V amplitude, 1 msec pulse width) except where otherwise stated

mg $tissue^{-1}$), respectively. DNP had a much more profound effect on both contractile force and Mg^{2+} efflux in the perfused rat-hearts. Contractile force decreased to 3.83 ± 0.95% of its original force with 10^{-4}M DNP and to 5.86 ± 0.53% with 5 x 10^{-5}M DNP. Following the decrease in contractile force, the tonic response in the heart was significantly reduced. A large and highly significant increase in net Mg^{2+} efflux accompanied the changes in contractile force. The metabolic inhibitors rotenone, KCN and oligomycin B inhibit ATP production at different locations on the electron transport chain, rotenone inhibits the NADH- Co-enzyme Q reductase, KCN inhibits the cytochrome-c oxidase and oligomycin inhibits the F_1-ATPase. DNP uncouples oxidation from phosphorylation, ATP production continues but is not linked to oxidation. During myocardial ischaemia, there is an imbalance in uptake of oxygen and cellular oxidation[4]. These results suggest that the influence of ATP on Mg^{2+} efflux is minimal, however, it would appear that the loss of the proton gradient is strongly associated with Mg^{2+} efflux.

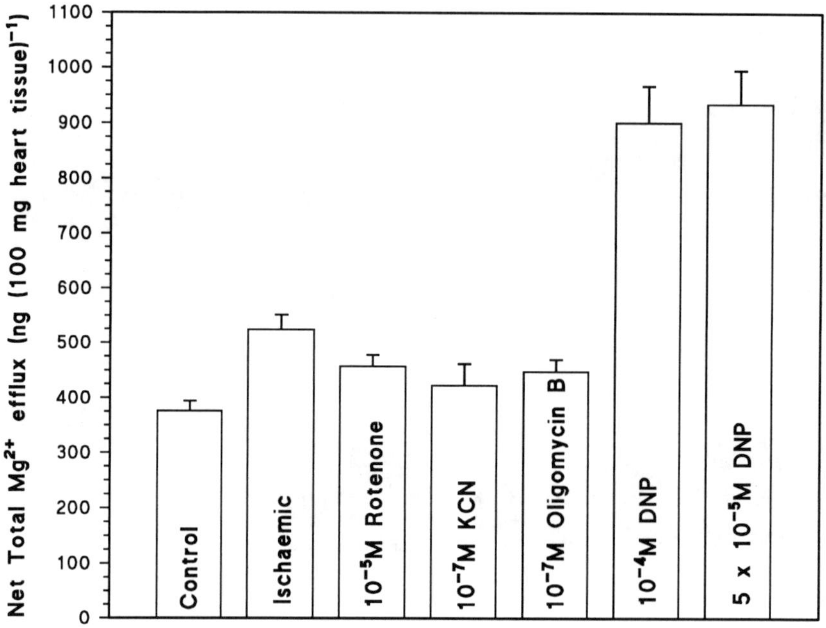

Figure 2. Bar-chart showing the net total Mg^{2+} efflux during perfusion of the isolated perfused heart for 10 min with Mg^{2+} free KH under control conditions, under ischaemic conditions (a flow rate of 1 ml min^{-1}), containing 10^{-5} M rotenone, 10^{-7} M KCN, 10^{-7}M oligomycin B, 10^{-4} M DNP and 5 x 10^{-5} M DNP. Data are mean ± SEM (n=3 to 7).

References

1. Whang, R., Hampton, E.M., Whang, D. (1994) Magnesium homeostasis and clinical disorders of magnesium deficiency. *Annals of Pharmacotherapy* **28,** 220 - 226.
2. Flatman, P.W. (1991) Mechanisms of magnesium transport *Ann Rev Physiol* **53,** 259-271.
3. Howarth, F.C., Waring J.J., Singh, J., Hustler, B.I. (1994), Effects of extracellular magnesium and β-adrenergic stimulation on contractile force and magnesium mobilization in the isolated rat heart. *Magnes. Res* **7,** 187-197.
4. Ferrari, R. (1995), Metabolic disturbances during myocardial ischemia and reperfusion. *Am. J. Cardiol* **76,** 17B-24B.

THE INFLUENCE OF Mg^{2+} AND Zn^{2+} ON VASCULAR EFFECT OF ANGIOTENSINE CONVERSION ENZYME INHIBITORS

M. NECHIFOR[1], A. MURARIU[1], G. DANILA[2], A. MIFTODE[3], C. FILIP[1], E. TESLARIU[1], A. VELEANU[1]

1.Dept. Pharmacology - Univ. Med. Pharm. "Gr. T. Popa", Iasi, Romania, 2.Dept. of Pharmaceutical Chemistry - Univ. Med. Pharm. "Gr. T. Popa", Iasi, Romania 3.Dept. of Inorganic Chemistry - Univ. Med. Pharm. "Gr. T. Popa", Iasi, Romania

1. Introduction

Magnesium and Zinc are two of the most important cations from both human and animal body [1]. These two metals have an important role in the enzymatic regulation of the smooth muscle function [2-4]. There are more than 200 enzymes which depends on Zinc [5] and there also are numerous enzyme with a Magnesium regulated activity [6,7].

Angiotensin II (Ang II) is one of the most potent endogenous vasoconstrictor factors. It is implicated in the vascular tonus regulation, in the vascular smooth muscle proliferation, in the hydroelectrolytic equilibrium regulation etc. An important role in the Ang II synthesis is that of the Ang I to Ang II converting enzyme (angiotensin converting enzyme - ACE). There have been realised many ACE inhibitors (ACEI), such as captopril, enalapril, lisinopril etc. All of these ACEI have become very useful drugs for the treatment of hypertension, of some forms of chronic heart failure etc.

In our work we searched the influence of Mg^{2+} and Zn^{2+} on the vasodepressor effect of lisinopril.

2. Material and Method

We searched the influence of Zinc Acexamate (ZnAc), Magnesium Acexamate (MgAc) and Magnesium Sulphate ($MgSO_4$) on the vascular effect of lisinopril (Sinopryl[R], Eczacibasi).

We worked on series of 6 male adult Wistar rats, weighing 180-200g and bred in the same laboratory conditions. They had got no food for 16 hours before the experiment, but they got water.

Series I received only lisinopril 1mg/kg i.p.
Series II received lisinopril 1 mg/kg i.p. and MgAc 0.5 mEq/kg i.p.
Series III received lisinopril 1 mg/kg i.p. and $MgSO_4$ 0.5 mEq/kg i.p.
Series IV received only MgAc 0.5 mEq/kg i.p.
Series V received lisinopril 1mg/kg i.p. and ZnAc 0.5 mEq/kg i.p.

T. Theophanides and J. Anastassopoulou (eds.), Magnesium: Current Status and New Developments, 231-234.

Series VI received only ZnAc 0.5 mEq/kg i.p.
Series VII received only Acexamic Acid (Aac) 0.5 mEq/kg i.p.
Series VIII received lisinopril 1 mg/kg i.p. and AAc 0.5 mEq/kg i.p.
All the substances were administrated in single dose.

The systolic blood pressure was measured with a non-invasive technique, using a Hugo Sachs Sphygmomanometer S-2 (with a cuff and a pulse sensor paced on the animal's tail). The blood pressure (BP) measurements were made initially (just before the substances administration) and at 1 hour, 3 hours and 6 hours after the substances administration. The obtained data were statistically interpreted by means of "t" test.

3. Results

The influence of MgAc, $MgSO_4$ ZnAc and AAc on the vascular effect of lisinopril is shown in Figures 1, 2, 3 and 4, respectively.

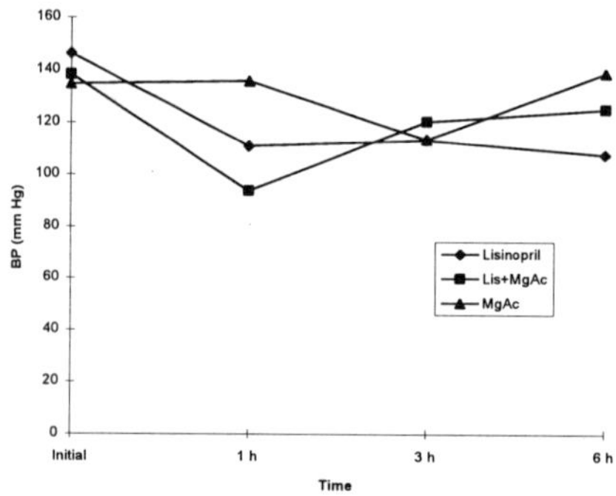

Figure 1. The Influence of MgAc on BP and on the Vascular Action of Lisinopril.

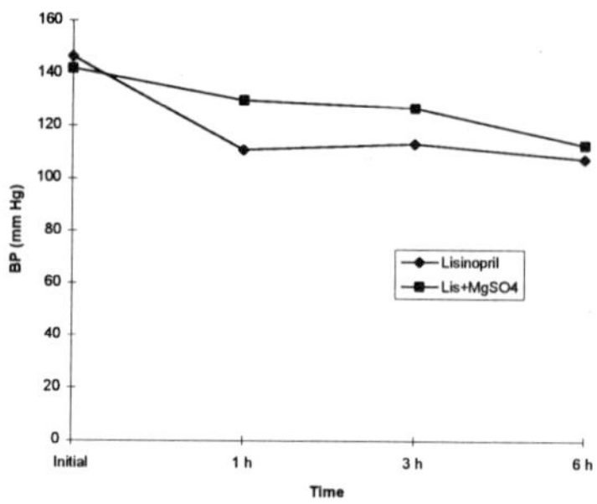

Figure 2. The Influence of MgSO⁴ on the Vascular Action of Lisinopril

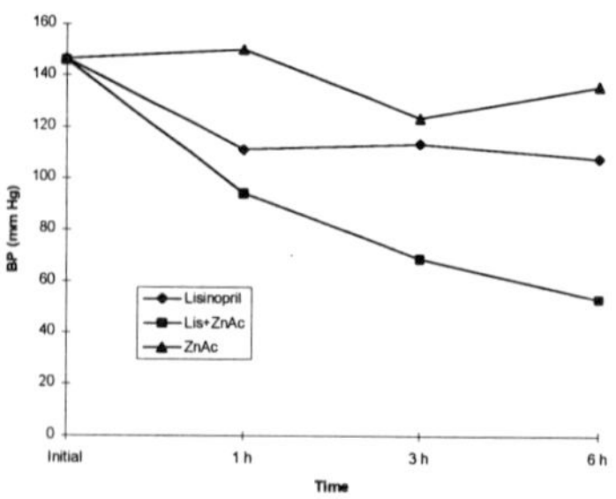

Figure 3. The Influence of ZnAc on BP and on the Vascular Action of Lisinopril.

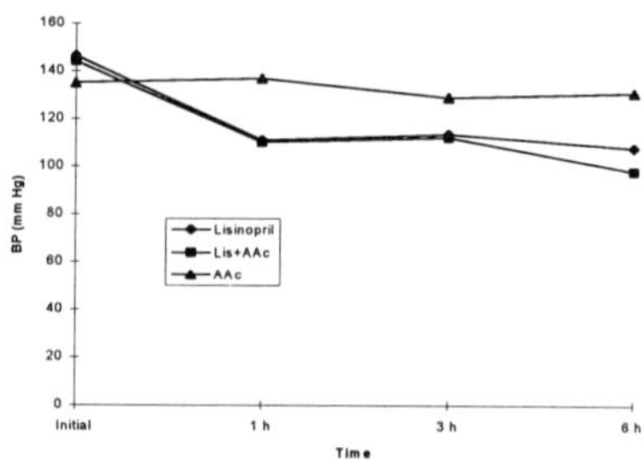

Figure 4. The Influence of Aac on BP and on the Vascular Action of Lisinopril

4. Discussions and Conclusions

Our data show that both Mg^{2+} and Zn^{2+} have an influence on the vasodepressor effect of lisinopril, but this action is different for those two elements MgAc significantly enhance the vasodepressor effect of lisinopril at 1 hour after drug administration (there is a BP decrease of 17% of the initial values in series I, which received only lisinopril and a decrease of 30% in series II, which received lisinopril and MgAc). At 3 and 6 hours neither MgAc nor MgSO4 have a significant influence on the vasodepressor effect of lisinopril.

Aac had no influence on BP when it was administered alone and it didn't influence the vascular effect of lisinopril. This thing allow us to sustain that all the influences on BP are due to Mg^{2+} and Zn^{2+} cations.

We consider that the shorter time effect of Mg^{2+} versus Zn^{2+} is due to some pharmacokinetic features of those two metals, Mg^{2+} being eliminated more rapidly.

Concerning the mechanism by which Mg^{2+} enhances the lisinopril vasodepressor effect, we consider there could be at least three explanations:

a) a local mechanism which implicates Mg^{2+} induced generation in the arterial endothelium of some vasodilator factors, such as PGI_2 and E prostaglandins (these eicosanoids antagonise some vasopressor factors actions)[8];
b) a central effect by which Mg^{2+} decrease the glutamate and aspartate releasing in CNS during the CNS injury [9]. Our experimental model implicates repeated immobilisations of the animal in order to measure the BP and this thing means a stress for the animal. In this context the decrease of the inhibiting amino acids from CNS could results in a decrease of BP;
c) the first discovered ACEI - captopril [10] determines a decrease of the central sympathetic tonus and of the presynaptic release of norepinephrine [11]. It could be possible that Mg^{2+} is able to increase this central sympathetic inhibition and thus it enhance the vasodepressor effect of ACEI.

Zn^{2+} has a significant vasodepressor effect at any moment of the BP measurement we chose and it also significantly enhance the vascular effect of lisinopril. We cannot state if the two vasodepressor substances have an additive or a synergistic effect.

The increasing interest for the ACEI therapeutic use [12] and the wide spread clinical use of both Mg^{2+} and Zn^{2+} compounds results in an useful association between the ACEI and the two metals' compounds for the treatment of arterial hypertension.

5. References:

1. Fabris N., Mocchegiani E. (1995) Zinc human diseases and aging, *Aging*, **7**, 77-93.
2. Paul R.J., Ruegg J.C. (1988) Role of magnesium in activation of smooth muscle, *Am. J. Physiol.*, **24**, C465-C472.
3. Moreland R.S., Ford G.D. (1981) The influence of magnesium on calcium activated vascular smooth muscle actomyosin ATP-ase activity, *Arch. Biochem. Biophys.*, **208**, 325-333.
4. Fullerton D.A., Hahn A.R., Agrafojo J., Sheridan B.C. McIntyre R.C.J. (1996) Magnesium is essential in mechanisms of pulmonary vasomotor control, *J. Surgical. Res.*, **63**, 93-97.
5. Berg J.M., Shi Y. (1996) The galvanization of biology: a growing appreciation for the roles of zinc, *Science*, , **271**, 1081-1085.

6. Gaussin V., Hue L., Stalmas W., Bollen M. (1996) Activation of hepatic acetyl-CoA carboxilase by glutamate and Mg2+ is mediated by protein phosphatase-2A, *Biochemical J.*, **316**, 217-224.
7. Sok D.E., Kim M.R. (1995) Involvment of both a Zn^{2+} site and an anonic binding site in the selective inhibition of a Zn^{2+} glicerophosphocolyne cholinephosphodiesterase by thiols and tellurites, *Neurochemical Res.*, **20**, 151-157.
8. McGiff J.C. (1980) Interactions of prostaglandins with the kallikrein-kininand renin-angiotensin systems, *Chemical Sci.*, **59**, 105s-116s.
9. Feldman Z., Gurevitch B., Artun A.A., Oppenheim E., Reichenthal E., Shapira Y. (1996) Effect of magnesium given one hour after head trauma on brain edema and neurogical outcome, *J. Neurosurg*, **85**, 131-137.
10. Cushman D.W., Cheung H.S., Sabo E.F., Ondetti M.A. (1977) Desing of potent competitive inhibitors of angiotensine converting enzyme carboxyalkylkanoyl and mecaptoalkanoyl amino acids, *Biochemistry*, **1**, 5484-5491.
11. Yasuda G., Shionoiri H., Kubo T., Misu Y. (1987) Presynaptic Angiotensin II Receptor and captopril-induced Adrenergic Transmision Failure Probably not via Converting Enzyme Inhibition in Guinea-Pig Pulmonary Arteries, *J. Hypertension*, **5** (2) S39-S45.
12. Timmermans P.BMWM, Wong P.C., Chin A.T., Herblin W.F., Benfield P., Carini D.J., Lee R.J., Wexler R.R. (1993) *Pharmacological Rev.*, **45**, 205-248.

A UNIQUE NON-INVASIVE INTRACELLULAR MAGNESIUM ASSAY CORRELATING WITH CARDIAC TISSUES, ARRHYTHMIAS, AND THERAPEUTIC INTERVENTION

BURTON B. SILVER[#], MARK C.P. HAIGNEY*, STEVEN P. SCHULMAN[^], GARY GERSTENBLITH[Δ], GORDON F. TOMASELLI[Δ], RONALD BERGER[Δ], HUGH CALKINS[Δ], AND SHAOKUI WEI[Δ]

*Δ The department of Medicin, Division of Cardiology, Johns Hopkins Medical Institutions, 600 M. Wolfe St. Baltimore, * The Department of Medicine, Division of Cardiology, Uniformed Services University of the Health Sciences, 4301 Jones Bridge Rd. Bethesda, # IntraCellular Diagnostics, Inc. 553 Pilgrim Dr (B) Foster City, CA*

1. Introduction

A novel technique using energy dispersive X-ray microanalysis, (EXA_{tm}), for non-invsive intracellular measurement of magnesium has now been accomplished and proven to be a valuable tool in multiple aspects of normal as well as pathological magnesium metabolism (U.S. patent No 4-717-826, B.S.). Over the last 15 years data has accumulated using the unique method which is now available to accurately evaluate tissue Mg levels.

Since only 1% of total body Mg^{2+} is found in the intravascular space, serum levels of Mg give little information about a patients overall status with respect to this essential mineral. Using the intracellular analysis it has been determined that Mg levels are significantly reduced in multiple physiological states which may lead to serious pathological conditions.[1] Description of the methodology and examples of data as well as potential applications will focus on intracellular $[Mg^{2+}]_i$ determinations obtained from subjects with cardiovascular disease syndromes related to Mg^{2+} deficiency.

Examples of the application of EXA_{tm} include examination of intracellular magnesium and other minerals in a wide spectrum of conditions which include: Effects of Microgravity, (NASA): cardiovascular conditions; arrthythmias; heart failure; myocardial infarction; and bypass surgery; standardization of normal controls (NASA)[2].

2. Rationale

Magnesium plays a protean role in intracellular metabolism, second only in potassium as the most common intracellular cation in human physiology. Multiple normal functions

T. Theophanides and J. Anastassopoulou (eds.), Magnesium: Current Status and New Developments, 235-240.

as well as pathological syndromes are affected by the cellular levels and availability of the magnesium stores within tissues. Sogically, measuring magnesium accurately and easily within tissues is of immense use to the understanding and treatment of a host of sideases in the human condition.

Cytosolic free magnesium acts as a cofactor for every ATPase and regulates calcium and potassium transport into tissues. Gecause of its many effects, Mg has been called the "chronic regulator" of cell metabolism. Murphy et al point out, though, that $Mg^{2+}i$ can only modulate intracellular function if $[Mg^{2+}]i$ changes in response to physiological stress[3].

Cellular magnesium activates many enzyme systems including those vital reactions which govern cardiac function. Regulated by mg levels, sodium-potassium ATPase maintains the normal intracellular concentrations of Na^+ and K^+ vital to normal cardiac function. Given the effect of a reduction in intracellular Mg on the action potential and the multiple channels and enzymes influenced by the Mg cation, a membrane depolarization would be expected due to cellular K^+ loss, Na^+ gain and Ca^{2+} gain[4].

3. Methodology

The measurement of intracellular $[Mg^{2+}]i$ was performed by x-ray dispersive microanalysis utilizing a specially configured electron microscope which images and irradiates cells with a focused electron beam. Sublingual epithelial cells were chosen as markers since they are easily accessible, are non-cornified, aerobic, have a high cytoplasm to nucleus ratio,turnover in less than 3 days, have long shelf life exhibit 99% viability, and show significant correlations with cardiac and muscle biopsies taken during bypass surgery. Excitation of cellular atoms displaces inner orbital electrons which are replaced by electrons from higher energy cells releasing fluoresced x-rays which allows quantitation of intracellular elements. EXA_{tm} units equals X-ray intensity (peak divided by background) divided by unit cell volume. Cardiac and muscle tissue biopsies were quick frozen in liquid nitrogen cooled with Freon and freeze dried tissue sections were then carbon coated for x-ray analysis.[5]

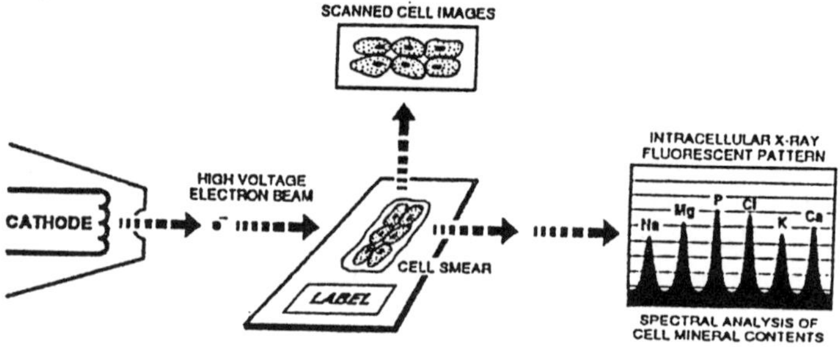

Figure 1 Intracellular analysis

EXA_{tm} units are converted to milliquivelants per liter by an conversion constant for each elementderived from reference standard of a highly stable matrix synthetic glass containing known amounts of the elements to be analyzed. The standard was certified by the National Bureau of Standards and Corning Diagnostics. The coefficient of variance of 40 determinations of X-ray emission energies for Mg with the reference standard was 0,82%. Applications of intracellular magnesium analysis will be reported in three investigations incorporating this unique method.

3.1 CORRELATION OF MG LEVELS BETWEEN SUBLINGUAL CELLS AND ATRIAL TISSUES TAKEN AT SURGICAL BYPASS AND THE EFFECT OF I.V. MG IN PATIENTS WITH ST ELEVATION AND ACUTE MYOCARDIAL INFARCTION

3.1.1 Results

We have found that energy despersive X-Ray analysis (EXA_{tm}) of sublingual epithelial cells correlate well with atrial specimens obtained at cardiopulmonary Bypass. Seventeen subjects, average age 65,7 years, had syvlingual smears made prior to atrial biopsies at the time of bypass surgery. Their sublingual levels were comapared to healthy volunteers. Atrial biopsy tissue frozen as described and elemental x-ray analysis was performed. Serum vs Intracellular $[Mg^{2+}]i$ in Surgical Patients: Mean serum Mg levels for the cardiac surgery patients was within normal range, 1.87 $\pm$ 0,06 mEq/L. Despite normal serum mg, intracellular $[Mg^{2+}]i$ levels in the sublingual epithelium of bypass patients were reduced when compared to healthy subjects despite normal serum levels. Linear regression comparing the individual values is shown in figure 2. A strong correlation exits between the sublingual and cardiac cells (R=0.68, p<0.002) Figure 2.

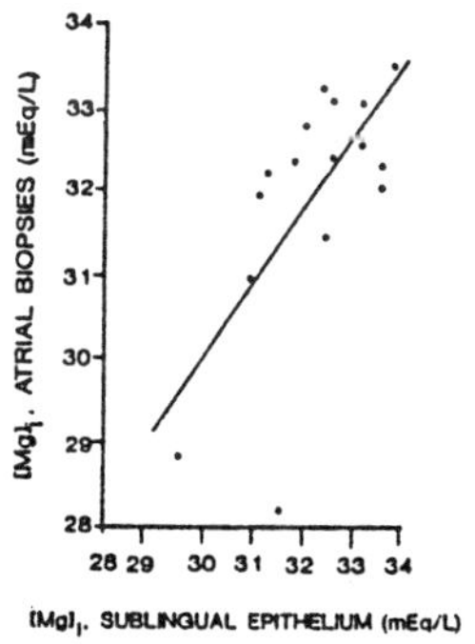

Figure 2 Scatterplot showing linear regression comparing [Mg] concentrations in atrial and sublingual cells in patients undergoing cardiopulmonary bypas (n=18, r=68, P<0.02

using the EXA_{tm} sublingual cell evaluation, magnesium intravenous intervention studies were done on 22 myocardial infarct patients who were compared to healthy controls and non-cardiac patients. Mean Mg in infarct patients was 30.7 $\pm$ 0.4 compared to 15 control subjects whose cellular Mg levels were 35.0$\pm$ 0.5, p<0.0001. Infarct patients receiced a mean dose of 36$\pm$ 6 mmol/24hrs of intravenous Mg_2SO_4. No Mg was given after the second 24 hours of the study. Intracellular [Mg $^{2+}$]i rose significantly in the infarct patients over the first 24 hours, mean sublingual magnesium continued to rise for 48 hours after the first 24 hours, mean sublingual magnesium continued to rise for 48 hours after the first dose of magnesium sulfate (from 30.7±0.4 to 35.2±0.6 mEq/L) suggesting that magnesium may be moving from the vascular space to the tissues over 48 hours (Figure 3).

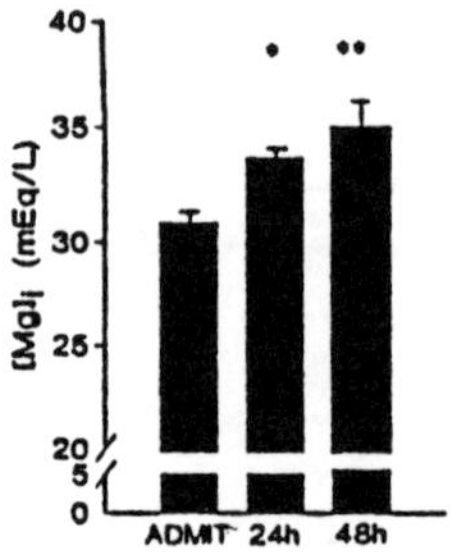

Figure 3 Bar graph comparing mean [Mg] concentration in the sublingual epithelium in AMI patients (n=22) on admission (ADMIT) to the CCU and 24 hours after admission. *P<0.05 vs admission and 48 hours.**P<0.05 vs admission and 24 hours.

3.2 ABNORMAL QT DISPERSION CORRELATES WITH TISSUE MAGNESIUM

Abnormal QT dispersion (QTd) on the 12 lead EKG, and magnesium deficiency have both been shown to predict a higher risk for sudden cardiac death, in patients with CHF, alcoholic liver disease, or requiring antiarrhythmic drugs. We tested the hypothesis that Mg deficiency predisposes to abnormal depolarization in the (QTd). In 37 subjects with malignant arrhythmias, Mg in serum and sublingual epithelial cells, were measured in a blinded fashion. Results were correlated with the EKG, (r=0,56, p=0.0001). Linear regression showed QTd correlated with depressed tissue Mg measured using energy dispersive x-ray analysis, (EXA_{tm}). Mean serum levels showed no such correlations.

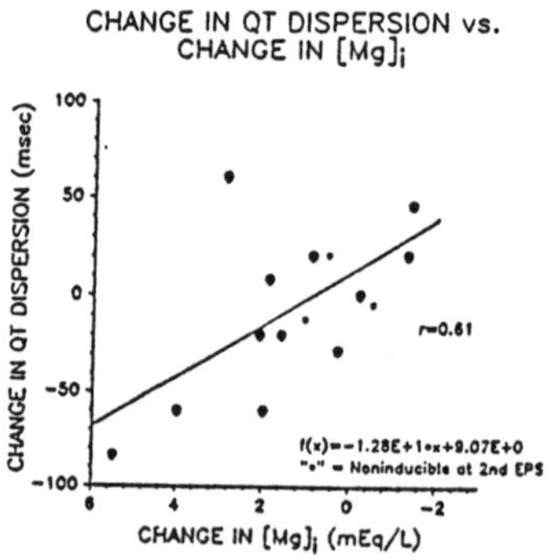

Figure 4 Linear regression comparing the change in QT dispression (y axis) and change in sublingual intracellular Mg in 12 patients having two measurements 3-7 days apart (r=0.61, p<0.05)

RepeatMg levels obtained in 12 treated patients 3-7 days later showed increase in tissue mg(1.5±-0.6 mEq/l (p<0.0001) and a decrease in QTd by 10±12 msecs (p<0.0001).

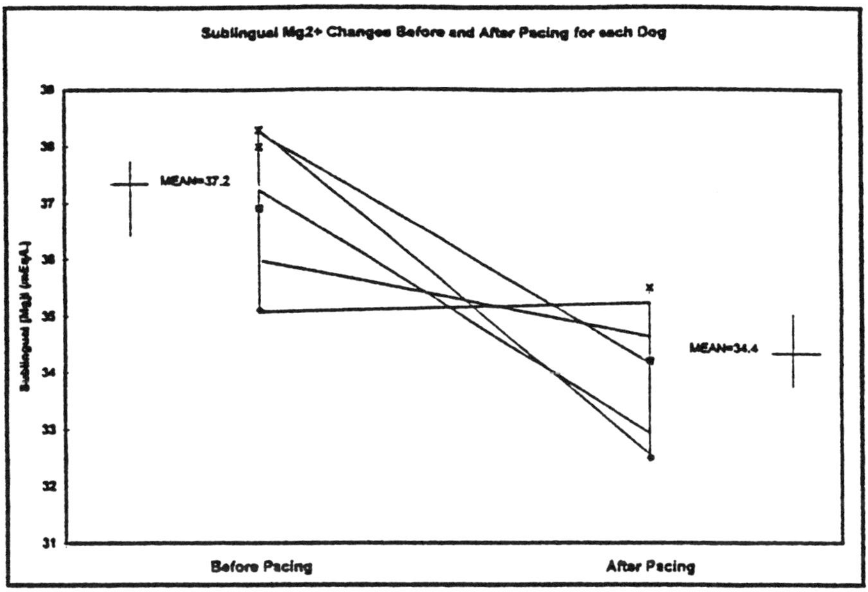
Sublingual Mg2+ Changes Before and After Pacing for each Dog
MEAN=37.2
MEAN=34.4
Before Pacing
After Pacing

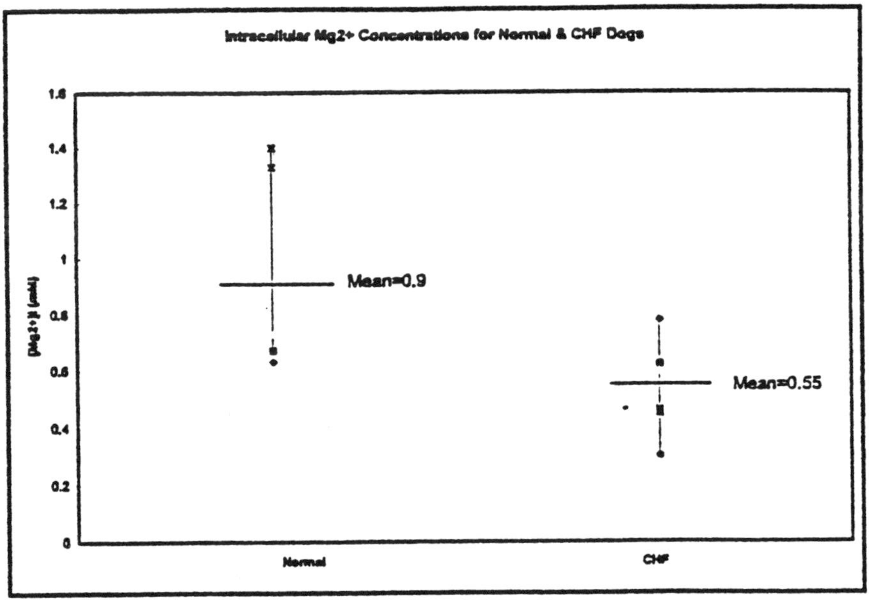
Intracellular Mg2+ Concentrations for Normal & CHF Dogs
Mean=0.9
Mean=0.55
Normal
CHF

Delta QTd [Mg] correlated at r=0.61, p<0.05. QT dispersion was shown to correlate with tissue Mg levels suggesting mg deficiency may be a risk factor for sudden cardiac by its effect on cardiac repolarization as indexed by QTd (Figure 4).

3.3 EFFECT OF EXPERIMENTAL HEART FAILURE ON SUBLINGUAL TISSUE MG AND CARDIAC IONIZED MG.

To test the hypothesis that CHF results in a rapid loss of tissue mg in the absence of confounding drugs or diseases, sublingual smears were obtained from normal dogs (n=4) which were exposed to 4 weeks of nentricular pacing at 250 bpm. All animals manifested bi-bentricular failure. Sublingual cell Mg was measured by energy dispersive x-ray analysis (EXA_{tm}). Cardiac myocytes were measured with the fluorescent probe Mg-Indo to measure $[Mg^{2+}]i$.

Sublingual cells demonstrated significantly lower $[Mg^{2+}]i$. in paced dogs, (34.8 mEq/L$\pm$0.8 vs. 38.7mEq/L$\pm$0.5, 2p<0.05). The fluorescence ratios indexing $[Mg^{2+}]i$. was also significantly reduced in the paced animals compared to the normal dogs $(0.573 \pm 0.007, n=49)_{paced}$ vs. normals $(0.749 \pm 0.029)_{normal}$, (n=18 cells, 2p<0.001). The reduction in total cellular Mg detected in the sublingual cells was (8%) yet was associated with a ~40% reduction in $[Mg^{2+}]i$ compared to controls. it seems that free Mg is not well buffered but fluctuates with changes in changes in total Mg. A small reduction in total cellular mg would cause a much large decrement in free mg which constitutes <5% of total stores. The data suggests that a reduction in tissue Mg results in a relatively large fall in free cardiac mg. Measures of sublingual Mg (EXA_{tm}) may be useful in future clinical investigation regarding the significance of Mg in heart failure. More importantly, EXA_{tm} testing will allow accutate serial measurement of Mg to guide Mg repletion therapy in a wide variety of syndromes in which Mg deficiency is a factor (Figures 5.6).

References

1. Eisenberg MJ (1992); Magnesium deficiency and sudden death. *Am. Heart J.* **124**: 544-549.
2. Haigney MCP, Berger R, Calkins H, Silverman HS, Tunin C, Haigney C, Silver B, Gerstenglith G, Schulman S, Tomaselli G(1992): QT dispersion forrelates wtihtissue magnesium concentration in humans. *Circ,* **86**: 1376-1382.
3. Murphy E, Freudenrich CC, Liegerman M. (1991): Cellular magnesium and Na/Mg exchange in heart cells. *Ann. Rev. Physiol.* **53**; 273-287.
4. Agus ZS, Morad M. (1991) Modulation of cardiac ion channels by magnesium. *Ann. Rev. Physiol.* 299-307.
5. Haigney MCP, Silver B, Tanglao E, Silverman HS, Hill JD, Shapiro JE, Gerstenglith G, Schulman SP (1995): Non-invasive measurement of tissue magnesium and correlation with cardiac levels. *Circ.* **92**; 2190-2197.

PHYSIOLOGICAL AND PHARMACOLOGICAL ROLE OF MAGNESIUM IN ACUTE MYOCARDIAL INFARCTION

P. Yang, G. Steurer, R. Smetana
University Clinics of Internal Medicine II and IV, University of Vienna Medical School, Waehringer Guertel 18-20, 1090 Vienna, Austria

Coronary artery disease, which is clinically expressed by reduced coronary blood flow followed by myocardial ischemia, hypoxemia, and myocardial dysfunction, is accompanied by increased concentrations of serum catecholamines. This increase is related to the severity of coronary artery disease and/or to the reduced performance of the left ventricle. The activation of the sympathetic nervous system has to be considered as a compensatory mechanism in order to increase the perfusion pressure to vital organs and to enhance myocardial contractility [1, 2]. Increased catecholamine output, however, requires increased magnesium supply to maintain intracellular, energy-bound metabolic processes. Therefore, magnesium requirements are increased during physical and psychological stress, which is particularly evident in acute myocardial infarction (AMI). A progressive decrease in the myocardial magnesium occurs under the influence of adrenergic overstimulation. Because stress and magnesium requirements/deficiency are mutually enhancing, a vicious circle may develop where the catecholamines cause movement of magnesium from the intra- to extracellular compartment [1, 2]. Consequently, the electric membrane becomes unstable and membrane permeability increases, followed by a further cellular loss of magnesium and potassium. The catecholamine-induced loss of myocardial magnesium is associated with the uptake of calcium that precedes cardiac damage by calcium accumulation [3-6].

In AMI, ischemia/reperfusion injury can compromise the benefits of early revascularization therapy. Cellular calcium overload is considered one of the essential pathomechanisms of ischemia/reperfusion injury [7, 8], in addition to oxygen radicals generated by the restoration of blood flow [9, 10], ischemia-induced activation of the complement system [11, 12], and complement-mediated damage of the myocardium. The development of the mechanical stunning of the heart is related to cytosolic calcium overload at the onset of reperfusion.

Magnesium is an activator of enzyme systems that are critical to cellular metabolism, especially those involved in the reactions with adenosine triphosphate. Magnesium regulates the transmembraneous electrolyte fluxes and is directly involved in the regulation of intracellular and extracellular homeostasis of potassium and calcium. Magnesium can be considered as "physiologic calcium blocker" which modulates the transmembrane shift and the transport of the calcium ion across the sarcoplasmic reticulum [13-15]. Since calcium overload is important in the pathogenesis of reperfusion injury, the channel blocking properties of magnesium may have important beneficial effects in patients with AMI [16].

T. Theophanides and J. Anastassopoulou (eds.), Magnesium: Current Status and New Developments, 241-244.

There is evidence that magnesium may be a suitable pharmacological agent to attenuate or to prevent myocardial reperfusion injury and to reduce infarct size. Many clinical trials have been performed to evaluate the effects of magnesium in patients with AMI [17-22, 24]. Meta-analyses on randomized trials indicate that intravenous magnesium significantly improved short-term (30 days) survival compared to controls and has beneficial effects on left ventricular function, infarct size, and arrhythmias [25-27]. The second Leicester Intravenous Magnesium Intervention Trial (LIMIT-2) [22] in which 2,316 patients with suspected AMI were given either intravenous magnesium or placebo has shown a significant reduction of mortality in the group of patients undergoing intravenous magnesium therapy.

This finding, however, was not confirmed by the later published forth International Study on Infarct Survival (ISIS-4) [24]. In a total of 58,050 patients with suspected acute myocardial infarction no definite beneficial effect of intravenous magnesium could be demonstrated on mortality following acute myocardial infarction. The discrepancy in the results of these two trials has mainly been attributed to the difference in the median time lapse prior to the initiation of intravenous magnesium administration [28, 29]. In LIMIT-2 patients were given intravenous magnesium a median of 3 hours from the onset of chest pain, while in ISIS-4 patients were treated a median of 8 hours after the onset of symptoms.

Current therapeutic regimen of AMI comprise coronary angioplasty or thrombolysis, intravenous nitrates, heparin, aspirin, beta-blockers, ACE-inhibitors and, optionally, calcium blockers. There is compelling evidence from experimental works which suggest that magnesium should be given before or in the early phase of reperfusion to achieve the protective effect. The decisive aspect, as pointed out in several critical reviews [25, 26, 29-36], leading to success or failure of magnesium application in AMI is the starting time of magnesium therapy. Similar to the guidelines of thrombolytic intervention, there is evidence that the early administration of intravenous magnesium, at the latest before myocardial reperfusion has started, leads to beneficial effects. These effects can be expressed by a reduced incidence of left ventricular failure during the acute phase of myocardial infarction and a reduction in mortality at the early stage of the post-infarction period.

Regarding our present knowledge, new clinical trials must be performed to point out future directions for the management of intravenous magnesium therapy in AMI with main focus on high-risk patients. These trials should determine the optimal dose and time schedule of intravenous administration of magnesium: an initial therapy of 8 to 10 mmol magnesium bolus, followed by an infusion of magnesium lasting over 12 hours at a dosage of 30 mmol; sustained magnesium infusion at a lower dosage of another 20 to 30 mmol for the next 12 hours in order to maintain the pharmaceutic effect on arrythmias and adrenergic overstimulation and to avoid the risk of hypotension and bradycardia.

Currently ongoing trials such as MAGIC (Magnesium In Coronaries; USA) and PICSO (Pressure-controlled Intermittent Coronary Sinus Occlusion; Austria) trials will provide further detailed concepts and guidelines for the treatment of ischemia/reperfusion injury in patients with AMI who are at high risk.

REFERENCES

1. Smetana R., Brichta A., Glogar D., Gottsauner-Wolf M., Weindlmayr-Goettel M., Meisinger V., Spona J.(1991)Stress and magnesium metabolism in coronary artery disease, *Magnesium Bulletin* 13(4), 125-7.

2. Smetana R., Huber K., Hartter E., Yang P., Meisinger V., Konnaris C., Gabriel H., Zehetgruber M., Klappacher G., Spona J., Vierhapper H. (1997) Coronary artery disease - magnesium and catecholamine interactions during stress burden, *Magnesium Research* Suppl I, 61-5.

3. Classen H.G. (1986) Systemic stress, magnesium status, and cardiovascular damage, *Magnesium* 5, 105-10.

4. Francis G.S. (1990) Interaction of the sympathetic nervous system and electrolytes in congestive heeart failure, *Am J Cardiol* 25(24E), 18-20.

5. Lasserre B. (1997) Magnesium and heart failure: an update., In R. Smetana (ed.), Advances in magnesium research - magnesium in cardiology, London: John Libbey & Company Ltd., pp. 12-20

6. Altura B.M., Altura B.T. (1997) Magnesium metabolism, atherogenesis and cardiovascular pathobiology, In R. Smetana (ed.), Advances in magnesium research - magnesium in cardiology, London: John Libbey & Company Ltd., pp. 28-37

7. Kusuoka H., Marban E. (1992) Cellular mechanisms of myocardial stunning, *Ann Rev Physiol* 54, 243-56.

8. DuToit E.F., Opie L.H. (1992) Modulation of severity of reperfusion stunning in the isolated rat heart by agents altering calcium flux at onset of reperfusion, *Circ Res* 70, 960-7.

9. Kloner R.A., Przyklenk K., Whittaker P. (1989) Deterious effects of oxygen radicals in ischemia/reperfusion: resolved and unresolved issues, *Circulation* 80, 1115-27.

10. Lucchesi B. (1994) Cardiac reperfusion injury, In B. Zikria (ed.), Reperfusion injuries and clinical capillary leak syndrome, Futura Publishing Company, Armonk, NY, pp. 171-202

11. Simpson P.J., Todd R.F.I., Fantone J.C., Mickelson J.K., Griffin J.D., Luccesi B.R. (1988) Reduction of experimental canine myocardial reperfusion injury by a monoclonal antibody (anti-Mo1, Anti-CD11b) that inhibits leucocyte adhesion, *J Clin Invest* 81, 624-9.

12. Homeister J.W., Satoh P.S., Kilgore K.S., Lucchesi B.R. (1993) Soluble complement receptor type 1 prevents human complement mediated damage of the rabbit isolated heart, *J Immunol* 150, 1055-64.

13. Iseri L.T., French J. (1984) Magnesium: nature's physiologic calcium blocker, *Am Heart J* 108, 188-93.

14. Krause S. (1991) Effect of increased free $[Mg^{2+}]i$ with myocardial stunning on sarcoplasmic reticulum Ca^{2+}-ATPase activity, *Am J Physiol* 261, H229-H235.

15. Shine K. (1979) Myocardial effects of magnesium, *Am J Physiol* 6, H413-H423.

16. Shechter M., Kaplinsky E., Rabinowitz B. (1992) The rationale of magnesium supplementation in acute myocardial infarction. A review of the literature, *Arch Intern Med* 152, 2189-96.

17. Feldstedt M., Bouchelouche P., Svenningsen A., Boesgaad S., Brooks L., Alderville J. (1988) Failing effect of magnesium-substitution in acute myocardial infarction, *Eur Heart J* 9, 226-32.

18. Shechter M., Hod H., Marks N., Behar S., Kaplinsky E., Rabinowitz B. (1990) Magnesium therapy and mortality in acute myocardial infarction, *Am J Cardiol* 66, 271-4.

19. Morton B.C., Smith F.M., J M.T., Nair R.C., Poznanski W.J. (1981) Magnesium therapy in acute myocardial infarction, *Magnesium Bulletin* 1a, 192-4.

20. Morton B.C., Smith F.M., Nair R.C., McKibbon T.J., Poznanski W.J. (1984) The clinical effects of magnesium sulphate treatment in acute myocardial infarction, *Magnesium Bulletin* 4, 133-6.

21. Rasmussen H.S., Gronbaek M., Cintin C., Balslov S., Norregard P., McNair P. (1988) One-year death rate in 270 patients with suspected acute myocardial infarction, initially treated with intravenous magnesium or placebo, *Clin Cardiol* 11, 377-81.

22. Woods K.L., Fletcher S., Roffe C., Haider Y. (1992) Intravenous magnesium sulphate in suspected acute myocardial infarction: results of the second Leicester Intravenous Magnesium Intervention Trial (LIMIT-2), *Lancet* 339, 1553-8.

23. Woods K.L., Fletcher S. (1994) Long-term outcome after intravenous magnesium sulphate in suspected acute myocardial infarction: the second Leicester Intravenous Magnesium Intervention Trial (LIMIT-2), *Lancet* 343, 816-9.

24. ISIS-4 (Fourth International Study on Infarct Survival) Collaborative Group (1995) ISIS-4: a randomised factorial trial assessing early oral captopril, oral mononitrate, and intravenous magnesium sulphate in 58,050 patients with suspected acute myocardial infarction, *Lancet* 345, 669-85.

25. Teo K.K., Yusuf S., Collins R., Held P.H., Peto R. (1991) Effects of intravenous magnesium in suspected acute myocardial infarction: overview of randomised trials, *BMJ* 303, 1499-503.

26. Yusuf S., Teo K., Woods K. (1993) Intravenous magnesium in acute myocardial infarction. An effective, safe, simple, and inexpensive intervention, *Circulation* 87, 2043-6.

27. Balkin J. (1997) The use of magnesium in critical coronary care patients., In R. Smetana (ed.), Advances in magnesium research - magnesium in cardiology, London: John Libbey & Company Ltd., pp. 21-7.

28. Antman E.M. (1995) Randomized trials of magnesium in acute myocardial infarction: big numbers do not tell the whole story, *Am J Cardiol* 75, 391-3.

29. Seelig M.S., Elin R.J. (1995) Reexamination of magnesium infusions in myocardial infarction, *Am J Cardiol* 76, 172-3.

30. Teo K.K., Yusuf S. (1993) Role of magnesium in reducing mortality in acute myocardial infarction. A review of the evidence, *Drugs* 46, 347-59.

31. Glogar D., Yang P., Steurer G. (1996) Management of acute myocardial infarction: Evaluating the past, practicing in the present, elaborating the future, *Am Heart J* 132, 465-70.

32. Smetana R. (1996) Magnesium in acute myocardial infarction, *Am Heart J* 132, 463-4.

33. Steurer G., Yang P., Rao V., Mohl W., Glogar D., Smetana R. (1996) Acute myocardial infarction, reperfusion injury, and intravenous magnesium therapy: Basic concepts and clinical implications, *Am Heart J* 132, 478-82.

34. Antman E. (1996) Magnesium in acute myocardial infarction: Overview of available evidence, *Am Heart J* 132, 487-95.

35. Seelig M., Elin R. (1996) Is there a place for magnesium in the treatment of acute myocardial infarction?, *Am Heart J* 132, 471-86.

36. Smetana R. (1997). Cardiovascular medicine - the importance of magnesium in coronary artery disease and acute myocardial infarction, In R. Smetana (ed.), Advances in magnesium research - magnesium in cardiology, London: John Libbey & Company Ltd., pp. 5-11.

MAGNESIUM IN ACUTE MYOCARDIAL INFARCTION (THE MAMI STUDY): EFFECT OF EARLY VERSUS LATE MAGNESIUM ADMINISTRATION ON CK-LEVELS.

STUEHLINGER HG, BEHRINGER W, HOEDL W, SEIDLER D, ZEINER A, SCHOERKHUBER W, LAGGNER AN.
Dept. of Emergency Medicine, Univ.Hospital, Vienna., Waehringer Guertel 18-20, 1090 Vienna, Austria

Abstract

Introduction: Two major studies have evaluated intravenous magnesium in patients with acute myocardial infarction. LIMIT-2 found a 24% reduction of mortality (day 28) in the magnesium treated group. However, ISIS-4 demonstrated no such benefit on mortality (day 35). In LIMIT-2, patients were started on intravenous magnesium immediately after admission, in contrast to ISIS-4, where magnesium was administered later.
Hypothesis: The beneficial effect of magnesium in acute myocardial infarction (AMI) depends upon early administration of the drug.
Methods: 102 patients undergoing thrombolysis (rt-PA) for acute myocardial infarction were randomised to receive magnesium at two different times. In group 1 the magnesium infusion was started concurrent with thrombolysis, group 2 received magnesium 3 hours later. Total dose of magnesium was 72 mmol (bolus of 8 mmol, followed by a continous infusion for 22.5 hours at a rate of 3.2 mmol/h).
Results: Maximum Ck levels (Ck max.) in group 1 were higher than in group 2 (1126 vs. 911 U/l) and reached the peak earlier (9.6 vs. 10.5 hours). In patients with anterior wall infarction these trends are more destinct: in group 1 (n=23) and 2 (n=21) Ck max. was 1598 U/l and 1241 U/l respectively. Group 1 reached these levels after 8.3 hours as opposed to 9 hours for group 2. A subgroup analysis was performed on patients with documented reperfusion (ECG: 70% reduction of initial ST-elevation after 3 hours). These patients had a Ck max. of 1694 U/l if magnesium had been started early (group 1) and 1209 U/l in group 2. These differences did not reach statistical significance.
Conclusion: Early administration of magnesium in AMI leads to higher and earlier maximum Ck levels which might indicate earlier and more pronounced reperfusion. The results of the MAMI study presented above might elucidate the divergent results of LIMIT-2 and ISIS-4.

T. Theophanides and J. Anastassopoulou (eds.), Magnesium: Current Status and New Developments, 245-249.

1.Introduction

Based on clinical experience and case reports magnesium sulfate has been used in patients with acute coronary syndromes for many years [1]. As early as 1935 effects of magnesium on the heart have been described [2]. In 1942 Elek and Katz recommended treatment of tachyarrhythmias due to ischemic events with magnesium sulfate [3]. These effects were confirmed by Boyd and Scherf [4]. In 1952 Hoffmann and Siegel observed that patients with coronary artery disease on oral magnesium therapy experienced less ischemia associated pain [5]. Malkiel-Shapiro could even document a reduction in mortality among patients with acute myocardial infarction treated with intravenous magnesium [6, 7]. Parson and coworkers confirmed a low mortality rate in a small study including 33 patients with myocardial infarction and parenteral magnesium therapy [8]. Experimental work published at the same time by Bajusz and Selye showed that parenteral magnesium effectively reduced the size of the necrotic area in experimental infarction [9, 10]

During the 1980s researchers started to test these interesting clinical and experimental observations in placebo-controlled studies [11 - 17]. Pooled results of these seven studies document 1301 patients with acute myocardial infarction. A meta-analysis shows a significant reduction of mortality of 55% in patients treated with magnesium as compared to controls. Six out of the seven studies found a lower incidence of ventricular arrhythmias in magnesium treated patients, in only two studies, however, did these differences reach statistical significance [18].

Despite a relatively low number of patients in all of these studies some important conclusions emerge. It seems to be crucial to administer magnesium sulfate early in the course of acute infarction. The most impressive effects on mortality were shown in the studies of Shechter, Ceremuzynski and Smith, who all started magnesium very early. While Abraham and coworkers started magnesium early too their study design allowed a relatively low systolic blood pressure (exclusion criterium: systolic blood pressure below 90 mm Hg!) which might explain the comparatively bad results. The highest mortality among the seven studies was described by Feldstedt who did not exclude patients until their blood pressure was below 80 mm Hg which is lower than in any of the other studies. A falsely set exclusion criterium in a study on a drug with known blood pressure lowering effects might explain the unfavorable results. All these considerations seem to underline the importance of early administration of magnesium and maintenance of coronary artery perfusion pressure.

Two recent trials currently dominate the scientific discussion on parenteral magnesium [19]. LIMIT-2 (Leicester Intravenous Magnesium Intervention Trial), published in 1992 in the Lancet studied the effects of magnesium in 2316 patients with acute myocardial infarction [20]. 28-day mortality in the magnesium treated group was 24% below that in the control group (7.8% versus 10.3%). There was no difference in the incidence of ventricular arrhythmias. In 1995, however, ISIS-4 (International Study of Infarct Survival) studied the effect of a long-acting nitrate, an ACE-blocker and of magnesium on the 35-day mortality of patients with AMI and could not demonstrate a beneficial effect of magnesium (7.28% versus 6.92% mortality) [21]. With regard to magnesium application both study designs differed considerably. While in LIMIT-2 patients received magnesium immediately after admission to the hospital, magnesium therapy in ISIS-4 was started up to 8 hours thereafter [22].

Experimental data have shown that calcium antagonistic agents such as magnesium have no effect if given after reperfusion [23, 24] These data support the hypothesis that the time of application of magnesium in acute myocardial infarction determines its effectiveness [25, 26]. The MAMI study was designed to test this hypothesis.

2. Methods

Intravenous magnesium sulfate infusion was started at two distinct time points: concurrent with begin of thrombolysis therapy (group1) or three hours after beginning of thrombolysis (group2).

Magnesium dosage followed the LIMIT protocol: a bolus of 8 mmol given over 6 minutes was followed by a continuous infusion of 72 mmol over 22.5 hours (= 3.2 mmol/h).

99 patients with acute myocardial infarction as defined by typical clinical signs (chest

TABLE 1. Exclusion criteria

Mean arterial blood pressure below 80 mmHg
Bradycardia (heart rate below 65/min)
II and III degree AV-block
Cardiogenic shock
Clinically apparent left heart failure
Catecholamine therapy
Renal insufficiency
Myasthenia gravis

TABLE 2. Criteria to stop magnesium therapy

Mean arterial blood pressure dropping below 80 mm Hg
Persistent bradycardia below 60 bpm
Development of AV-block
Development of left heart failure requiring therapy
Serum magnesium level above 3 mmol/l

TABLE 3. Standard therapy for all study patient

Thrombolysis therapy with rt-PA
Inhibition of thrombocyte aggregation with 500 mg of salicylic acid
i.v. Heparin, dosage according to PTT
Analgesia with i.v. morphin
Sedation
i.v.nitrates, i.v.beta-blocking agents, diuretics and ACE-inhibitors with diazepam orally
According to clinical necessity

pain), ECG changes (ST-segment elevation) and creatin kinase (Ck) kinetics were included into the study. Ck on admission had to be negative. Exclusion criteria are summarized in table 1, table 2 lists criteria for termination of magnesium therapy. All patients received thrombolysis therapy with rt-PA according to the Neuhaus protocol (standard treatment see table 3).

Mortality was analyzed after 7 and 28 days, as well as after 6 and 12 months. The incidence of arrhythmias was documented by Holter monitoring over 12 hours. Ck kinetics and myocardial reperfusion reflected by clinical signs and ST-segment dynamics were analyzed.

3. Results

99 patients were included into the study, 50 patients in the early magnesium group (group1) and 49 in the late group (group2).

Maximum Ck levels (Ck max.) in group 1 were higher than in group 2: 1126 +/-767 U/l vs. 911 +/-650 U/l and reached their peak earlier: 9.6 vs. 10.5 hours.

In patients with anterior wall infarction these trends are more destinct: in group 1 (n=23) and 2 (n=21) Ck max. were 1598 +/-816 U/l and 1241 +/-796 U/l respectively. Group 1 reached these levels after 8.3 hours as opposed to 9 hours for group 2.

A subgroup analysis was performed on patients with documented reperfusion (ECG criterium: 70% reduction of initial ST-elevation within 3 hours). Reperfusion could be achieved in 71% of patients in group 1 vs. 57% in group 2. Patients with reperfusion had a Ck max. of 1694 U/l (+/- 839) at 8,7 hours if magnesium had been started early (group 1, n=15) and 1209 U/l (+/-865) at 9,1 hours in group 2 (n=11).

These differences did not reach statistical significance.

4. Conclusion

1.There is a trend towards earlier and higher Ck-peak levels with early application of intravenous magnesium in acute myocardial infarction which might be due to earlier and more efficient reperfusion.

2. The differences in Ck kinetics resemble the results from plazebo controlled studies with values in the late magnesium group equal to data with plazebo.

3. The results of the MAMI study might explain some of the different results of LIMIT-2 and ISIS-4.

References

1. Heesch CM, Eichhorn EJ. (1994); Magnesium in acute myocardial infarction. *Ann Emerg Med,* **24**: 1154-1160.
2. Zwillinger L (1935). Über die Magnesiumwirkung auf das Herz. *Klin Wochenschr*; **14**: 1429-1433.
3. Elek SR, Katz LN. (1942) Further observations on the action of drugs on coronary vessel caliber: paredrine, angiotonin, renin, quinidine, insulin, magnesium sulphate, morphine, acid and alkali. *J Pharm Exptl Therap*; **75**: 178.
4. Boyd LJ, Scherf D. (1943) Magnesium sulfate in paroxysmal tachycardia. *Am J Med Sci*; **206**: 43.
5. Hoffmann L, Siegel R. (1952); Die Behandlung von Kreislaufstörungen mit Magnesiumverbindungen. *Hippocrates* **23**: 387-390.
6. Malkiel-Shapiro B, Bersohn I, Terner PE. (1956) Parenteral magnesium sulphate therapy. Med Proc; **2:** 455-462.
7. Malkiel-Shapiro B. (1958) Further observations on parenteral magnesium sulphate therapy in coronary heart disease: A clinical appraisal. *South Afr Med J*; **32**: 1211-1215.
8. Parsons RS, Butler TL, Sellars EP. (1960) Coronary artery disease. *Med Proc*; **6**: 479.
9. Bajusz E, Selye H. (1959) Conditioning factors for cardiac necroses. *Trans NY Acad Sci*; **21**: 659-667.
10. Bajusz E, Selye H. (1960) The chemical prevention of cardiac necroses following occlusion of coronary vessels. *Can Med Assoc J*; **82**: 212-213.

11. Morton BC, Nair RC, Smith FM, McKibbon TG, Poznanski WJ. (1984) Magnesium therapy in acute myocardial infarction - a double blind study. *Magnesium*; **3**: 346-352.
12. Smith LF, Heagerty AM, Bing RF, Barnett DB. (1986) Intravenous infusion of magnesium sulfate after myocardial infarction: effects on arrhythmias and mortality. *Int J Cardiol*; **12**: 175-180.
13. Abraham AS, Rosenmann D, Kramer M, Balkin J, Zion MM, Farbstien H: (1987) Magnesium in the prevention of lethal arrhythmias in acute myocardial infarction. *Arch Intern Med*; 147: 753-755.
14. Rasmussen HS, Gronbaek M, Cintin C, Balslov S, Norregard P, McNair P. (1988) One-year death rate in 270 patients with suspected acute myocardial infarction, initially treated with intravenous magnesium or placebo. *Clin Cardiol*; **11**: 377-381.
15. Ceremuzynski L, Jurgiel R, Kulakoswski P, Gebalska J. (1989) Threatening arrhythmias in acute myocardial infarction are prevented by intravenous magnesium sulfate. *Am Heart J*; 118: 1333-1334.
16. Shechter M, Hod H, Marks N, Behar S, Kaplinsky E, Rabinowitz B. (1990) Magnesium therapy and mortality in acute myocardial infarction. *Am J Cardiol*; **66**: 271-274.
17. Feldstedt M, Boesgaard S, Bouchelouche P, Svenningsen A, Brooks L, Lech Y, Aldershvile J, Skagen K, Godtfredsen J. (1991) Magnesium-substitution in acute ischaemic heart syndroms. *Eur Heart J*; **12**: 1215-1218.
18. Teo KK, Yusuf S, Collins R, Held PH, Peto R. (1991) Effects of intravenous magnesium in suspected acute myocardial infarction: overview of randomised trials. *Br Med J*; **303**: 1499-1503
19. Antman EM. (1995) Randomized Trials of Magnesium in Acute Myocardial Infarction: Big Numbers Do Not Tell the Whole Story. *Am J Cardiol*; **75**: 391-393.
20. Woods KL, Fletcher S, Roffe Ch, Haider Y. (1992) Intravenous magnesium sulphate in suspected acute myocardial infarction: result of the second Leicester Intravenous Magnesium Intervention Trial (LIMIT-2). *Lancet*; **339**: 1553-1558.
21. ISIS-4 (1995) Collaborative Group. A randomised factorial trial assessing early oral captopril, oral nitrate, and intravenous magnesium sulphate in 58.050 patients with suspected acute myocardial infarction. *Lancet*; **345**: 669-682.
22. Beuckelmann DJ. Editorial: (1994) Magnesiumtherapie bei Myocardinfarkt: Bei differenzierter Indikation sinnvoll. *Münch Med Wschr*; **136**: 34.
23. du Toit EF, Opie LH. (1992) Modulation of severity of reperfusion stunning in the isolated rat heart by agents altering calcium flux at reperfusion. *Circ Res*; **70**: 960-967.
24. Opie LH. (1994) Myocardial stunning - are calcium antagonists useful? *Cardiovasc Drugs Ther*; **8**: 533-541
25. Lüderitz B, Manz M. (1994) Die Bedeutung von Magnesium in der Intensivmedizin. *Z Kardiol;* **Suppl 6**: 121-124.
26. Seelig MS, Elin RJ. (1995) Reexamination of magnesium infusions in myocardial infarction. *Am J Cardiol*; **76**: 172-173.

CHANGES IN PLATELET, ERYTHROCYTE AND PLASMA MAGNESIUM LEVELS IN NORMOTENSIVE AND HYPERTENSIVE OBESES DURING ORAL GLUCOSE TOLERANCE TEST.

F. CORICA, A. ALLEGRA, R. IENTILE*, M. BUEMI, A. CORSONELLO, S. BONANZINGA, R. D'ANGELO*, T. DE GREGORIO, D.CERUSO

Department of Internal Medicine. University of Messina, Italy.

**Institute of Biochemical Sciences and Clinical Biochemistry. University of Messina, Italy.*

Previous studies have provided evidence that insulin can induce important modifications in plasma and erythrocyte magnesium levels (1), and that insulin-resistance could modify the ionophoric action exerted by insulin on erythrocytes (2). No data are reported in literature as regards the modifications of platelet magnesium concentrations during oral glucose tolerance test (OGTT). The aim of this study was therefore to evaluate the behavior of plasma, erythrocyte and platelet magnesium concentrations, measured by direct current spectrometry, in normotensive and hypertensive obeses, and in a group of age- and sex-matched healthy subjects, during OGTT. Plasma glucose, insulin and norepinephrine were also evaluated.

Table 1. *General characteristics of the subjects studied.*

	Controls (N=12)	Normotensive Obeses (N=18)	Hypertensive Obeses (N=15)
Age (years)	34.3 ± 5.0	39.3 ± 10.3	38.3 ± 5.1
Sex (F/M)	8/4	13/6	8/7
B.M.I. (kg/m^2)	22.1 ± 2.5	36.3 ± 6.9*	36.8 ± 11.0*
Waist circumference (cm)	74.6 ± 9.8	107.1 ± 15.6*	113.6 ± 13.1*
Waist-to-Hip Ratio	0.79 ± 0.06	0.89 ± 0.1**	0.97 ± 0.1**††
Plasma cholesterol (mg/dl)	168.6 ± 19.4	193.0 ± 30.6	212.5 ± 59.5***
S.B.P. (mmHg)	120.5 ± 4.6	123.9 ± 9.1	158.1 ± 25.1**†
D.B.P. (mmHg)	77.2 ± 3.6	77.9 ± 8.9	90.6 ± 11.5**†
M.B.P. (mmHg)	91.7 ± 3.4	93.2 ± 7.9	113.1 ± 14.9**†

Data are means ± SD.
$^*p < 0.001$, $^{**}p < 0.01$, $^{***}p < 0.05$ vs controls; $^{\dagger}p < 0.01$, $^{\dagger\dagger}p < 0.05$ vs normotensive obeses (Student-Newman-Keuls test for multiple comparisons)

In the control group, a significant reduction in plasma Mg at 30' (p<0.05), 60' (p<0.05), 90' (p<0.001) and 120' (p<0.01) was observed; erythrocyte Mg concentrations were significantly higher at 30' (p<0.05), 60' (p<0.05) and 90' (p<0.05) with respect to baseline; a significant increase in platelet Mg levels at 60' (p<0.001) and 90' (p<0.05) was found. In obese subjects plasma Mg levels were significantly reduced at 30' (p<0.05), 60' (p<0.01), 90' (p<0.01), 120' (p<0.001), 150' (p<0.01) and 180' (p<0.01), while a significant reduction in erythrocyte Mg concentrations at 60' (p<0.05), 90' (p<0.01) and 120' (p<0.05) was found; platelet Mg concentrations were significantly lower at 30' (p<0.05), 60' (p<0.01), 90' (p<0.001), 120' (p<0.01), 150' (p<0.01) and 180' (p<0.05) when compared to baseline. In hypertensive obeses plasma Mg levels were significantly reduced at 120' (p<0.01), 150' (p<0.05) and 180' (p<0.05), while a significant decrease in erythrocyte Mg at 90' (p<0.01) was found; a significant reduction

T. Theophanides and J. Anastassopoulou (eds.), Magnesium: Current Status and New Developments, 251-252.

of platelet Mg at 120' ($p<0.01$), 150' ($p<0.05$) and 180' ($p<0.05$) was observed. As regards plasma norepinephrine and plasma, erythrocyte and platelet Mg levels during OGTT, the differences among groups are shown in Figure 1. The fall in plasma Mg levels, occurring in each group of subjects studied, could be a consequence either of the activation of the adrenergic system induced by insulin, by means of lypolisis (3), or of magnesium uptake within the cells (e.g. adipocytes) (4). In the control group, the fall in plasma Mg levels could also be a consequence of the increase in erythrocyte and platelet Mg concentrations, which follows the increased insulinaemia. In both normotensive and hypertensive obese patients, insulin-resistance seems to reduce the ionophoric activity of insulin in erythrocytes and especially in platelets, where insulin receptors, similar to those found in other cell types, were also identified. Moreover, the increased norepinephrine levels found in obeses, could be involved in the lowering of platelet magnesium content by receptorial mechanisms that are still unknown (α_2-adrenoceptor)(5). In conclusion, our results seem to confirm the hypothesis of the existence of an impaired ionophoric action of insulin in obeses, probably related to insulin-resistance.

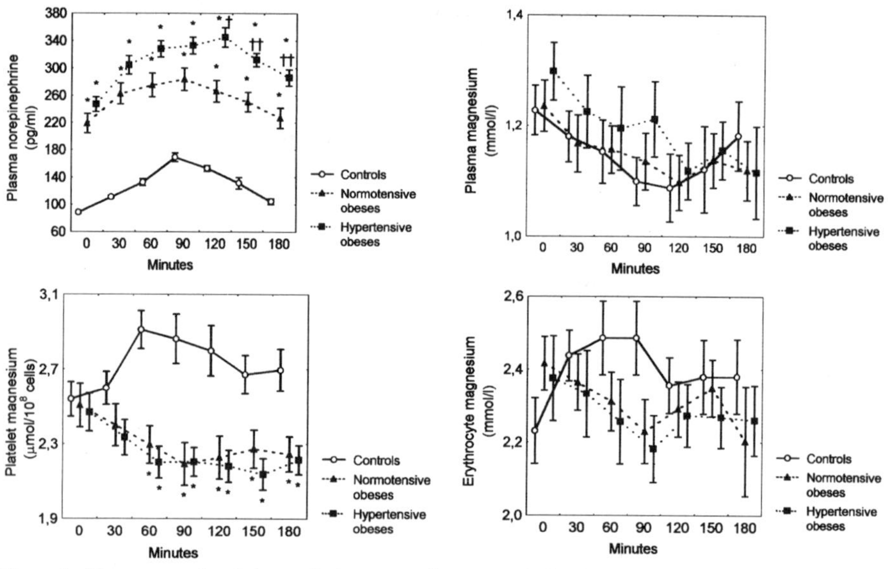

Figure 1. *Plasma norepinephrine and plasma, erythrocyte and platelet magnesium levels during OGTT in the groups studied.* * $p < 0.01$, ** $p < 0.05$ *vs controls;* † $p < 0.01$, †† $p < 0.05$ *vs normotensive obeses.*

References

1. Durlach, J. and Rayssiguier, Y. (1983) Magnésium et glucides I. Données cliniques et thérapeutiques. *Magnésium* **2**, 192-224.
2. Paolisso, G., Sgambato, S., Giugliano, D., Torella, R., Varricchio, M,, Scheen, A.J., D'Onofrio, F. and Lefèbvre, P.J. (1988) Impaired insulin-induced erythrocyte magnesium accumulation is correlated to impaired insulin-mediated glucose disposal in Type 2 (non-insulin-dependent) diabetic patients. *Diabetologia* **31**, 910-915.
3. Rayssiguier, Y. (1977) Hypomagnesemia resulting from adrenaline infusion in ewes: Its relation to lipolysis. *Horm. Metab. Res.* **9**, 309-314.
4. Elliot, D.A. and Rizac, M.A. (1974) Epinephrine and ACTH stimulated Mg accumulation in adipocytes and their plasma membranes. *J. Biol. Chem.* **249**, 3985-3990.
5. Kahn, N.N. and Sinha, A.K. (1992) Down regulation of α_2-adrenergic receptor numbers in platelets by insulin. *Biochim. Biophys. Acta* **1134**, 292-296.

THE ROLE OF MAGNESIUM: A NATURAL CALCIUM ANTAGONIST IN THE TREATMENT OF ACUTE ISCHAEMIC STROKE

Th. GALEAS[1], G. ZIOGAS[1], B. VALOTASIOU[1], B. SINTZERMA[1], AND F. KARAPANOS[2]
[1] B' Internal Medicine Clinic, General Hospital of Trikala, [2] Medical Center of Trikala, 42 100 Trikala, Greece.

1. INTRODUCTION

In the ischaemic stroke, the infarct formed, is either generated by an embolus distal from the brain (atrial fibrilation, valvular heart disease, carotid atherosclerosis) or locally, in the endothelium of the cerebral vessels, gradually, as a symptom of a generalized vascular disease like diabetes mellitus.
However, in both cases, when clinical symptomatology is present, blood flow in the ischaemic brain area is lower or equal to 20 ml/100 gr/min. When flow is between 20 and 15 ml/100 mg/min there is only a reduction of cell energy, manifested with subfunction of the cell membrane pumps, which sustain the ion difference on either side. When blood flow falls below 10 ml/100 mg/min there is an uncontrolled in flux of ions in the neuronal cells resulting in the inevitable death. [1].
The purpose of the work is to show the contribution of Mg a natural Ca antagonist in the treatment of acute ischaemic stroke because of its specific inhibitory action against damage mechanism, on cellular lever.

2. MATERIAL - METHOD

Material of the study were all the cases with the indication of acute ischaemic stroke admitted in the Internal Medicine Clinics of the General Hospital of Trikala, during the period of one year.
510 patients, men and women aged 72,26 ± 10,91 were hospitalized, of whom 70 died, aged 78,82 ± 8,75. We separated the patients in two groups one of 233 patients aged 71,25 ± 10,95 who received a supplement of magnesium together with the usual medication and the other of 277 patients aged 73,27 ± 10,87 without the supplement. Those in the first group who had azotemia or chronic renal insufficiency were not given Mg. Diagnosis was set with clinical and laboratory criteria. Brain CT was necessary.

T. Theophanides and J. Anastassopoulou (eds.), Magnesium: Current Status and New Developments, 253-255.

3. RESULTS

The administration of Mg significantly improved the clinical and laboratory picture of the patients in the first group and rehabilitation was more rapid in the transient strokes.

4. DISCUSSION

Reduction of blood flow in the ischaemic cerebral area is followed by reduction of oxygen supply to the tissues of this area, reduction of oxidative phosphorolysis and ATP production, so ion and especially Ca^{++} influx into the cell becomes uncontrollable resulting in activation of destructive lipases, proteases and endonucleases and finally cell damage. Free radicals which cause contraction of the precapillary sphincters and transient ischaemia with subsequent dilation of precapillary sphincters are generated during this process [2]. In this way larger quantity of serum comes through the capillary wall which results in the development of aseptic inflammatory edema [3]

In order to secure the necessary ATP, neuronal cells use anaerobic metabolism and this results in an increased H^+ production which are extruded by Na^+ /H^+ pump. Na^+ ion influx leads to cellular edema when the other Na^+ pumps subfunction. So edema is due to neuronal cell damage (part of ischaemic cascade) as well as damage of the area vessels (inflammatory edema). Besides, increased intake of hexose by polymorphonuclears (eg D-glucose) and influx of calcium ions is stimulated [4,5] while mobility of intracellular calcium is stimulated by LTB, promoting increased transportation of calcium ions and D-glucose to the edema area. Finally, in the area around the infarct and parallel to the neuronall cell edema the initial inflammatory edema is developed which in a short time is *'flooded'* with constrictive substances, like hydroxyl products of arachidonic acid (HETES). Txs and free radicals, as well as increased concentrations of calcium ions and D-glucose which promote ischaemic cascade and intracellular edema because of the anaerobic formation of ATP. It is also noted that although blood glucose is at normal rate, however, it is comparatively increased in the ischaemic area. Also an increase in extracellular glutamate is observed which is probably caused by the uncontrolled flux of calcium ions into the preconjugated sites, resulting in the retaking of glutamate from the neurons and astrocytes with further ATP consumption . It is suggested that Mg ions have an antiplatelet action as well, by blocking platelet Ca^{++} channels. [6]

4. CONCLUSION

It seems that administration of Mg could be mostly indicated in ischaemic stroke because it activates vasodilation, prohibits reduction of blood flow in the ischaemic area and activation of the ischaemic cascade. However, if ischaemia does happen, neuronal cells are sufficiently protected from the uncontrolled influx of calcium ions which leads them to irreversible death.

Also extension of inflammatory edema is reduced because vasodilation caused antagonizes vasoconstrictive action of arachidonic acid, Txs and serotonin. Finally Mg a natural Ca antagonist must be administered as a supplement in the treatment of acute ischaemic stroke as it improves clinical and laboratory picture of the patients.

References

1. Astrup, J. ,Siesjo, B.K and Symon, L. (1981) The state of penumbra in the ischaemic brain: viable and lethal threshold in cerebral ischaemia. *Stroke*; **12**: 273-25.
2. Schmidley, J.W. (1990) Free radicals in central nervous system ischemia. *Stroke*; **25**: 7-12.
3. Kandreviotis, N. (1983) *Elements of Internal Medicine and Pathology*, 3rd edition, Thessaloniki; pp. 182-189, 190-196, 201-208.
4. Bass, D.A, O'Flaherty, J.T., Goetzly, E.J et al. (in press) Arachidonic acid and hexose transport in human polymorphonuclear leucocytes. *Prog Lipid Res.*
5. Naccache, P.H, Borgeat, P., Goetzl, E.J. et al. (in press) Mono and di-hydroxy eicosatetraenoic acids after calsium homeostasis in rabbit neutrophils. *J Clin Invest.*
6. Galeas, Th. and Hatziioannidis, B. (1993) Ischaemic stroke. Biochemical mechanism. The role of Ca^{++} inhibitors in the treatment. *Health Review*; **4**, 23.

DOES CARDIAC ISCHEMIA INDUCE LOWERING OF BLOOD IONIZED MAGNESIUM LEVEL IN PATIENTS?

MASAKO HADARA, KENJI UESHIMA, MASASHI SHIBATA AND KATSUHIKO HIRAMORI
Second Department of Internal Medicine, Iwate Medical University, Uchimaru 19-1, Marioka, Japan

1. INTRODUCTION

It is considered that magnesium relates in the pathophysiology of cardiovascular disease. A various state of ischemic heart disease is associated with a deficiency of plasma and tissue ionized magnesium, but it is not well known whether low ionized magnesium levels are the cause or result of myocardial ischemia.
Recently the instrument (NOVA 8; NOVA Biomedical Co, USA.) for measuring ionized magnesium was developed. We investigated the correlation between plasma ionized magnesium and myocardial ischemia by measuring the plasma ionized magnesium levels in patients who underwent percutaneous transluminal coronary angioplasty (PTCA).

2. MATERIALS AND METHOD

The study population consisted of 42 pts. With stable angina pectoris. PTCA to the left anterior descending arteries were performed in 37 pts (20 men, 17 women; mean age 63 years), and coronary angiography in 5 pts (4 men, 1 woman; mean age 68 years).

Whole blood was taken from a femoral artery from all patients. The blood samples were obtained in standard green top vacuutainer tubes (Teruno, Japan) containing 15 units of sodium heparin per ml of blood sample using a 7 ml sample. All blood samples were collected and maintained under anaerobic conditions and processed within 1-2 hours. We measured plasma ionized magnesium levels using NOVA 8. In the PTCA group, we monitored the total balloon inflation time and ECG during PTCA.

3. STATISTICAL ANALYSIS

Data are presented as mean ± SD. Comparison of ionized magnesium levels, before and after PTCA, were made by use of paired t-test. Then, in PTCA group, we divided

T. Theophanides and J. Anastassopoulou (eds.), Magnesium: Current Status and New Developments, 257-258.

into tow groups by median Δ ionized magnesium levels and dose parameter were made by use of unpaired t-test.

4. RESULTS

Plasma ionized magnesium levels were significantly decreased in patients after PTCA compared to the values before PTCA (0.64±0.001 mmol/l vs. 0.50±0.002 mmol/l, p<0.01). And whole blood ionized magnesium did not significantly change with CAG (0.49± 0.02 mmol/l vs. 0.49±0.01 mmol/l). During PTCA, there was a positive correlation between Δ ionized magnesium levels and the balloon inflation times (r=0.47, p<).01). And during PTCA, maximum ST deviation and the total balloon inflation times in patients with high decreases in ionized magnesium levels (Δ iMg≥0.03 mmol/l, n=17) were greater than with only a slight decrease in ionized magnesium levels (Δ iMg <0.03 mmol/l, n=20) (643.5±357.2 sec vs 387.5±172.1 sec, 3.8±0.4 mm vs. 1.6±0.5 mm)

5. CONCLUSION

1. Plasma ionized magnesium levels did not significantly change in the control group.
2. Cardiac ischemia, provoked by TPCA, induced a decrease of plasma ionized magnesium levels in patients.
3. Myocardial ischemia results in lower ionized magnesium levels.

6. SUMMARY

Cardiac ischemia may induce a decrease of plasma ionized magnesium levels in patients with IHD. The decreased degree of plasma ionized magnesium depends on the severity of myocardial of myocardial ischemia.

EFFECT OF OUABAIN ON THE BEHAVIOR OF ^{28}Mg IN THE MICE MYOCARDIUM

TATSUO IDO[1], REN IWATA[1], AND MIEKO KAWAMURA[2]
1Radiopharmaceutical Chemistry, Cyclotron and Radioisotope Center, Tohoku University, Sendai 980-77, JAPAN
2Department of Applied Biochemistry, Faculty of Agriculture, Tohoku University, Sendai 981, JAPAN

Key words: Mg uptake, ^{28}Mg, Ouabain, myocardium, Mg transport.

1. INTRODUCTION

Magnesium is essential for maintaining ATPase reaction which is responsible for hydrolyzing ATP to ADP. Under normal conditions the intracellular magnesium ion concentration is 8 times higher than the extracellular concentration [1]. Ouabain, a specific pump inhibitor, is known to inhibits the ATPase activity. The purpose of this study was to observe the effect of ouabain on magnesium behavior in the mouse myocardium.

2. MATERIALS AND METHODS:

Ouabain(0.147mg/animal) was intraperitonealy administered to ddY male mice (6 weeks old; 25-30g)[2]. After the administration of ouabain at intervals of 5, 10, 30, 60, and 120 min., each experimental mouse received a venous injection of ^{28}Mg (10^5-10^6 cpm/animal) via the tail[3]. The mice were sacrificed 10 min. after the injection of ^{28}Mg. The uptake of ^{28}Mg in the blood, heart, liver, small intestine, muscle, and bone were then measured.

Autoradiography of ^{28}Mg concentrations in the heart was also performed. ^{28}Mg was injected 10 min. after the administration of ouabain. The mice were sacrificed 10 min. after the injection of ^{28}Mg and myocardial uptake levels were then measured (BAS-3000 system, Fuji film Corp, Japan).

3. RESULT AND DISCUSSION:

The relative uptake ratio of ^{28}Mg in the heart, small intestine, muscle, bone, liver and blood at different time after ouabain injection are shown in Table 1. As shown, the administration of ouabain affected the uptake of ^{28}Mg in the heart, small intestine, muscle, bone and liver. Compared to the control, the accumulation of ^{28}Mg in heart and small intestine showed a substantial decrease at 10min. and 30 min. after the administration of ouabain. However, after 60 min., the effect of ouabain could not be observed and ^{28}Mg uptake in the organs showed no significant differences. Low

T. Theophanides and J. Anastassopoulou (eds.), Magnesium: Current Status and New Developments, 259-260.

accumulation of ^{28}Mg in the skeletal muscle and bone was observed initially but gradually recovered by 120 min.. In contrast to the other organs, the uptake level in the liver was approximately 1.7 times higher than control at 60min. but returned to normal by 120 min..

TABLE 1: Effect of ouabain on ^{28}Mg uptake in Mice

ORGANS	5min	10min	30min	60min	120min
HEART	0.69	0.50	0.59	1.02	1.07
S.INTESTINE	0.75	0.36	0.56	1.06	1.15
MUSCLE	0.61	0.70	0.91	0.88	1.29
BONE	0.54	0.70	0.79	0.87	1.24
BLOOD	0.91	0.73	0.47	0.98	1.17
LIVER	0.95	0.66	0.69	1.76	0.80

*time(min) : after injection of ouabain
*data is shown as relative uptake ratio
*uptake ratio = DAR of treated animal / DAR of control group

Autoradiographical analysis revealed that, after administration of ouabain, the accumulation of ^{28}Mg in the myocardium was 0.54 times lower (average) at 10 min..

3. CONCLUSION:

Ouabain was shown to deduce the uptake of Mg in mouse myocardial tissue *in vivo*. Moreover, this effect was reversible.

The mechanism of Mg^{2+} transfer into the cell may be depended on the active transport of ATPase (perhaps the Mg pump) and may have caused the higher intracellular concentration of Mg^{2+}.

References:

1. Smith E.L. et al (1982). *Principles of biochemistry: mammalian biochemistry*, 7th ed., McGraw-Hill Book Co. NY.
2. Navararnam S. et al. (1989) Influence of the diabetic state on digitalis-induced cardiac arrhythmias in rat, *Arch. Int. Pharmacodyn. Ther.*, **301** 151-64.
3. Iwata R., Ido T et al. (1992) Chromatographic purification of no-carrier-added magnesium-28 for biological studies, *J. Radioanal. Chem. Article,* **159,** 233-237

EXPERIMENTAL AND CLINICAL CONTROVERSIES IN THE ANTIISCHEMIC AND ANTIARRHYTHMIC EFFECTS OF MAGNESIUM SULFATE

Z. KISS
Department of Medicine, University oteaching Hospital, H-6901-Mako, P.O. Box 72, Hungary

Deprivation of oxygen supply implies a defect of energy resources necessary for the active ion transfer across the membrane, extrusion of sodium and calcium and intrusion of potassium. Depolarization is the consequence of an accumulation of sodium and calcium in the cell, which is associated with a depletion in potassium [1] and high extracellular potassium level. Intracellular sodium accumulation with water leads to cell swelling. The effect of cell swelling is important since during a period of myocardial ishaemia/ or low blood flow as occurs, one of the cell's responses is to increase the intracellular volume, which at this time contains comparatively less osmotically active particles [2]. The concentration of potassium, in the extracellular space, rises from 4 to 8 mmol/l, within the first minutes of ischemia [3]. Calcium antagonists, including magnesium, the nature's physiologic calcium channel blocker, tend to control both calcium overload and potassium depletion [1,4] and consequently, ischemic depolarization, responsible for the enhancement of excitability and slowing of conduction, leading to fibrillation.

Experimental data showed that magnesium restores high extracellular potassium, induced partial depolarization and inactivation of the fast sodium channel in the ventricular muscle [5], even if magnesium ions are not directly taking part in the ion fluxes associated with the normal cardiac potential [6]. Potassium loss from the cardiac myocytes occurs almost instantaneously during ischemia and hypoxia [7] and the first minutes is due to increased K^+-efflux [8]. It, has been showed recently that 9-15 min of global myocardial ischemia results in a 3-6 fold increase in intracellular free magnesium ion concentration and it remains elevated on repercussion and may contribute to myocardial stunning decreasing the calcium sensitivity of cardiac cells.

Several authors described a transient decrease of total serum magnesium during the first 24 hours after acute myocardial infarction which reached normal values 6 to 14 days later [9]. The fall of serum magnesium may be explained by increased lipolysis due to stress-induced catecholamine release. The free fatty acids bind Mg^{2+}. Thus the free fatty acids in adipocytes reduce the concentration of intracellular free Mg^{2+} in these cells followed by Mg^{2+} uptake in adipocytes [10]. The changes of ionized K^+ are closely correlated to the changes of ionized Mg^{2+}. A close relationship was also shown for total Mg and ionized K^+ in serum [11].

T. Theophanides and J. Anastassopoulou (eds.), Magnesium: Current Status and New Developments, 261-264.

The use of intravenous magnesium sulfate ($MgSO_4$) in human acute myocardial infarction (AMI) also remains controversial, although the probable cause of different efficacy of the drug in LIMIT-2 and ISIS-4 trial is the timing of $MgSO_4$ administration, i.e. administering before reperfussion resulted in substantial myocardial salvage, but it has no significant effect when magnesium was given after reperfussion had occurred [12,13]. The LIMIT-2 [14] found a 24% reduction in 28 day-mortality, whereas the ISIS-4 [15] showed no benefit. In latter trial magnesium infussion was typically administered after initiation of the fibrinolytic infussion, and magnesium levels may not have been adequately elevated at the time of reperfussion, possibly invalidating the results of this trial as to the question of magnesium efficacy in reducing reperfussion injury [16]. In the subgroup of inferior myocardial infarction ISIS-2 showed that mortality after streptokinase treatment alone was 7.2 %, compared with 8.8 % (NS) after conventional treatment without aspirin [17].
Before thrombolytic era, in our 366 patients suffering from AMI and treated by intravenous $MgSO_4$, the two week hospital mortality was half than in the control 108 pts. [18]. The difference of arrhythmia incidence was not significant. Other recent results also indicate that $MgSO_4$ infussion may have a beneficial effect on mortality in pts with AMI not receiving thrombolytic therapy; while clinical signs of spontaneous reperfussion-induced ventricular arrhythmias were observed [19]. Shechter et. al. [20], randomly assigned patients, who were considered unsuitable for thrombolysis, to receive magnesium infussion or placebo. 194 patients were recruited and the average time between onset of chest pain and magnesium therapy was 7 h (an average of 5 h earlier than the equivalent group in ISIS-4). In-hospital mortality was 4.2% in their treatment group compared with 17.3% in controls, and most of the benefit was seen in terms of lower frequencies of cardiogenic shock and congestive cardiac failure. We have reported the effectiveness of intravenous $MgSO_4$ in the treatment of paroxysmal supraventricular tachyarrythmias in 7 out of 14 pts. In the remainer, $MgSO_4$ proved to be effective in combination with verapamil, digitalis, or KCl. Sinus tachycardia was abolished by intravenous $MgSO_4$ in a further four cases [21]. According to published data, the predominant site of $MgSO_4$ effect is the atrioventricular node and has similar electrophysiologic properties to those of calcium channel blockers; slowing of AV-nodal conduction (and sinus node activity without affecting infranodal conduction) may be mediated by alteration in slow calcium channel kinetics.

Conclusions

Animal model reperfussion data demonstrate a myocardial protective effect of $MgSO_4$ [22] and data from multiple small trials of ~13,000 pts also suggested a significant mortality benefit [23], meanwhile is difficult to measure the direct effect of $MgSO_4$, because magnesium acts at molecular and cellular level [24]. Possible beneficial effects of magnesium included enchanced collateral flow, decreased free radical formation, afterload reduction, antiarrhythmic effects, and reduced repercussion injury because of impeded calcium influx after relief of ischemia.

As an indirect sign, after magnesium pretreatment during early ischemia in the *in vivo* human heart, the serum magnesium level was elevated (1.69±0.07 versus 0.6±0.03 mmol/l in placebo group) and epicardial monophasic action potential was significantly shorter, than in placebo group [25].

As long as it remains controversial whether a general Mg^{2+} infusion of all acute myocardial infarction patients is beneficial. Bertschat et al. [26] suggested that ionized Mg^{2+} should be measured and Mg^{2+} should be infussed only in acute myocardial infarction patients with low ionized Mg^{2+}. When it is not possible to measure ionized Mg^{2+}, hypokalaemia may be taken as an indicator for low ionized Mg^{2+} [26]. Concerning magnesium homeostasis, decreased renal magnesium excretion, <1 mmol/day, may be a good indicator of magnesium deficiency [27,28].

From literature and our data it is concluding that clinicians may elect to administer $MgSO_4$ intravenously as adjunctive therapy for AMI. Relative contraindications of intravenous $MgSO_4$ include second-degree av-block, sinus-bradycardia, hypotension and significant renal impairment.

References

1. Blake, K. and Clusin, W.T. (1986), Effect of diltiazem on ischemic myocardial depolarization and extracellular K+ accumulation, *Eur. J. Pharmacol.*, **127**, 261-265
2. Rees, S.A. (1996), Academic research, K+ channels and an afair of the heart, *TiPS* **17**, 53-55
3. Gettes, L.S., Cascio, W.E., Johnson, T. and Fleet, W.F. (1991), Local myocardial biochemical and ionic alterations during myocardial ischemia and reperfusion. *Drugs*, **42 Suppl. 1**, 7-13
4. Heijnis, J.B., Coronel, R. and Van Zwieten, P.A. (1991), The effects of calcium antagonists on extracellular potasium accumulation during global ischemia in isolated perfuced rat hearts. *Cardiovasc. Drugs Ther.* **5,** 1035-1041
5. Kiyosue, T. and Arita, (1982), M. Magnesium restores high K-induced inactivation of fast Na channel in guinea pig verticular muscle. Pflugers Arch. 395, 78-80
6. Shine, K. (1979), Myocardial effects of Magnesium, *Am. J. Physiol.* **237**, 4, H413
7. Kleber, A.G. (1983), Resting membrane potential, extracellular potassium activity, and intracellular sodium activity during acute global ischemia in isolated perfused guinea pig hearts. *Circ. Res.* **52**, 442-450
8. Gassr, R.N.A., Vaughan-Jones, R.D. (1990), Mechanism of potassium efflux and action potential shortening during ischemia in isolated mammalian cardiac muscle, *J. Physiol. (London)* **431**, 713-741
9. Rasmussen, H.S., Aurop, P., Hojberg, S. Kehn-Jensen, E., McNair, P. (1986), Magnesium and acute myocardial infarction. Transient hypomagnesemia not induced by renal magnesium loss in patients with acute myocardial infarction. *Arch. Intern. Med.* **146**, 872-874
10. Vormann, J., Foerster, R., Gunther, T., Ebel, H. (1983), Lipolysis-induced magnesium uptake into fat cells. *Mag. Bull.* **5**, 39-41
11. Landmark, K., Urdal, P., Basmo, G.M. (1988), Changes in serum K and Mg during acute myocardial infarction (abstract) *Eur. Heart J.* **9, Suppl. 1**, 227, 1260
12. Christensen, C.W., Rieder, M.A., Silverstein, E.L., Gencheff, N.E. (1995), Magnesium sulfate reduces myocardial infarct size when administrered before but not after coronary reperfusion in a canine model, *Circulation*, **92**, 2617-2621
13. Herzog, W.R., Schlossberg, M.L., MacMurdy, K.S. (1995), Timing of magnesium therapy effects experimental infarct size, *Circulation,* **92**, 2622-2626
14. Woods, K.L., Fletcher, S., Roffe, C., Haider, Y. (1992), Intravenous magnesium sulfate in suspected myocardial infarction: results of the second Leicester Intravenous Magnesium Intervention Trial, LIMIT-2, *Lancet*, **339**, 1553-1558
15. ISIS-4, Fourth international study of infarct survival (Collaborative group). (1995), ISIS-4: a randomised factorial trial assessing ealry oral captopril, oral manonitrate, and intravenous magnesium sulfate in 58,050 pts with suspected acute myocardial infarction. *Lancet*, 345, 669-685.

16. Antman, E.M. (1995), Magnesium in acute MI. *Circulation*, **92**, 2367-2372.
17. ISIS-2, Second international study of infarct survival, Collaborative group, (1988), Randomised trial of intravenous streptokinase, oral aspirin, both, or neither among 17,187 cases of suspected acute myocardial infarction, *ISIS-2, Lancet ii*, 349-360.
18. Kiss, Z., Bardos, J., Domokos, M.,, Toth, M., Sepp., J. (1992), Beneficial effect of intravenous magnesium sulfate in acute myocardial infarction, *Magnes. Res.*, **5**, 242-243.
19. Thoegersen, A.M., Johnson, O., Wester, P.O. (1993), Effect of magnesium infusion on thrombolytic and non-thrombolytic treated patients with acute myocardial infarction. *Inter. J. Cardiol.* **39**, 13-22.
20. Shechter, M., Hod, H., Chouraqui, P., Kaplinsky, E., Rabinowitz, B. (1995), Magnesium therapy in acute acute myocardial infarction when patients are not candidates for thrombolytic therapy. *Am. J. Cardiol.* **75**, 391-393.
21. Kiss, Z., Barna, A., Gaspar, A. (1989), Efficacy of intravenous MgSO4 in the treatment of paroxysmal supraventricular tachycardias. *Magnes. Res.* **2**, 228.
22. Hearse, D.J. Stewart, D.A., Braimbridge, M.H.V. (1978), Myocardial protection during ischemic cardiac arrest. J. Thorac. *Cardiovasc. Surg.* **75**, 877-885.
23. Theo, K.K., Yusuf, S., Collins, R., et al. (1991). Effects of intravenous magnesium in suspected acute myocardial infarction: overwiew of randomised triala. BMJ 303, 1499-1503.
24. Smetana, R., (1991). Magnesium in acute myocardial infarction. *Am. Heart J.* **132**, 463-464.
25. Redwood, S.R., Taggart, P.I., Sutton, P.M., Bygrave, A., Bashir, Y., Purkayastha, D., Camm, A.J., Treasure, T. (1996), Effect of magnesium on monophasic action potential during early ischemia in the in vivo human heart. *J. Am. Coll. Cardiol.* **28**, 1765-1769.
26. Bertschat, F., Ising, H., Gunther, T., Jeremias, A. Jeremias, E. (1995) Changes of ionized magnesium and free fatty acids in serum after acute myocardial infarction. *Eur. J. Clin. Chem. Clin. Biochem.* **33**, 553-558.
27. Kiss, Z. (1994). *Magnesium deficiency syndrome, Magnesium in biological systems- Environmental and biomedical aspects.* Akademiai Kiado, Budapest.
28. Kiss, Z., Bardos, J. (1995), Asymptomatic hypomagnesaemia in diabetes melittus, *Magnes. Res.* **8**, 109.

INFLUENCE OF MAGNESIUM ON THE AORTIC VASORELAXATION IN ADULT AND OLD RATS

Magnesium age and vasorelaxation

P. LAURANT, A. BERTHELOT
Laboratoire Physiologie, Pharmacologie et Nutrition Préventive Expérimentale
UFR Médecine et Pharmacie, Place St-Jacques, 25030 Besançon Cedex, France.

1. Introduction

Changes in extracellular magnesium (Mg) concentration can affect blood flow, vascular reactivity and blood pressure. Mg modifies blood vessels responsiveness by modulating agonist-receptor affinity, calcium flux across the vascular smooth muscle cell membrane, and calcium release from intracellular storage sites (Altura et al., 1987). Mg differentially affects the vascular reactivity of isolated blood vessels obtained from rats with various cardiovascular diseases such as hypertension in which the vascular function is altered (Altura and Altura, 1981; 1983; Laurant and Berthelot, 1996). Aging alters the vascular function and the increased peripheral resistance observed with aging is believed to be due to a reduced vasodilator responsiveness (Vanhoutte, 1988). Since the structural and functional alterations observed with aging are similar to those described in hypertension (Laurent et al., 1996), it is speculated that, with aging, Mg may differentially modulate the relaxant response of isolated blood vessels. The present study was designed to investigate the effects of varying Mg concentrations on vasorelaxant response of isolated aortae from adult and old rats.

2. Methods

Thoracic aortae from adult (4-month-old) and old (24-month-old) male Sprague-Dawley rats were excised, cleaned of connective tissue and cut into rings. All rings (with or without endothelium) were suspended between two L-shaped stainless steel hooks, placed in 10 ml water-jacketed tissue bath filled with physiological salt solution (PSS) and stretched to obtain a passive force of 2 g. Rings were incubated in PSS containing normal (1.2 mM), high (4.8 mM), or low (0.1 mM) Mg concentrations for 60 min. They were exposed to a concentration of norepinephrine inducing 70-80% of maximal contraction. In rings with endothelium, a cumulative concentration-response curve to acetylcholine was generated and in rings without endothelium, a cumulative

T. Theophanides and J. Anastassopoulou (eds.), Magnesium: Current Status and New Developments, 265-268.

concentration-response curve to sodium nitroprusside or isoproterenol was generated. Some rings with and without endothelium were incubated in Mg-free PSS for 60 min. A cumulative concentration-response curve to $MgSO_4$ was then performed after norepinephrine contraction.

3. Results

In PSS containing normal Mg concentration, acetylcholine-induced relaxation was significantly less pronounced in the aortic rings from old rats than in those from adult rats ($p < 0.001$). Low Mg did not modify relaxant response to acetylcholine in the aortic rings from either adult or old rats. High Mg significantly increased relaxation to acetylcholine in the aortae from both adult and old rats (Emax values: 91.1 ± 1.3 vs 85.8 ± 2.2% in adult rats, $p < 0.05$ and 65.7 ± 3.4 vs 44.8 ± 3.7% in old rats, $p < 0.01$). The increasing effect of high Mg on acetylcholine-induced relaxation was significantly greater in the rings from old rats than in those from adult rats (2-way-ANOVA, $p < 0.02$) (Fig 1). In normal PSS, sodium nitroprusside-induced relaxation was similar in the aortae from adult and old rats. Low Mg did not modify relaxant response to sodium nitroprusside in the aortic rings from either adult or old rats. High Mg significantly increased relaxation to sodium nitroprusside in the aortae from both adult and old rats (pIC_{50} values: 7.92 ± 0.15 vs 7.32 ± 0.12 in adult rats, $p < 0.05$ and 7.67 ± 0.04 vs 7.27 ± 0.07 in old rats, $p < 0.05$) (Fig 1).

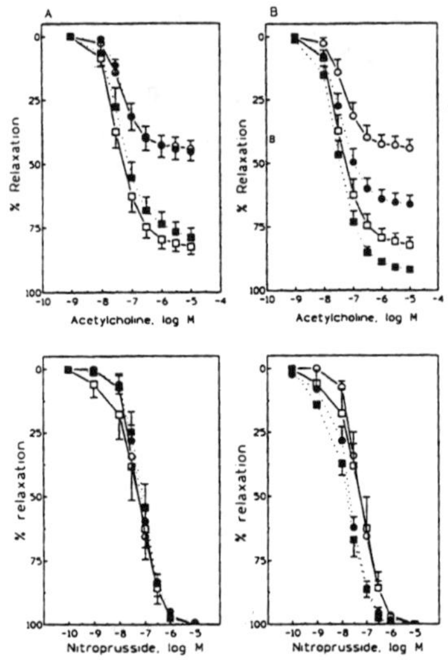

Figure 1 Concentration-response curves to acetylcholine and sodium nitroprusside in isolated aortae from adult (□, ■) and old (○, ●) rats bathed in PSS containing (A) 1.2 mM (□—□, ○—○) or 0.1 mM (■•••■, ●•••●) Mg and, (B) 1.2 mM (□—□, ○—○) or 4.8 mM (■•••■, ●•••●) Mg. Magnitudes of relaxation are expressed as percentage of contraction induced by norepinephrine. n = 6-10 rats per group. Symbols indicate means ± s.e.mean.

In normal PSS, isoproterenol-induced relaxation was significantly less pronounced in the endothelium-denuded aortic rings from old rats than in those from adult rats. Low Mg significantly decreased relaxation to isoproterenol in the aortae from adult (pIC_{50} values: 6.85 ± 0.06 vs 7.30 ± 0.11, $p < 0.05$) but not from old rats (6.38 ± 0.05 vs 6.42 ± 0.11). High Mg significantly increased relaxation to isoproterenol in the isolated aortae from both adult and old rats (pIC_{50} values: 7.83 ± 0.10 vs 7.30 ± 0.11 in adult rats, $p < 0.05$; 6.85 ± 0.12 vs 6.42 ± 0.11 in old rats, $p < 0.05$).

Addition of $MgSO_4$ caused concentration-dependent relaxation of the isolated aortae with intact endothelium which was significantly lower in the aortae from old rats than in those from adult rats ($p < 0.01$). When endothelium was disrupted, Mg-induced relaxation was significantly attenuated in the isolated aortae from both adult and old rats ($p < 0.01$). Mg-induced relaxation was, however, similar in adult and old rats when endothelium was disrupted. Thus, aging significantly interfered with endothelium on Mg-induced relaxation (2-way-ANOVA, $p < 0.01$) (Fig. 2).

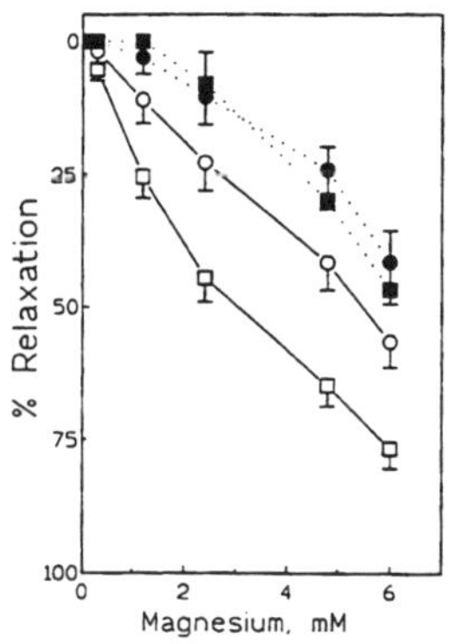

Figure 2 Concentration-response curves to $MgSO_4$ in isolated aortae with intact (□, ○) or disrupted (■, ●) endothelium from adult (□—□, ■•••■) and old (○—○, ●•••●) rats. Magnitudes of relaxation are expressed as percentage of contraction induced by norepinephrine in magnesium-free PSS. n = 6-10 rats per group. Symbols indicate means ± s.e.mean.

4. Discussion

Changes in Mg concentration differentially affected vasorelaxant response of the isolated aortae from old rats compared to adult rats. The differential modulatory effect of Mg may be due to the altered vascular and endothelial function observed in aortae from old rats.

The present study showed that response to β-adrenergic agonists (isoproterenol), which activate adenylyl cyclase via G-proteins, was impaired with aging (for references see Vanhoutte, 1988). High Mg potentiates isoproterenol-induced relaxation, while lowering extracellular Mg attenuates it. Mg is an activator of adenylyl cyclase. It is an essential component in the coupling of β-adrenoceptors to G-proteins and is required for hormonal activation of G-protein. In aortae from old rats, high Mg potentiated isoproterenol-induced relaxation suggesting that Mg may improve the altered β-adrenergic vascular response observed with aging. Low Mg did not decrease isoproterenol-induced relaxation in aortae from old rats, as it did in aortae from adult rats. Impairment of isoproterenol-mediated adenylyl cyclase activation of old blood vessels may be responsible. High Mg increased acetylcholine- and nitroprusside-induced relaxation of aortae from both adult and old rats suggesting that Mg might activate cellular events leading to increases endothelium nitric oxide (NO) synthase and/or smooth muscle guanylyl cyclase activities. The relationship between extracellular Mg and NO activity is controversial and not well understood and suggest that there are differential effects of Mg on agonist-induced NO release (Altura and Altura, 1987; Farago et al., 1991; Ku and Ann, 1991). Nevertheless, because Mg inhibits NO synthase activity without altering NO release in endothelial cells (Howard et al., 1995), it could be suggested that Mg probably stimulates NO diffusion in vascular smooth muscle cells and/or activate guanylyl cyclase activity of smooth muscle cells. The increased relaxation induced by exogenous NO (nitroprusside) of aortae in response to high Mg would confirm this assumption.. Endothelium- and NO-dependent relaxation mediated by acetylcholine is reduced with aging while endothelium-independent relaxation mediated by nitrovasodilators such as nitroprusside is not modified (Vanhoutte, 1988). The impaired acetycholine-induced relaxation may be due to a dysfunction of endothelium rather than smooth muscle. In the present study, the increasing effect of high Mg on acetylcholine-induced relaxation was greater in old than in adult rats whereas it was similar on nitroprusside-induced relaxation. These findings suggest that Mg may be effective in aortae from old rats in improving some endothelial events which increase the blunted endothelium-dependent relaxation when stimulated by acetylcholine.

Relaxation induced by increasing extracellular Mg was greatly inhibited in aortae with disrupted endothelium bathed in Mg-free PSS, suggesting that Mg-induced relaxation is mediated, at least in part, by endothelium-dependent mechanisms in which the release of endothelium-dependent relaxing factors would be implicated. However, relaxation induced by increasing Mg was markedly attenuated in aortae with intact endothelium from old rats compared to adult rats, whereas it was similar in adult and old rats when endothelium was disrupted. These findings suggest that endothelium-dependent relaxation mediated by the increase of Mg was blunted in old rats and these defects may be due to an endothelial dysfunction observed with aging.

In conclusion, the altered aortic responsiveness observed with aging modifies the modulatory role of Mg on vasorelaxation. The fact that high Mg improved both the endothelium-, NO-dependent relaxation to acetylcholine and the adenylyl cyclase-dependent relaxation to isoproterenol in aortae from old rats, suggests that Mg has a beneficial effect on age-related alterations on vascular function. These results support the recommendation that Mg administration be increased in the elderly to prevent cardiovascular diseases.

5. References

Altura BM, Altura BT (1981): Magnesium ions and contraction of vascular smooth muscle: relationship to some vascular diseases. *Fed. Proc.* **40**, 2672-2679.

Altura BM, Altura BT (1983): Influence of magnesium on vascular smooth muscle and serum biochemical parameters from diabetic and hypertensive rats. *Magnesium* **2**, 253-266.

Altura BT, Altura BM (1987): Endothelium-dependent relaxation in coronary arteries requires magnesium ions. *Br. J. Pharmacol.* **91**, 449-451.

Altura BM, Altura BT, Carella A, Gebrewold A, Murakawa T, Nishio A (1987): Mg^{2+}-Ca^{2+} interaction in contractility of vascular smooth muscle: Mg^{2+} versus organic calcium channel blockers on myogenic tone and agonist-induced responsiveness of blood vessels. *Can. J. Physiol. Pharmacol.* **65**, 729-745.

Farago M, Szabo C, Dora E, Horvath I, Kovach AGB (1991): Contractile and endothelium-dependent dilatory responses of cerebral arteries at various extracellular magnesium concentrations. *J. Cereb. Blood Flow Metab.* **11**, 161-164.

Howard AB, Alexander RW, Taylor WR (1995): Effect of magnesium on nitric oxide synthase activity in endothelial cells. *Am. J. Physiol.* **269**, C612-C618.

Ku DD, Ann HS (1991): Differential effects of magnesium on basal and agonist-induced EDRF relaxation in canine coronary arteries. *J. Cardiovasc. Pharmacol.* **17**, 999-1006.

Laurant P., Berthelot A (1996): Endothelin-1-induced contraction in isolated aortae from normotensive and DOCA-salt hypertensive rats: effect of magnesium. *Br. J. Pharmacol.* **119**, 1367-1374.

Laurent S, Lacolley P, Girerd X, Boutouyrie P, Bezie Y, Safar M (1996): Arterial stiffening: opposing effect of age- and hypertension-associated structural changes. *Can. J. Physiol. Pharmacol.* **74**, 842-849.

Vanhoutte PM (1988): Aging and vascular responsiveness. *J. Cardiovasc. Pharmacol.* **12** (suppl 8), S11-S18.

THE USE OF MAGNESIUM PREPARATIONS – KORMAGNESIN AND MAGNEROT IN CARDIOLOGICAL PRACTICE

VERTKIN A.L., GORODETSKIY V.V., DUHANINA I.V., STOVBUR O.V., AVILOVA O.A.
Moscow Medical Stomatological Institute Moscow, Russian Federation

The variety of biochemical and physiological effects of magnesium in the organism provides the abundance of indications for its appliance in the treatment practice. With it all, the widely spread in the population general shortage of magnesium in the organism does not seem to play a decisive role in the clinical effectiveness of magnesium preparations.

It is well known that whenever magnesium shortage manifests itself in such symptoms and syndromes as the syndrome of chronical fatigue, headaches, arterial hypotension, heart arrhythmia, inclination to thrombosis, cramps of skeleton muscles, spastic contractions of smooth muscles including bronchus and vessels and others, the introduction of magnesium ions into the organism can either decrease or completely curb the clinical symptoms. Whereas the effectiveness of magnesium preparations is indicated even in case the magnesium shortage in the organism is not proved, the shortage being evaluated by its concentration in the blood serum and erythrocytes. A test with magnesium load is considered to be a more informative way of magnesium shortage revelation. The case is diagnosed as magnesium shortage if during the test 3400 mg of magnesium are introduced into the organism within 24 hours but less than 70% or less than 1780 mg are exuded with the urine.

This work studied the effectiveness of intravenous "Kormagnesin" in cases of heart arrhythmia, and oral "Magnerot" in cases of angina pectoris and postinfarction cardiosclerosis with the patients with normal volume of magnesium in the blood serum. "Kormagnesin" is a solution of sulfate magnesia (200 and 400 mg in the ampoule). The tablets of Magnerot contain 0.5 gr. of magnesium orotat (32.8% of magnesium).

Besides determination of the magnesium volume in the blood serum all the patients were subjected to the test with magnesium load.

Among the 67 patients with the heart arrhythmia 38 ones had extrasystole arrhythmia as a complication after the acute infarction of myocard whereas 29 ones had parochsisms of atrium fibrillation which developed in the background of ischemia. "Kormagnesin" had been injected for 7–10 minutes in the dosage corresponding to 600 mg of ionized

T. Theophanides and J. Anastassopoulou (eds.), Magnesium: Current Status and New Developments, 269-271.

magnesium with the further increase of the dosage up to 2400 mg necessary for carrying out the test with the magnesium load.

All the patients who took Magnerot had angina pectoris of II–III functional class, postinfarction cardiosclerosis and extrasystole, during the Holter–monitoring ECG episodes of silent myocard ischemia were found 87% of the patients.

The test with magnesium load was carried out according to the classical Protocol. The presence of magnesium shortage in the patients′ organisms was fixed before the Magnerot treatment with the help of physiological test. For this purpose 40.0 ml of intravenous Kormagnesin 200 (880 mg of magnesium in 100 ml of 5% solution of glucose) was introduced every 8 hours within 24 hours. Before the first introduction the patient relieved his bladder, after that the 24 hour urine was collected and the volume of magnesium was fixed in it (a coloured reaction with magon). The volume of magnesium in the 24-hour urine was fixed before the test as well. The case of magnesium shortage was diagnosed in case more than 30% of introduced magnesium, i.e. more than 800 mg was held in the organism. The concentration of magnesium in the blood serum was also fixed before and after the magnerot treatment. Clinical manifestations of magnesium shortage such as muscle cramps, paresthesia, depressive conditions, cardiovascular malfunctions (tachycardia, extrasystole, arrhythmia of different etiology, frequent overfalls of arterial hypotension), headaches were evaluated. Magnerot was prescribed in the dosage of 3.0 gr. per 24 hours during the first week and 1.5 gr. during the next 5 weeks.

As a result of Kormagnesin application its high effectiveness was revealed as an anti - arrhythmic drug in cases of supra ventricular arrhythmia and extrasystole arrhythmia which complicated the current of infarction of myocard. In cases of fibrillation arrhythmia, though the sinusoid rhythm did not restore, in the background of ¨Kormagnesin¨ introduction, the normosystole form of fibrillation without pulse deficit and gemodynamic disorder was achieved. The achieved effects were kept without further application of additional anti–arrhythmia drugs, including the patients subjected to electroimpulse therapy.

The intravenus injection of magnesium salt did not cause the decrease of arterial blood pressure lower than the physiological rate even if its initial level was not high.
Like in the other researches we paid attention to the sedative and analgetic effects of Kormagnesin which manifested itself in the decrease of angina pains recurrences and demand for narcotic analgetics.

Good results were also achieved while treating 15 patients with ischemia by Magnerot. These were patients with angina pectoris of II – III functional classes, postinfarction cardiosclerosis and extrasystole, with that subjective improvement of general condition was marked in all the patients starting with the third week of the Magnerot treatment, as well as considerable decrease of angina pectoris attacks per 24 hours (by 45%) and the quantity of taken tablets of nitroglycerin.

Besides, repeatedly held Holter – monitoring ECG showed the decrease of ventricular extrasystole frequency by more than twice as much, the supraventricular extrasystole has become 40% more seldom, and what is most important from the point of view of infarction of myocard and sudden death prophylaxis, the frequency of silent ischemia of myocard decreased by 25%.

Dynamic veloergometric loads showed a tendency for the increase of tolerance to load. No side effects appeared in the process of Magnerot usage.

While comparing the clinical effectiveness of magnesium containing drugs – Kormagnesin and Magnerot with the results of the test with the magnesium load no direct connection between the magnesium shortage sings in the organism and the treatment results was revealed.

Thus, the magnesium containing drugs – Kormagnesin and Magnerot are indicated for the patients with different forms of ischemia not depending on presence or absence of the signs of general magnesium shortage in the organism of the patient.

SERUM AND ERYTHROCYTE MAGNESIUM IN PATIENTS WITH UNSTABLE ANGINA PECTORIS DURING AND IN THE ABSENCE OF CRISIS

MARIANA VLAD[1], G. UZA[1], I. DRAGATOIU[2], I. SORTAN[3], CRISTINA IGA[1], P.J. PORR[1]

[1]Institute of Public Health, Str. Pasteur 6, 3400 Cluj, [2] Clinic of Endocrinology, Cluj, [3]Heart Institute, Cluj, Romania

Eighteen patients with unstable angina pectoris were studied. Serum and erythrocyte Mg concentrations were determined during a crisis and also in the absence of a crisis. In addition, serum and erythrocyte Mg was assessed in a control group of 22 normotensive patients without angina pectoris. The patients' age ranged between 41 - 67 years, mean age 52.3 $\pm$ 8.7. In controls age range was 37 - 62, mean age 43.5 $\pm$ 7.2. Serum and erythrocyte Mg levels were determined using the AAS Perkin-Elmer model 300. In controls the average concentration of serum Mg was 0.88 $\pm$ 0.05 mmol/l, and of erythrocyte Mg 0.67 $\pm$ 0.04 mmol/l. In patients with unstable angina pectoris average concentration of serum Mg on admission to the hospital was 0.78 $\pm$ 0.06 mmol/l, and of erythrocyte Mg 0.65 $\pm$ 0.05 mmol/l. During the crisis the average serum Mg decreased to 0.69 $\pm$ 0.03 mmol/l ($p<0.01$), and erythrocyte Mg to 0.61 $\pm$ 0.07 mmol/l as compared to the average values found in controls and in patients without a crisis ($p<0.01$).

The average of serum and erythrocyte Mg values for the patients in which the levels decreased during a crisis was smaller than the average values calculated for the whole group of patients during a crisis.

We consider that a more marked decrease of serum Mg as compared to erythrocytic Mg favours coronary spasm and angina crisis.

1. INTRODUCTION

High blood pressure, smoking and dyslipidemia are major risk factors for atherosclerotic cardiovascular diseases. Minerals and trace elements have long been suspected to be potential risk factors for cardiovascular diseases.

Low serum magnesium (Mg) is a risk factor for early death in people with acute myocardial infarction, but the impact of low-serum Mg on long-term prognosis is not well known. In a recent study we have demonstrated that the abdominal aortic Mg levels in patients who died from coronary ischaemic heart disease and from acute myocardial infarction are significantly lower than in controls. At the same time, the calcium (Ca) level in the abdominal aorta was very high as compared to controls. These

T. Theophanides and J. Anastassopoulou (eds.), Magnesium: Current Status and New Developments, 273-276.

results suggest that the Mg deficit in the aortic tissue leads to impairment of the main mechanism and determining removal of Ca from the arterial wall cells.
The aim of this study is to determine the Mg content in the serum and erythrocytes of patients with unstable angina pectoris during and in the absence of crisis, based on the hypothesis that pain and anxiety - powerful stress factors - induce adrenergic activity and cause Mg depletion.

2. MATERIAL AND METHODS

The study included 18 patients (10 women and 8 men) aged between 41 and 67 years (mean age 52.3 $\pm$ 8.7) with unstable angina pectoris observed at rest, with ST-segment depression and chest pain despite conventional therapy (beta-blockers, heparin, nitrates and acetylsalicylic acid). Diagnosis was established according to the WHO criteria.
The control group included 22 normotensive patients without angina pectoris (14 women and 8 men), aged between 37 - 62 years (mean age 43.5 $\pm$ 7.2).
Blood samples for Mg determination were collected in a 10-ml mineral- and additive-free BD-vacutainer tube and allowed to clot at room temperature. The serum was then separated and frozen at 0^{0}C until time for analysis. Samples for erythrocytic Mg were collected in a 5 ml sodium-heparinized BD-vacutainer tube, separated and frozen at 0^{0}C until time for ananlysis. In preparation for analysis blood samples were thawed and prepared in duplicate. Serum samples were diluted 1:40 in 0.5% lanthanum oxide and 10% HCl. Analysis for Mg concentrations was performed by atomic absorption spectrophotometry using a model 300 Perkin-Elmer (Norwalk, CT).
Statistical analysis was performed with Student's unpaired t-test.

3. RESULTS

In controls, the average concentration of serum Mg was 0.88 $\pm$ 0.05 mmol/l and of erythrocyte Mg 0.67 $\pm$ 0.04 mmol/l. In patients with unstable angina pectoris the average concentration of serum Mg at admission to the hospital was 0.78 $\pm$ 0.06 mmol/l, and of Mg in erythrocytes 0.65 $\pm$ 0.05 mmol/l (Table 1).

TABLE 1. Mean serum and erythrocyte Mg levels in study and control groups

	Control (n=22)	Patients (n=18)	
		On admission	During crisis
Serum (mmol/l)	0.88 ±0.05	0.78±0.06	
		0.69±0.03*	
Erythrocyte (mmol/l)	0.67±0.04	0.65±0.05	
		0.61±0.07*	

* $p < 0.01$
All values are expressed as mean $\pm$ SD and refer to Mg concentration

During the crisis the mean serum Mg level decreased to 0.69 $\pm$ 0.03 mmol/l ($p<0.01$) and erythrocyte Mg to 0.61 $\pm$ 0.07 mmol/l as compared with the average values recorded in controls and in patients not in crisis ($p<0.01$) (Table 1).
The mean serum (0.65 $\pm$ 0.04) and erythrocyte (0.59 $\pm$ 0.03) M g values, calculated only in the patients in which this concentration decreased during a crisis, was less than the mean values calculated for the whole group of patients during a crisis.

4. DISCUSSION

In this study we measured serum and erythrocyte Mg levels in patients with unstable angina pectoris during a crisis and free of crisis. and found a significant decrease of serum Mg levels during crisis.
Patients with angina pectoris and risk of myocardial infarction have a deficiency of the total body Mg due to a combination of factors, including inadequate diet, disease and treatment induced loss, old age, and prior stressful conditions that may speed Mg loss.
Pain and anxiety promote release of catecholamines, which induce a sharp rise of free fatty acid levels. Catecholamines also induce movement of Mg between intra- and extra-cellular compartments. As a consequence, the electric membrane becomes unstable, followed by increased loss of Mg and K. Hypomagnesemia noted in patients with unstable angina pectoris during a crisis was developed in the course of repetitive angina attacks, accompanied by high catecholamine secretion. In conclusion it was shown that i) in patients with unstable angina pectoris serum and erythrocyte Mg levels are lower than in controls ii) during angina crisis these levels drop significantly lower than in the patients with angina pectoris without a crisis iii) the decrease of Mg levels may be a consequence of stress, and also the deficiency can be considered a risk factor for coronary spasm, which results in a *circulus viciosum*.

References

1. Altura BM, Altura BT, Carela A, Prasad DMV (1981) Hypomagnesemia and vasoconstriction: possible relationships to etiology of sudden death, ischemic heart disease and hypertensive vascular disease. *Artery*, **9**, 213-234.
2. Balkin J (1995) The use of magnesium in critical coronary care patients, management of cardiac arrhythmias. *Advances in Magnesium Research*: 1 Magnesium in Cardiology. Fifth European Magnesium Congress, Vienna, pp. 21-27.
3. Elliott M Antman (1996) Magnesium in acute myocardial infarction: Overview of available evidence. *Am Heart J*, **132,** 487-494.
4. Reunanen A, Knekt P, Marniem IJ, Naki J, Maatela J, Aromaa A (1996) Serum calcium, magnesium, copper and zinc and risk of cardiovascular death. *European J Clin Nutr*, 50, 431-457.
5. Seelig MS (1989) Cardiovascular consequence of magnesium deficiency and loss: pathogenesis, prevalence and manifestations-magnesium and chloride loss in refractory potassium repletion. *Am J Cardiol*, **63,** 4G-21G.
6. Seelig MS, Preuss HG (1994) Magnesium metabolism and perturbations in the elderly. *Geriatr Nephrol Urol*, **4**, 101-141.
7. Seelig MS (1994) Consequences of magnesium deficiency on the enhancement of stress reactions; preventive and therapeutic implications. *J Am Coll Nutr*, **13**, 429-446.

8. Smetana R *et al.* Coronary artery disease - magnesium and catecholamine interactions during stress burden.*Magnes. Res.* **8, suppl.1**.
9. Uza G, Pavel O, Uza D, Vlaicu R (1989) Hypomagnesemia in patients with essential arterial hypertension *Magn Bull* **9**, 172-180.
10. Vlad M, Uza G, Caseanu E, Petrescu M, Porr PJ, Albu A (1995) Magnesium and calcium concentrations in the abdominal aortas of patients who have died from coronary heart disease. *Magnes.Res.*.**8, suppl.1**.
11. Wang R (1982) Hypomagnesemia and hypokalemia in 1000 treated ambulatoring patients. *Am Coll Nutr*, **1**, 317-322.

RELATIONSHIP BETWEEN MAGNESIUM PARAMETERS, LIPID PROFILE AND PEROXIDATION LEVELS IN HEALTHY CHILDREN

LAIRES, M.J.*, MONTEIRO, C.P.*, SARDINHA, L.**, BICHO, M.***, LLOBET, S.****, MARQUES VIDAL, P.****, MONTEIRO, C.****, HALPERN, M.J.****
** Laboratório de Bioquímica, Faculdade de Motricidade Humana, U.T.L.; ** Núcleo de Exercício e Saúde, Faculdade de Motricidade Humana, U.T.L.; *** Laboratório de Genética, Faculdade de Medicina de Lisboa, U.L.; **** Instituto Superior de Ciências da Saúde*

1. Introduction

Several studies have provided evidence that the ingestion of a diet deficient in magnesium may be a factor in the aetiology of cardiovascular disease [1].

A major factor of cardiovascular diseases is atherosclerosis. In the atherosclerosis process may be implicated several factors such as genetic, lipid profile, state of nutrition and hypertension [2].

A crucial step in the pathogenesis of atherosclerosis is believed to be the oxidative modification of low density lipoprotein (LDL). The oxidation of LDL is a free radical driven process and the aldehyde products of the breakdown of the resulting lipid hidroperoxides are responsible for the modification of the LDL apoproteins [3].

Modified Apo B have altered receptor affinity, causing it to be scavenged by macrophages with the development of foam cells and the initiation of the atherosclerotic lesion [4]. Recent animal-experiments have shown that magnesium deficiency enhances susceptibility to free radicals

T. Theophanides and J. Anastassopoulou (eds.), Magnesium: Current Status and New Developments, 277-280.

mediated injury and may increase tissue damage [5].
It is important to evaluate the state of nutrition, the mineral status and lipid profile from childhood in order to prevent or to reduce the risk of further cardiovascular diseases.
The aim of this study was to determine, in healthy children, the relationship between magnesium status, state of nutrition, lipid profile, peroxidation levels, (TBARS and LDL TBARS), and anthropometric parameters.

2. Population and Methods

We performed a study in 171 children from Lisbon, 76 boys with a mean age of 13.0 ± 1.7 years and 95 girls with a mean age of 12.9 ± 1.7 years.
The permission for the study was previously obtained from their parents.
Anthropometric evaluation included weight and height. Evaluation of state of nutrition included body mass indice (BMI) and total fat percentage (T Fat %).
A blood sample was collected after a 12 hours fast to evaluate: plasma and red blood cells magnesium by atomic absorption, plasma ionised magnesium by selective electrodes; the lipid profile, which included total cholesterol (TC), high (HDL-Col) and low (LDL-Col) density lipoproteins cholesterol, triglycerides (TG) by commercial kits and apolipoproteins A_1 (Apo A_1) and B (Apo B) by imunonepholometry; and peroxidation indices namely plasma TBARS by a thiobarbituric acid assay and *in vitro* LDL peroxidation induced by phenilhydrazine detected by the same thiobarbituric acid assay (LDL-TBARS).
Statistical analysis included t-test to compare the mean values.

3. Results

In this study we analysed the results of the studied population (Stud. Pop.) according with sexes and compared them with other populations (Ref. Pop.). RBC-Mg values were compared with those of a young adult Lisbon population [6]; the state of nutrition and the lipid profile with

those of a rural population from the northern Portugal [7]; and peroxidation indices with an age matched sports girls control [8].
Analysing the results according with sexes, we observed that boys had the mean values of Tot fat % and of Apo B significantly lower and of HDL--Col significantly higher.
When compared with a population of young adults, red blood cells magnesium mean values were lower (4.0 ± 0.5 *vs* 5.9 ± 0.8 mg/dl).
In table 1 are presented the mean values of the state of nutrition and lipid profile parameters of the studied population and of the reference populations, and the statistical significance of their comparison by the t-test (only significant values).

Table 1 - Comparison of the state of nutrition and lipid profile of the studied population with those of the reference populations.

Parameter	Boys			Girls		
	Stud. Pop.	Ref. Pop	p	Stud. Pop.	Ref. Pop	p
Tot fat % (%)	21.7 ± 10.5	17.6 ± 5.5	0.004	30,3 ± 7.5	21.6 ± 7.3	0.001
BMI (kg/m^2)	20.4 ± 3.3	18.1 ± 2.8	0.001	20.7 ± 3.2	19.9 ± 3.1	0.001
TC (mg/dl)	172.9 ± 30.6	150.0 ± 28.3	0.000	175.8 ± 28.2	139.4 ± 22.6	0.000
HDL-Col. (mg/dl)	60.4 ± 14.4	53.0 ± 13.0	0.018	55.7 ± 12.9	46.0 ± 8.8	0.001
Apo A1 (g/l)	1.4 ± 0.2	1.6 ± 0.3	0.000	1.3 ± 0.2	1.5 ± 0.2	0.000

LDL TBARS mean values were lower then that of a age matched sports-girls (22.3 ± 15.3 *vs* 76.3 ± 22.3 μM).

4. Discussion

The significant differences observed for Apo B between sexes can be explained by a more precocious sexual maturity of the girls.
When compared with a population of young adults, red blood cells magnesium mean values were lower. This is in accordance with another study in a portuguese young population [9].
Nevertheless there is no relationship between this lower levels and the lipid profile witch is also in accordance with another study [10].
LDL TBARS mean values were lower then that of a age matched sports girls which could be explained by the effect of exercise promoting peroxidation through oxygen metabolism activation, hypoxic xanthine

oxidase activation or inflammatory reaction [8].
The differences of the state of nutrition and lipid profile between the Lisbon population and the rural population could be explained by the different food intake.

4. References

1 Rayssiguier, Y., Gueux, E., Bussière, L., Durlach, J., and Mazur, A. Dietary magnesium affects susceptibility of lipoproteins and tissues to peroxidation in rats. *Journal of American College of Nutrition*, **12** (1993), 133-137.

2 Durlach, J.: Antioxidant dietary status and genetic cardiovascular risk, or how an adequate intake of α-tocopherol, selenium, taurine, magnisium and various other natural antioxidants may overcome the deleterious metabolic consequences related to the E4-4 type of apolipoprotein E, *Magnesium Research*, **9** (1996), 139-141.

3 Steinberg, D., Parthasarathy, S., Carew, T.E., Khoo, J.C., and Witztum, J.L.: Beyond cholesterol: modifications of low-density lipoprotein that increase its atherogenicity, *N. Engl. J. Med.*, **320** (1990), 915-924.

4 Esterbauer, H., Wäg, G., and Puhl, H.: Lipid peroxidation and its role in atherosclerosis. *Br. Med. Bulletin*, **49** (1993), 566-576.

5 Freedman, A.M., Cassidy, M.M., and Weglicki, W.B.: Magnesium-deficient myocardium demonstrates an increased susceptibility to an in vivo oxidative stress, *Magnesium Research*, **4** (1991), 185-189.

6 Laires, M.J., and Cunha Monteiro, A.L.: Magnésio em patologia: estudo prospectivo e prática clínica, *Revista Portuguesa de Bioquímica*, **III** (1980), 149-152.

7 Guerra, A.J.M., Castro, E.M.B., Carvalho Guerra, F., Rego, C., Monteiro, C., Silva, D., Teixeira Santos, N., and Bicho, M.: Associação entre ėstado de nutrição, composição corporal, lípidos e apolipoproteínas séricas e fenótipos da haptoglobina numa população dos 9 aos 12 anos, *Rev. Port. Cardiol.*, in press.

8 Guerra, A., Rego, C., Laires, M.J., Castro, E., Silva, D., Lourenço, S., Monteiro, C.P. Silva, Z., Carvalho Guerra, F., Bicho, M. and Teixeira Santos, N.: LDL susceptibility to peroxidation in high performance rhyrhmic female teenagers gymnasts (Abstract), *SFRR Europe Summer Meeting*, (1997); 228-229.

9 Laires, M.J., Monteiro, C.P., Sérgio, J., Madeira, F., Sainhas, J., Palmeira, A., Felisberto, G.M. and Vaz, C.: Content of trace elements in blood before and after three standard efforts, in P. Collery, L.A. Poirier, N.A. Littlefield and J.-C. Etienne (eds.), *Metal Ions in Biology and Medicine*, Vol. 3, John Libbey Eurotext, Paris, (1994), pp. 557-563.

10 Laires, M.J., Pereira, T.G. and Halpern, M.J.: Magnesium and Hyperlipemias, *Magnesium*, **2** (1983), 250-252,.

ASSOCIATION BETWEEN MAGNESIUM AND HYPERTENSION: SPECULATION OR FACT?

T. D. MOUNTOKALAKIS, M.D.
3rd Department of Medicine, Medical School, University of Athens, Sotiria General Hospital, 152 Mesogion Avenue, Athens, Greece

Abstract

Experimental evidence suggests a role for Mg in blood pressure regulation. In animal models, Mg deficiency leads to potentiation of the vasoconstrictive effect of catecholamines and angiotensin, and to increased intracellular Ca concentration which may also result to vasoconstriction. Observational studies indicating an inverse correlation between water hardness and cardiovascular mortality have led to the investigation of the relation between Mg intake and blood pressure. However, because Mg exists in common foods together with other cations, such as K and Ca, which may also be involved in the genesis of hypertension, it has been difficult to distinguish the individual effect of each cation on blood pressure. In one international study (Intersalt), salt intake was found to have a significant although weak effect on blood pressure and K seemed to play a role in limiting the rise in blood pressure associated with Na-rich diets, but there was no evidence that Mg intake might significantly affect blood pressure. On the other hand, results of most of the clinical trials on the effect of dietary Mg supplementation on blood pressure have been hard to interpret because of methodologic reasons. Historically controlled trials have shown a hypotensive effect of oral Mg, but controlled trials have provided little evidence that Mg supplementation can lower blood pressure. Data from some studies suggest that a hypotensive effect of Mg supplementation can be expected only in K- or Mg-deficient patients or with higher doses of Mg which may have a true pharmacological effect on blood pressure. Uncertainty concerning the association between Mg and hypertension will persist until the results of further large, carefully designed clinical trials become available.

T. Theophanides and J. Anastassopoulou (eds.), Magnesium: Current Status and New Developments, 281-289.

1. Introduction

Essential hypertension is recognized as an important risk factor for cardiovascular disease. Among nonpharmacological measures proposed for the control of hypertension in the community, nutritional interventions have consistently kept a prominent place. Experimental as well as observational studies have indicated a role for Mg in blood pressure regulation. However, the effectiveness of Mg supplementation in reducing blood pressure remains uncertain.

2. Experimental Evidence

Experimental findings suggest that Mg is important in the control of vascular tone. Thus, in canine coronary arteries, a reduced extracellular Mg concentration has been shown to result in increased intracellular Ca by allowing excess entry of this cation (Altura and Altura, 1982). In addition, a low concentration of Mg in blood and tissue has been found to potentiate the vasoconstrictor effects of catecholamines and angiotensin. These and other similar findings have led to the characterization of Mg as the nature's own calcium blocker (Iseri and French, 1984). Further to its effect on vascular tone, Mg has been implicated in the pathogenesis of arterial hypertension and of atherosclerosis. Thus, experimental studies in rats have shown that a low Mg diet is associated with the development of hypertension, platelet activation, atherogenic alterations in plasma lipids, and vascular lesions (Seelig, 1989; Durlach et al., 1992; Hwang et al., 1992).

3. Observational Studies

Interest concerning a possible role of Mg in human cardiovascular disease has been stimulated by observational studies indicating an inverse correlation between water hardness and cardiovascular mortality. To explain this correlation, a "cardiac water factor" has been speculated. Presumably, such a factor would be either something beneficial in hard water or something harmful in soft water. Since the bulk of water hardness is made up of Ca and Mg, these two ions have been the most likely candidates for the "cardiac water factor". Mg is much more abundant intracellular ion than Ca, and a considerably greater proportion of the daily Mg intake, as compared with the daily Ca intake, comes from drinking water. Coupled with the known arrhythmogenic potential of Mg depletion, these facts had led to the hypothesis that Mg is the cardioprotective factor contained in hard water (Anderson et al., 1980).

In epidemiological studies in Britain (Stitt et al., 1973) and in USA (Dawson et al., 1978), drinking water Mg was found to be inversely correlated with hypertension and hypertensive heart disease. However, negative results from other simple correlation studies, as well as from multivariate statistical analysis of registry data, in association with the demonstration that only a small proportion of the daily Mg intake comes from drinking water, have cast doubt on the possible physiological significance of water hardness (Hammer and Heyden, 1980). On the other hand, advocates of the "water factor" have adopted a somewhat different approach which emphasizes the ratio of Ca:Mg intake rather than water hardness as the important determinant (Karppanen, 1981; Marier, 1981).

Community-based studies that have investigated the relation between Mg intake and blood pressure have reported conflicting findings. In a Sweedish study, no difference in Mg intake measured by biochemical analysis of duplicate portions of ingested food, was found between normotensive and hypertensive women (Thulin et al., 1980). In contrast, in a cohort of Japanese men with no prevalent cardiovascular disease, living in Honolulu, Mg intake measured by dietary recall, was found to be associated inversely with blood pressure (Joffres et al., 1987).

Urinary micronutrient excretion has been repeatedly used as a measure of the corresponding nutrient dietary intake. In most studies, no relation was found between 24-hour urinary Mg excretion and blood pressure (Whelton and Klag, 1989) (Table 1). In the Intersalt study (Intersalt Cooperative Research Group, 1988), an interpopulation study involving 10 079 men and women in 52 centers from 32 countries, blood pressure was found to be positively related to urinary Na excretion, and negatively related to urinary K excretion. The relation of blood pressure to the ratio of urinary Na:K was closer than to either electrolyte alone. Of the 52 centers of the study, Tianjin (People's Republic of China) had the highest recorded Na:K excretion ratio. In a subsequent cross-sectional study of a random sample of the Tianjin population, dietary Na, as well as urinary Na and Na:K ratio was found to be positively correlated with blood pressure. However, no significant associations were found between blood pressure and either intake or urinary excretion of Ca and Mg (Tian et al., 1995). Finally, several clinical studies that have examined the relation between serum or intra-erythrocyte Mg and blood pressure have reported conflicting findings, mainly because of the lack of adjustements for important covariables (Whelton and Klag, 1989).

4. Clinical Trials

Mg was used in the management of eclamptic convultions and also as a potent antiarrhythmic and vasodilator agent long before it was recognized as a nutrient. Furthermore, several randomized trials have examined its vasodilatory effect in acute myocardial infarction, in recent years. It is clear that the characterization of Mg as a drug or as a nutrient lies on the distinction between pharmacological action and replenishment of deficient stores. When there is a proven Mg deficiency, the patient is expected to be completely cured as soon as the deficit is corrected. On the other hand, the potency and duration of action of parenteral Mg salts will depend on Mg concentration in plasma, and their effect will be the same in each case of repeated administration.

TABLE 1. Studies on the relationship between urinary Mg (UMg) excretion and blood pressure (BP)

Authors (year)	Population studied	Results
Dawson et al. (1978)	24 West Texas communities	Inverse correlation between UMg and hypertension mortality
Staessen et al. (1983)	2 Belgian towns	No correlation between UMg and BP
Kesteloot (1984)	Belgian army troops	Inverse correlation between UMg and diastolic BP
M' Buyamba-Kabangu et al. (1986)	Urban Bantus	No correlation between UMg and BP
Kesteloot et al. (1987)	Korean people	No correlation between UMg and BP
Zhu et al. (1987)	Chinese boys	No correlation between UMg and BP
Yamori et al. (1994)	Japanese people	Inverse correlation between UMg and risk of high BP
Tian et al. (1995)	Chinese people	No correlation between UMg and BP

Several clinical trials have examined the effect of Mg supplementation on blood pressure. However, interpretation of their results has been difficult because of considerable inadequacies in study design. Thus, in several trials Mg supplementation was given concurrently with diuretics which are known to increase urine Mg excretion and as a consequence, to reduce serum Mg, particularly when administered in high doses and for long periods of time. On the other hand, in most studies, the sample size was too small and the duration of treatment did not exceed a period of 4 weeks. Finally, the doses used in different studies varied largely (from 12.5 to 40.0 mmol per day) and only a few studies included a control group and even fewer used a double-blind assessment (Table 2).

TABLE 2. Differences in patient selection and oral Mg dose in randomized controlled clinical trials on the effect of oral Mg on blood pressure

Authors (year)	Study design	Use of diuretics	Oral Mg dose (mmol)	Mean decrease in BP (mmHg)
Dyckner and Wester (1983)	Open parallel	Yes	15.0	12/8
Reyes A.J. et al. (1984)	Double blind parallel	Yes	15.0	11/7
Cappucio et al. (1985)	Double blind cross over	No	15.0	No effect
Henderson et al. (1986)	Double blind parallel	Yes	12.5	No effect
Lind et al. (1991)	Double blind parallel	No	15.0	No effect
Widman et al. (1993)	Double blind cross over	No	15.0-40.0	7.9/8.2
Witteman et al. (1994)	Double blind parallel	No	20.0	7.6/3.3

In 1989, Whelton and Klag (1989) reviewed eight trials of Mg supplementation published until 1988. Half of them were historically controlled trials and provided evidence that oral Mg can exert a hypotensive effect. In contrast, only one of 4 randomized controlled trials showed a statistically significant reduction in blood pressure after oral Mg. Several controlled trials published since then provided little evidence of a hypotensive response to oral Mg administration.

It has been postulated that a reduction of blood pressure by oral Mg can be expected only in hypertensive patients on long-term diuretic treatment. In such patients, restoration of the diuretic-induced K and Mg deficiency would potentiate the hypotensive effect of diuretic treatment. In fact, in a randomized controlled trial, Dyckner and Wester (1983) reported a mean reduction in systolic blood pressure of 12 mmHg, and a mean reduction in diastolic blood pressure of 8 mmHg, in 18 patients with essential hypertension and in 2 patients with congestive heart failure treated with diuretics, after 6 months of oral Mg aspartate in the daily dose of 15 mmol. However, a subsequent controlled study failed to show any effects of Mg treatment, in the daily dose of 12.5 mmol, on blood pressure, in hypertensive patients treated with diuretics (Henderson et al., 1986). It is interesting to note that in two double-blind studies with positive results, the daily dose of Mg supplementation was 20.0 mmol and 30.0-40.0 mmol (Widman et al., 1993; Witteman et al., 1994), i.e. higher than the dose of 13.0 and 15.0 mmol, which is the RDA for adult women and men, respectively. Thus, a true pharmacological effect of Mg in these studies cannot be ruled out.

Mg supplementation has also been used in intervention studies addressing multiple changes in mineral intake. In a study by Karppanen et al. (1984), usual table salt was substituted by a K- and Mg- enriched salt mixture (Na:K:Mg 65%:25%:10% or 45%:35%:20%). A statistically significant reduction in blood pressure was observed, but the lack of a control group make difficult the interpretation of the results. In a later randomized double-blind placebo controlled trial in older people with untreated mild to moderate hypertension, the intervention group received a mineral salt (Na:K:Mg 8:6:1) for 24 weeks. Systolic blood pressure fell by 7.6 mmHg and diastolic blood pressure by 3.3 mmHg in the mineral salt group compared with the control group (Geleijnse et al., 1994). Finally, in a recent randomized, parallel-group, double-blind trial, 125 patients with mild or borderline hypertension were randomly allocated to receive one of the following 4 regimens: K 60 mmol and Ca 25 mmol, K 60 mmol and Mg 15 mmol, Ca and Mg, or placebo. No significant effect on blood pressure was produced by either of the 3 dietary cation combinations (Sacks et al., 1995).

5. Comment

Despite a considerably high number of publications, the quality of evidence supporting the argument that Mg deficiency plays an important role in hypertension is rather poor. Studies on Mg intake measured either by dietary recall or by biochemical analysis of duplicate portions of food, are difficult to interpret because Mg exists in common foods together with other cations, such as K and Ca, which may also be involved in the pathogenesis of hypertension. On the other hand, it could be that other nutrients, such as fiber, vitamins or proteins, contained in fruits or vegetables have the true hypotensive effect (Sacks and Kass, 1988). Conflicting results of interventional trials may be explained, at least in part, by the divergence of the different study protocols.

The interesting hypothesis that Mg supplementation is effective in reducing blood pressure only in patients with pre-existing Mg deficiency, as a result of long-term diuretic treatment, is abated by the fact that there is no sound evidence that diuretic treatment in the doses used in hypertension, is capable to reduce intracellular Mg and therefore, to cause a true Mg deficiency state (McInnes et al., 1992). On the other hand, the response of blood pressure to oral Mg in daily doses higher than 15.0 mmol could be due to a true pharmacological effect. It is obvious that uncertainty concerning the association between Mg and hypertension will persist until the results of further large, carefully designed clinical trials become available.

6. References

Altura, B.M., and Altura, B.T. (1982) Mg, Na, and K interactions and coronary heart disease. *Magnesium* **1,** 241-265.

Anderson, T.W., Leriche, W.H., Hewitt, D., Neri, L.C.(1980) Magnesium, water hardness, and heart disease, in M. Cantin and M.S. Seelig (eds.), *Magnesium in Health and Disease,* New York, pp. 565-571.

Cappucio, F.P., Markandu, N.D., Beynon, G.W., Shore, A.C., Sampson, B., MacGregor, G.A.(1985) Lack of effect of oral magnesium on high blood pressure. A double blind study. *Br. Med. J.* **291,** 235-238.

Dawson, E.B., Frey, M.J., Moore, T.D., McGanity, W.J. (1978) Relationship of metal metabolism to vascular disease mortality rates in Texas. *Am. J. Clin. Nutr.* **31,** 1188-1197.

Durlach, J., Durlach, V., Rayssiguier, Y., Bara, M., Guiet-Bara, A. (1992) Magnesium and blood pressure II. Clinical studies. *Magn. Res.* **5,** 147-153.

Dyckner, T., and Wester, P.O. (1983) Effect of magnesium on blood presure. *Br. Med. J.* **286,** 1847-1849.

Geleijnse, J.M., Witteman, J.C.M., Bak, A.A.A., den Breeijen, J.H., Grobbee, D.E. (1994) Reduction in blood pressure with a low sodium, high potassium, high magnesium salt in older subjects with mild to moderate hypertension. *Br. Med. J.* **309,** 436-440.

Hammer, D.I., and Heyden, S. (1980) Water hardness and cardiovascular mortality. An idea that has served its purpose. *J.A.M.A.* **243,** 2399-2400.

Henderson, D.G., Schierup, J., Schodt, T. (1986) Effect of magnesium supplementation on blood pressure and electrolyte concentration in hypertensive patients receiving long-term diuretic treatment. *Br. Med. J.* **293,** 664-665.

Hwang, D.L., Yen, C.F., Nadler, J.L. (1992) Effect of extracellular magnesium on platelet activation and intracellular calcium mobilization. *Am. J. Hypertens.* **5,** 700-706.

Intersalt Cooperative Research Group (1988) Intersalt: an international study of electrolyte excretion and blood pressure. Results for 24 hour urinary sodium and potassium excretion. *Br. Med. J.* **297,** 319-328.

Iseri, LF, and French, J.H. (1984) Magnesium: nature's physiologic calcium blocker. *Am. Heart J.* **108,** 188-193.

Joffres, M.R., Reed, D.M., Yano, K. (1987) Relationship of magnesium intake and other dietary factors to blood pressure. The Honolulu Heart Study. *Am. J. Clin. Nutr.* **45,** 469-475.

Karppanen, H. (1981) Epidemiological studies on the relationship between magnesium intake and cardiovascular diseases. *Artery* 9, 190-199.

Karppanen, H., Tanskanen, A., Tuomilehto, J., Puska, P., Vuori, J., Jantti, V., Seppanen, M.L. (1984) Safety and effects of potassium- and magnesium-containing low sodium salt mixtures. *J. Cardiovasc. Pharmacol.* **6,** S236-S243.

Kesteloot, H. (1984) Urinary cations and blood pressure. Population studies. *Ann. Clin. Res.* **16** (suppl. 43), 72-80.

Kesteloot, H., Huang, D.X., Li, Y.-L., Geboers, J., Joossen, J.V. (1987) The relationship between cations and blood pressure in the People's Republic of China. *Hypertension* **9,** 654 659.

Lind, L., Lithell, H., Pollare, T., Ljunghall, S. (1991) Blood pressure response during long-term treatment with magnesium is dependent on magnesium status. A double-blind, placebo-controlled study in essential hypertension and in subjects with high-normal blood pressure. *Am. J. Hypertens.* **4,** 674-679.

Marier, J.R.(1981) Water hardness and heart disease. *J.A.M.A.* **245,** 1316-1326.

M' Buyamba-Kabangu, J.R., Fagard, R., Lignen, P., Mbuy wa Mbuy, R., Staessen, J., Amery A. (1986) Blood pressure and urinary cations in urban Bantu of Zaire. *Am. J. Epidemiol.* **124**, 957-968.

McInnes, G.T., Yeo, W.W., Ramsay, L.E., Moser, M. (1992) Cardiotoxicity and diuretics: much speculation-little substance. *J. Hypertens.* **10,** 317-335.

Reyes, A.J., Leary, W.P., Acosta-Barrios, T.N., Davis, W.H. (1994) Magnesium supplementation treated with hydrochlorothiazide. *Curr. Ther. Res.* **36,** 332-340.

Sacks, F.M., and Kass, H.E. (1988) Low blood pressure in vegeterians: effects of specific foods and nutrients. *Am. J. Clin. Nutr.* **48,** 795-800.

Sacks, F.M, Brown, L.E., Appel, L., Borhani, N.O., Evans, D., Whelton, P. (1995) Combinations of potassium, calcium, and magnesium supplements in hypertension. *Hypertension* **26,** 950-956.

Seelig, M.S. (1989) Cardiovascular consequences of magnesium deficiency and loss in refractory potassium repletion. *Am. J. Cardiol.* **63,** 4G-21G.

Staessen, J., Bulpitt, C., Fagard, R., Joosens, J.V., Lijnen, P., Amery, A. (1983) Four urinary cations and blood pressure. A population study in two Belgian towns. *Am. J. Epidemiol.* **117,** 676-687.

Stitt, F.W., Clayton, D.G., Crawford, M.D., Morris, J.N. (1973) Clinical and biochemical indicators of cardiovascular disease among men living in hard and soft water areas. *Lancet* **1,** 122-126.

Thulin, T., Abdulla, M., Dencker, I., Jargestad, M., Melander, A., Norden, A., Schersten, B., Akesson B. (1980) Comparison of energy and nutrient intakes in women with high and low blood pressure levels. *Acta Med. Scand.* **208,** 367-373.

Tian, H.-G., Nan, Y., Shao, R.-C., Dong, Q.-N., Hu, G., Pietinen, P, Nissinen, A. (1995) Association between blood pressure and dietary intake and urinary excretion of electrolytes in a Chinese population. *J. Hypertens.* **13,** 49-56.

Whelton, P.K., Klag, M.J. (1989) Magnesium and blood pressure: review of the epidemiologic and clinical trial experience. *Am. J. Cardiol.* **63,** 26G-30G.

Widman, L., Wester, P.O., Stegmayr, B.K., Wirell, M. (1993) The dose-dependent reduction in blood pressure throug administration of magnesium. A double blind placebo controlled cross-over study. *Am. J. Hypertens.* **6,** 41-45.

Witteman, J.C.M., Grobbee, D.E., Derkx, F.H.M., Bouillon, R., De Bruijn, A.M., Hofman, A. (1994) Reduction of blood pressure with oral magnesium supplementation in women with mild to moderate hypertension. *Am. J. Clin. Nutr.* **60,** 129-135.

Yamori, Y., Mizushima, S., Ueda, M., Sawamura, M., Nara, Y. (1994) Inverse relationships of urinary magnesium excretion to blood pressure: WHO-CARDIAC study. *J. Hypertens.* **12** (suppl. 3), 561.

Zhu, K., He, S., Pan, X., Zheng, X., Gu, Y. (1987) The relation of urinary cations to blood pressure in the People's Republic of China. *Hypertension* **9,** 654-659.

DOES SERUM MAGNESIUM CONCENTRATION PREVENT MITRAL ANNULAR CALCIFICATION IN HEMODIALYSIS PATIENTS ?

J. TZANAKIS, N. KALLIVRETAKIS, A. PRAS,
K. VIRVIDAKIS, D. MAYOPOULOU-SYMVOULIDOU,
and T. MOUNTOKALAKIS
Departments of Nephrology and Cardiology, General Hospital of Chania, Crete, and 3rd Department of Medicine, Medical School, University of Athens, Sotiria General Hospital, Athens, Greece

1. Introduction

Cardiac valve calcifications are common in patients on chronic hemodialysis. Magnesium (Mg) status is one of several factors implicated in the pathogenesis of this complication (Meema et al., 1987; Rostand et al., 1988). The present study was scheduled to study the relationship between mitral annular calcification and various biochemical, hormonal, and demographic parameters, namely serum Mg, calcium (Ca), phosphate (P), CaXP product, parathyroid (PTH) levels, age and duration of dialysis therapy, in patients on chronic hemodialysis.

2. Subjects and Methods

Fifty-six patients, 34 men and 22 women, treated with chronic hemodialysis, and 122 control subjects with normal renal function were studied. Patients were on hemodialysis treatment for more than 8 months and had a CaXP product not higher than 60 for at least once a year, in regular monthly monitoring. Control subjects were recruited from the Cardiology outpatient Clinic of the same hospital after being age- and sex- matched with hemodialyzed patients. Serum Mg, Ca, CaXP and intact PTH levels from the patients' monthly follow-up were used for analysis. Mitral annular calcifications were detected by M-mode and two dimensional and Doppler echocardiography, in patients and control subjects.

T. Theophanides and J. Anastassopoulou (eds.), Magnesium: Current Status and New Developments, 291-292.

3. Results

The incidence of mitral annular calcifications in hemodialysis patients was found to be 41.1%. This incidence was significantly higher than that found in control subjects (4.1%, $\div^{2}=36.58$, $p<0.05$). Patients who were found to have mitral annular calcifications were significantly ($p<0.05$) older than those who did not (67.5±9.6 vs 59.4±9.9 years, respectively). Patients with serum Mg higher than 3 mg/ml were twice as likely to develop mitral annular calcifications than those with lower serum Mg values ($\div^{2}=7.0$, $p<0.008$). The rest biochemical parameters did not differ significantly between hemodialysis patients with and without mitral annular calcifications. Multiple logistic regression analysis revealed that serum Mg could predict successfully (overall accuracy 86%) mitral annular calcifications when age and serum Ca and P were taken into account.

4. Comment

The above findings are consistent with previous evidence indicating a protective role of hypermagnesemia on cardiovascular calcification in hemodialysis patients (Meema et al., 1987; Rostand et al., 1988). The hypothesis that this action might be mediated through a suppressive effect of Mg on PTH secretion is not supported by the present finding of a lack of significant difference in serum PTH between patients with and without mitral annular calcifications. Further prospective studies in larger samples of patients are needed to clarify this issue.

5. References

Meema, H.E., Oreopoulos, D.G., Rapaport A. (1987) Serum magnesium levels and arterial calcification in end-stage renal disease. *Kidney Int.* **32,** 388-394.

Rostand, S.G., Sanders, C., Kirk, K.A., Rutsky E.A., Fraser R.G. (1988) Myocardial calcification and cardiac dysfunction in chronic renal failure. *Am. J. Med.* **85,** 651-657.

FUROSEMIDE-LIKE NATRIURETIC ACTIONS OF GENISTEIN WITHOUT EFFECTS ON MAGNESIURESIS

I. GIMENEZ, J.M. LOU, J.O. ALDA & J.F. ESCANERO.

Dpt. of Pharmacology and Physiology, School of Medicine, c/Domingo Miral s/n, 50009 Zaragoza, Spain.

1. Introduction

Loop-diuretics, as furosemide, increase magnesiuresis; prolonged treatment with these drugs may cause magnesium deficiency[1]. Magnesium renal reabsorption depends on the work of Na-K-Cl cotransport, which is inhibited by loop diuretics. Magnesium renal reabsorption is modulated also by other factors including aldosterone, parathormone and calcemia. Recently, we reported that isoflavonoids, natural compounds with potential benefits in cancer, cardiovascular diseases or osteoporosis[2] inhibited Na-K-Cl cotransport with a potency only 2-3 times lesser than that of furosemide[3]. Present work was addressed to investigate if genistein, an isoflavonoid, has a magnesiuretic effect like furosemide.

2. Material and Methods

Isolated perfused rat kidney technique was carried out following previously detailed procedures[4]. After a first ten min basal period, cumulative dosis of genistein (dissolved in DMSO) were administered in the perfusate during 4 further consecutive periods of 10 min each. Urinary volume, Na, K and Mg were measured for each one of the five 10 min periods. In control experiments, same volume of DMSO did not change urinary excretion rates of water and electrolytes, which remained constant for more than 90 min.

T. Theophanides and J. Anastassopoulou (eds.), Magnesium: Current Status and New Developments, 293-298.

To measure genistein effects in vivo, fourteen rats were used to live in metabolic cages. In the day of the experiment, 7 rats received a subcutaneous injection of 15 mg/kg genistein (dissolved in DMSO and suspended in 1 mL of 0.2% metanol/saline) and a second group of 7 rats received the same volume of vehicle alone. Urinary sodium, potassium and magnesium contents were measured by flame photometry.

3. Results

3.1. STIMULATION OF SALIDURESIS IN ISOLATED PERFUSED RAT KIDNEY.

Table 1 shows maximal increments in water, sodium, potassium and magnesium urinary excretion induced by administration of genistein (308 μM) or furosemide to the isolated perfused rat kidney, compared with those of an equivalent dose of furosemide (100 μM). Genistein was as good diuretic and natriuretic as furosemide. This drug induced a higher kaliuresis than genistein. Neither furosemide nor genistein were able to cause a statistically significant increase in magnesiuresis.

TABLE 1. Effect of genistein/furosemide administration to the isolated perfused rat kidney. Relative increment in water and electrolyte urinary excretion.

Variable	ΔGenistein 310 μM	ΔFurosemide 100 μM
Diuresis	3,20±0,38*	3,47±0,27*
Natriuresis	5,80±1,25*	6,62±1,30*
Kaliuresis	2,29±0,39*	3,44±0,13*
Magnesiuresis	1,62±0,33	1,45±0,13

Δ = relative increment (rate between values after and before addition of drug/vehicle).
Data are given as mean±S.E.M. (n=4)
* = $p<0.05$, drug *vs* vehicle (ANOVA followed by unpaired t-test)

Figure 1 shows dose-dependency of natriuresis and magnesiuresis stimulation by genistein. It can be seen that, differently from sodium excretion, genistein-stimulated magnesiuresis was not statistically different from basal magnesiuresis, and that it reaches its maximum at a smaller genistein concentration (205 μM).

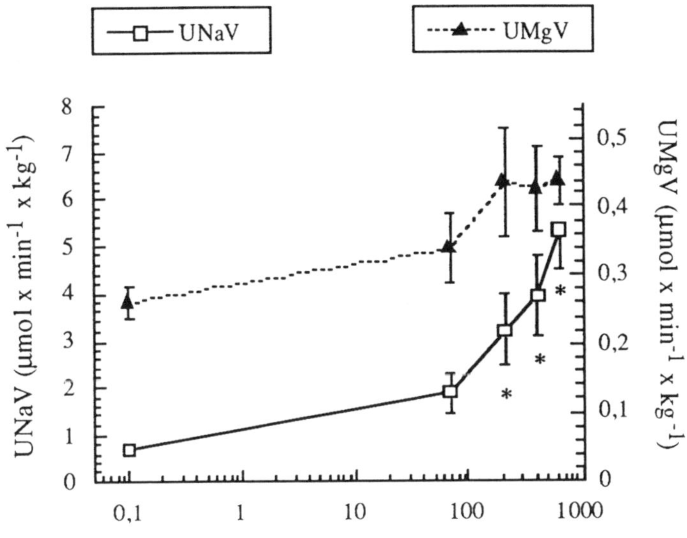

Figure 1. Cumulative dose-response curve for urinary excretion rate of sodium (UNaV) and magnesium (UMgV) in the isolated perfused rat kidney. After a ten min basal period, genistein was administered subsequently to reach concentrations of 70, 205, 410, and 616 µmol/L. Administration of vehicle alone did not cause any effect. Data are given as mean±S.E.M. (n=4). * = $p<0.05$ (ANOVA followed by paired t-test). kg – kidney weight in grams.

3.2. RENAL EFFECTS OF GENISTEIN IN VIVO.

Figure 2 shows the effects of genistein subcutaneous administration on urinary excretion of water, sodium and magnesium. It can be seen that both diuresis and natriuresis were higher in rats treated with genistein, in the first six hours after isoflavonoid administration. Magnesiuresis (Fig. 2) and kaliuresis (data not shown) did not change in genistein group compared to vehicle group.

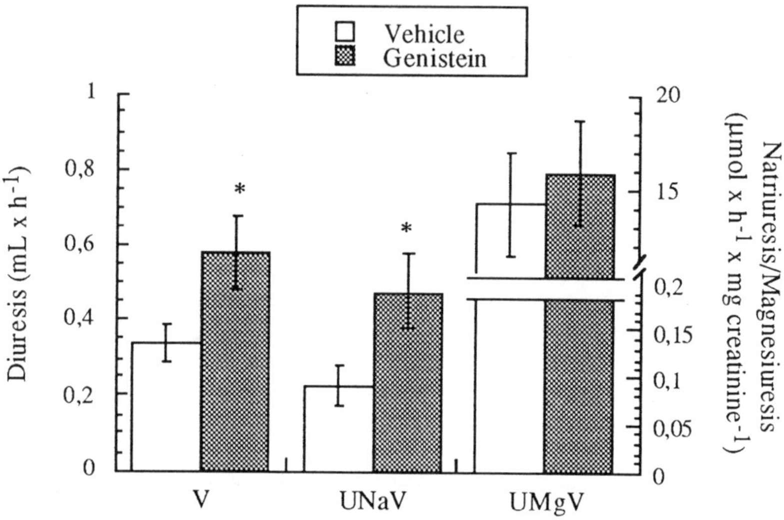

Figure 2. Urinary excretion rate of water (V), sodium (UNaV) and magnesium (UMgV) during the first 6 hours following subcutaneous administration of genistein 15 mg/Kg (solid bars) or vehicle alone (clear bars). Data are given as mean±S.E.M. (n=4). * = $p<0.05$ (ANOVA followed by unpaired t-test).

4. Discussion

In LLC-PK1 cells, genistein inhibited Na-K-Cl cotransport with a potency approximately three times lesser than that of furosemide[3]. In the present work, we showed that, in the isolated perfused rat kidney, genistein 308 µM was able to induce similar stimulation of diuresis, natriuresis and kaliuresis than furosemide 100 µM. However, neither furosemide nor genistein increased significantly magnesium renal excretion, supporting the idea that stimulation of magnesiuresis by loop-diuretics requires extra-renal participation besides cotransport inhibition.

In vivo, genistein retained its diuretic and natriuretic properties, but was unable to induce a significant increase in magnesiuresis. A similar result was obtained by Crippa *et al* [5], who demonstrated that muzolimine, nevertheless inducing a similar diuresis and natriuresis than other loop-diuretics, did not increase magnesium renal excretion. They suggested that this difference might indicate distinct interferences with humoral

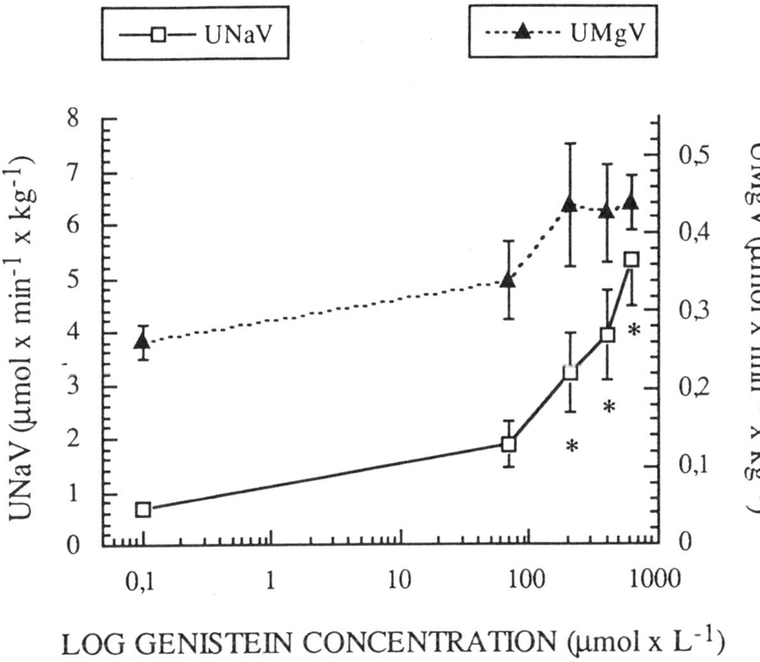

UNaV
UMgV
UNaV (µmol x min-1 x kg-1)
UMgV (µmol x min-1 x kg-1)
LOG GENISTEIN CONCENTRATION (µmol x L-1)

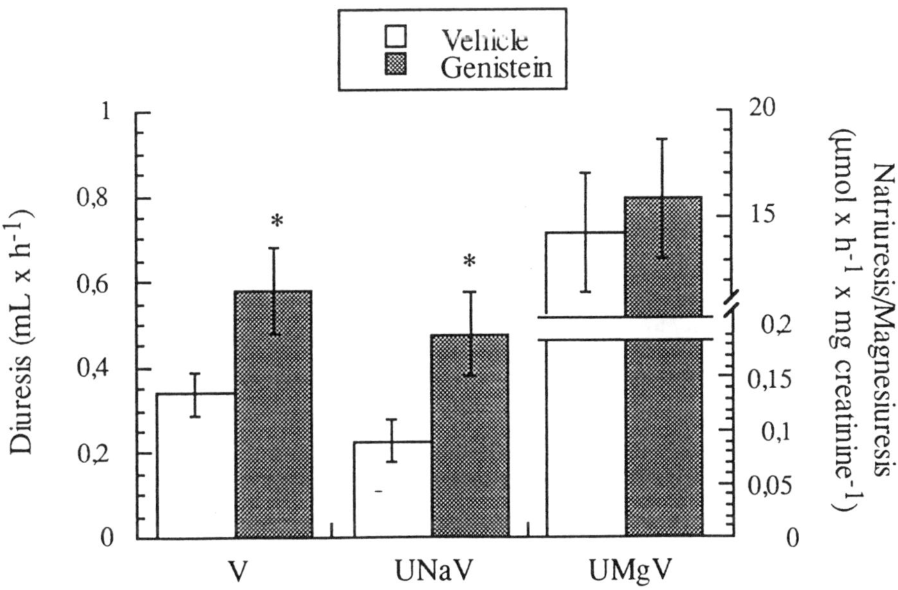

Vehicle
Genistein
Diuresis (mL x h-1)
Natriuresis/Magnesiuresis (µmol x h-1 x mg creatinine-1)
V
UNaV
UMgV

regulators of magnesium renal excretion, such as aldosterone, PTH or calcium. This could also be the case with genistein, which has been shown to possess a miriad of bioactivities, including modulation of steroid metabolism. It is possible that, besides its effects at the Henle's loop, genistein inhibits directly aldosterone secretion[6]. This would eliminate the rebound effect seen with traditional loop-diuretics and the increment in potassium and magnesium renal excretion. More research is needed to elucidate genistein effects on renal functions, specially now that genistein is being involved in prevention of cardiovascular diseases like hypertension[2].

5. References

1. Reyes, A.J. (1996) Magnesium and diuretics: 10 years of progress. In: *Current Resesarch in Magnesium*, eds. M.J. Halpern & J. Durlach, pp. 293-295. London: John Libbey.
2. Knight, D.C. and Eden, J.A. (1996) A review of the clinical effects of phytoestrogens. *Obstet Gynecol* **87**, 897-904
3. Martinez, R.M., Gimenez, I., Lou, J.M., Mayoral, J.A. and Alda, J.O. (1997) Isoflavonoids possess biological activities of loop-diuretics. *Am J Clin Nutr* (in press)
4. Maack, T. (1980) Physiologycal evaluation of the isolated perfused rat kidney. *Am J Physiol.***238**, F71-F78
5. Crippa, G., Giorgi-Pierfranceschi, M., Sverzellati, E., Castelli, A. and Cararra, G.C. (1996) Magnesiuretic action of very high formulations of loop diuretics in healthy subjects. In: *Current Resesarch in Magnesium*, eds. M.J. Halpern & J. Durlach, pp. 293-295. London: John Libbey.
6. Bodart, V., Ong, H. and De Lean, A. (1995) A role for protein tyrosine kinase in the steroidogenic pathway of angiotensin II in bovine zona glomerulosa cells. *J Steroid Biochem Mol Biol* **54**, 55-62

PLATELET, ERYTHROCYTE AND PLASMA MAGNESIUM LEVELS IN RENAL TRANSPLANT RECIPIENTS DURING CYCLOSPORIN OR AZATHIOPRINE TREATMENT

A. ALLEGRA, F. CORICA, R. IENTILE*, A. CORSONELLO, V. SPARACINO**, R. D'ANGELO*, F. RUBINO, L. CASTAGNA, F. CAPUTO**, M. BUEMI.

*Department of Internal Medicine, * Institute of Biochemical Sciences and Clinical Biochemistry, University of Messina, Italy. **Division of Nephrology and Transplantations, Ospedale Civico, Palermo, Italy.*

Abstract

Using direct current spectrometry, the authors evaluated magnesium concentrations in plasma, erythrocytes and platelets from two groups of subjects who had received a renal transplant and were treated either with cyclosporin, azathioprine and prednisolone or with azathioprine and prednisolone, and in a group of healthy subjects. The plasma, erythrocyte and platelet concentrations of magnesium in the two groups of subjects treated with immunosuppressors were lower than in the control group, while no difference was found between the two groups of transplant recipients for magnesium status. Magnesium fractional excretion was higher in the subjects who had undergone transplantation than in the control group, whereas no difference was found between patients and controls for urinary ion excretion over 24 hours. The depletion of intraplatelet magnesium induced by immunosuppressive drugs may have been responsible for the platelet hyperaggregability found following treatment with cyclosporin and azathioprine.

Key words: Platelet magnesium, Cyclosporin, Azathioprine, Renal transplantation.

Introduction

It is known that cyclosporin treatment is accompanied by hypomagnesemia (1-3). However, the numerous studies in literature on the capacity of cyclosporin to modify the body content of magnesium in the different compartments report contradictory findings (4). Cyclosporin administration in rats not only triggers a reduction in serum magnesium, but also appears to be accompanied by an increase in the ion content in muscle, renal and hepatic cells (5-8), whereas the drug does not appear to modify the ion content in muscle in humans (9). Epstein et al. report a reduced magnesium content in the bone after cyclosporin treatment (10). On the other hand, Bley et al. report that cyclosporin-induced modifications in magnesium status, in the different compartments, depend on the different cell type studied. In fact, serum magnesium levels show an early reduction during cyclosporin treatment, while a reduction in lymphocyte and erythrocyte magnesium appears later (11). Of greater interest, however, is the suggestion made by different authors that many of the collateral effects that can follow the administration of cyclosporin (neurotoxicity, hypertension and nephrotoxicity in particular), may depend on a magnesium depletion induced by cyclosporin itself (12-16). As different authors have demonstrated that both cyclosporin and other immunosuppressors, such as azathioprine, can cause platelet function disturbances (17-

T. Theophanides and J. Anastassopoulou (eds.), Magnesium: Current Status and New Developments, 299-302.

19), it is known that a reduction in the platelet magnesium content is accompanied by a modification in platelet aggregation (20). The aim of our study was to evaluate the effects of the administration of cyclosporin and of other immunosuppressors on the content of magnesium in platelets.

Materials and Methods

The study was conducted on two groups of transplant recipients. A group of eight subjects underwent treatment with cyclosporin, azathioprine and prednisolone (group CAP), while the second group was given azathioprine and prednisolone only (group AP). Ten healthy age and sex matched subjects served as controls. The time after transplant and type of transplant (from a cadaver or living donor), renal function, incidence of hypertension, prednisolone and azathioprine doses were not significantly different in the two groups of patients treated. None of the patients had taken diuretics or other drugs that could have interfered with magnesium metabolism (gentamycin, aminoglycosides, tobramycin, theophillin, phenobarbital, etc) in the two months prior to the beginning of the study. Criteria for exclusion from the study were: obesity, diabetes mellitus, heart failure or cancer and/or the presence of diarrhea. After obtaining their fully informed consent, a blood sample was taken from all patients in the morning after an overnight fast, and plasma, erythrocyte and platelet magnesium levels were assayed using direct current spectrometry. To evaluate platelet magnesium we used a modified version of the technique described by Touyz and Milne (21). Plasma cyclosporin was measured by radioimmunoassay using the specific monoclonal antibody. In all subjects, moreover, urinary magnesium excretion over 24 hours and magnesium fractional excretion were evaluated, the latter being calculated dividing the clearance of magnesium by the clearance of creatinine. The statistical analysis was made utilizing the Kruskal-Wallis test, values of $p < 0.05$ being considered statistically significant. The study of correlations was made using Spearman's correlation coefficient.

Results

In both the groups of transplant recipients, plasma, erythrocyte and platelet magnesium

Table 1. *General characteristics and magnesium levels in the groups studied.*

	Healthy subjects (N=10)	Group AP (N=13)	Group CAP (N=8)
Age (years)	40 (31-45)	41 (29-47)	44 (30-52)
Sex (M/F)	7/3	8/5	5/3
B.M.I. (kg/m^2)	23.7 (22.6-25.7)	24.7 (22.8-25.5)	23.1 (22.4-25.4)
Serum creatinine (mg/dl)	1.0 (0.8-1.1)	1.5 (1.0-2.1)*	1.1 (0.9-1.4)
Creatinine clearance (ml/min.)	85.4 (78.2-104.2)	53.1 (31.8-74.1)*	74.3 (55.6-81.4)*
Time after transplantation (months)	---	102 (4.2-142.2)	54.5 (27.2-74.5)
Cadaveric/living donor	---	12/1	7/1
Prednisolone dose (mg/die)	---	15 (11-15)	10 (10-15)
Azathioprine dose (mg/die)	---	100 (50-112)	75 (75-100)
Cyclosporine dose (mg/die)	---	---	255 (250-313)
Blood cyclosporine (ng/ml)	---	---	184 (102-218)
Hypertension (yes/no)	0/10	7/1	10/3
Urinary magnesium (mmol/24h)	4.4 (3.7-5.1)	3.8 (3.0-5.0)	4.4 (3.8-5.8)
Magnesium fractional excretion (%)	3.5 (2.8-4.1)	18.6 (11.1-22.8)***	12.8 (8.4-16.3)***
Plasma magnesium (mmol/l)	0.86 (0.81-0.96)	0.62 (0.45-0.74)**	0.61 (0.55-0.78)**
Erythrocyte magnesium (mmol/l)	2.56 (2.44-2.76)	2.30 (1.98-2.43)*	2.18 (1.94-2.47)*
Platelet magnesium (μmol/10^8 cells)	2.84 (2.75-2.95)	1.75 (1.52-2.42)***	1.75 (1.71-1.90)***

Data are medians with 25th and 75th percentile in parentheses.
* $p < 0.05$, ** $p < 0.01$, *** $p < 0.001$ vs healthy subjects.

concentrations were significantly lower than those in the control group, whereas no difference was found between the two groups of transplanted patients for magnesium status (Table 1). No difference was found between the groups studied as regards urinary 24 hour magnesium, whereas values for fractional magnesium excretion were significantly higher in patients who had received a transplant than in controls (Table 1). Moreover, in the whole group of transplant recipients (N=21), an inverse correlation was found between the levels of platelet magnesium and urinary magnesium ($r_s = - 0.54$, $p < 0.05$) and time after transplantation showed a negative correlation with platelet magnesium levels ($r_s = - 0.73$, $p < 0.001$), and a direct correlation with fractional magnesium excretion ($r_s = 0.53$, $p < 0.05$).

Discussion

Our data appcar to confirm that there is an alteration in the compartmentalization of magnesium in patients treated with cyclosporin. An analogous effect would appear to be determined also by the administration of other immunosuppressors, such as azathioprine and prednisolone. The increase in the fractional excretion of magnesium found in both groups of patients treated with immunosuppressors, and the finding of a direct correlation between time after transplantation (i.e. immunosuppressive treatment duration) and magnesium fractional excretion, appears to explain the mechanism inducing the magnesium deficit present in these patients. It is also known that, in the kidney, cyclosporin causes an intra-adrenal vasoconstriction with hypoperfusion and chronic medullary ischemia that can cause a tubular alteration and modify renal ion metabolism (11). On the other hand, the finding of normal 24 hour urinary magnesium levels in the treated patients is not surprising because this parameter is influenced to an important extent by numerous variables, diet above all. The most important result of our study is the reduction found in intraplatelet magnesium concentrations in patients on immunosupprcssors, which may explain complications due to an alteration in the platelet aggregation found in these subjects (18, 19). Cyclosporin appears to have an important impact on platelet aggregation. It has, in fact, been demonstrated that cyclosporin increases thromboxane synthesis, and reduces the synthesis of prostacyclin by the endothelium. In experimental models, moreover, it can trigger an increase in the platelet aggregation to an extent ranging from 163 to 253% with respect to control values (17). As it has been demonstrated that a reduced intraplatelet content of magnesium causes platelet hyperaggregability (20), it cannot be ruled out that the administration of immunosuppressors may give rise to a condition analogous to that in women on estrogen therapy (22, 23), in whom reduced magnesium concentrations increase the risk from platelet hyperaggregability. Although these patients eventually become susceptible to hemorrhages, it cannot be ruled out that inverse conditions may appear in patients with a particular predisposition or exposure to other risk factors (24). In conclusion, the reduction in the total body content of magnesium found in patients treated with immunosuppressors such as cyclosporin, azathioprine and prednisolone is probably due to the increased renal excretion of ions induced by these drugs and, in particular, immunosuppressive therapy appears capable of significantly modifying platelet magnesium content.

References

1) Barton, C.H., Vaziri, N.D., Martin, D.C., Choi S. and Alikhani S. (1987) Hypomagnesemia and renal magnesium wasting in renal transplant recipients receiving cyclosporin. *Am. J. Med.* **83**, 693-9
2) Kaiser, W., Biesenbach, G., Kramer, E. and Zazgornik, J. (1989) Magnesium after renal transplantation comparison between cyclosporin A and conventional immunosoppression. *Adv. Exp. Med. Biol.* **252**, 297-

302

3) June, C.H., Thompson, C.B., Kennedy, M.S., Nims, J. and Thomas, E.D. (1985) Profound hypomagnesemia and renal magnesium wasting associated with the use of cyclosporin for marrow transplantation. *Transplantation* **39**, 620-4.
4) McDiarmid, S.V., Colonna, J.O., Shoked, A., Ament, M.E. and Busuttil, R.W. (1993) A comparison of renal function in cyclosporin and FK-506 treated patients after primary orthotopic liver transplantation. *Transplantation* **56**, 847-53.
5) Nozue, T., Kobayashi, A., Saki, A., Satoh, T., Kodama, T., Yamazaki, H., Kurosawa, M., Vemasu, F., Endoh, H. and Tokagy, Y. (1993) Evidence that cyclosporine causes both intracellular migration and inappropriate urinary excretion of magnesium in rats. *Transplantation* **55**, 346-9.
6) Markell, M.S., Altura, B.T., Barbour, R.L. and Altura, B.M. (1993) Ionized and total magnesium levels in cyclosporin-treated renal transplant recipients: relationship with cholesterol and cyclosporin levels. *Clin. Sci. Colch.* **85**, 315-8.
7) Ramos, E.L., Barry, Y.M., Kubilis, P., Peterson, J.C., Pfaff, W.W., Howard, R.J., Parris, C.J., Patton, R.R. and Karlix, J.L. (1995) Hypomagnesemia in renal transplant patients improvement overtime and association with hypertension and cyclosporine levels. *Clin. Transplant.* **9**, 185-9.
8) Barton, C.H., Vaziri, N.D., Mina-Araghi, S., Crosby, S. and Seo, M.I. (1989) Effects of cyclosporin on magnesium metabolism. *J. Lab. Clin. Med.* **114**, 232-236.
9) Frost, L., Danielsen, H., Dorup, I., Kjaer, T. and Pedersen, E.B. (1993) Skeletal muscle magnesium content during cyclosporin and azathioprine treatment in renal transplant recipients. *Nephrol. Dial. Transplant.* **8**, 79-93.
10) Epstein, S., Scholsberg, M., Fallon, M., Thomas, S., Movsowitz, C. and Ismail, F. (1990) 1,25 dihydroxyvitamin D3 modifies cyclosporin-induced bone loss. *Calcif. Tissue Int.* **47**, 152-7.
11) Bley, N., Dick, K., Rob, P.M., Rohwer, J. and Sack, K. (1996) Follow up measurements of magnesium concentrations in serum, lymphocytes and red blood cells after renal transplantation and cyclosporin treatment - with or without oral magnesium supplementation. *Magnes. Res.* **9**, 74.
12) June, C.H., Thompson, C.B., Kennedy, M.S. and Loughran, T.P. (1985) Correlation of hypomagnesemia with the onset of cyclosporin associated hypertension in marrow transplant patients. *Transplantation* **41**, 47-51.
13) Kaiser, W., Biesenbach, G., Kramer, E. and Zazgornik, J. (1992) Hypomagnesemia. - A feature of cyclosporin tubulotoxicity in renal transplant recipients. *Magnesium B.* **14**, 130-2.
14) Thompson, C.B., Sullivan, K.M., June, H.C. and Thomas, E.D. (1984) Association between cyclosporin neurotoxicity and hypomagnesemia. *Lancet* **2**, 1116-20.
15) Palestine, A.G., Austin, H.A. and Nussenblatt, R.B. (1985) Cyclosporin-induced nephrotoxicity in patients with autoimmune uveitis. *Transplant. Proc.* **17**, 209-14.
16) Forslund, T., Hannonen, P., Reitano, S. and Fyhrqvist, F. (1995) Hypertension in cyclosporin A-treated patients is independent of circulating endothelin levels. *J. Intern. Med.* **238**, 71-5.
17) Gonzalez-Correa, J.A., De La Cruz, J.P., Lucena, M.I. and Sanchez de la Cuesta, F. (1996) Effect of cyclosporin A on platelet aggregation and thromboxane/prostacyclin balance in a model of extrahepatic cholostasis in the rat. *Thromb. Res.* **81**, 367-81.
18) Malyszko, J.S. (1996) Platelet aggregation and peripheral serotonergic system in kidney transplant recipients treated with cyclosporin. *Transplant. Proc.* **28**, 1954-7.
19) Malyszko, J.S. (1996): Effects of Azathioprine and mizoribine on aggregation od narmal and uremic platelets. *Transplant. Proc.* **28**, 3488-9.
20) Nadler, J.L., Malayan, S., Luong, H., Shaw, S., Natarajan, D. and Rude, R.K. (1992) Intracellular free magnesium deficiency plays a key role in increased platelet reactivity in type II diabetes mellitus. *Diabetes Care* **15**, 835-41.
21) Touyz, R.M. and Milne, F.J. (1991) A method for determining the total magnesium, calcium, sodium and potassium contents of human platelets. *Miner. Electrolyte Metab.* **17**, 173-8.
22) Durlach, J. (1970) Pilula et thrombose; ou des plaquettes, des estrogenes et du magnesium. *Revue Franc. Endocrinol. Clin.* **11**, 45-54.
23) Goldsmith, N.F. and Johnston, S.O. () Mg estrogen hypothesis: thromboembolism and mineralization ratios. in: M. Cantin and M.S. Seelig (eds.), *Magnesium in health and disease.* Spectrum Press NT, London, pp. 313-324.
24) Gallardo, D., Ferré, C., Berlanga, J.J., Banda, E.D., Ponce, C., Saler, A., Alonso, E., Espannol, I., Riu, C. and Grenena, H. (1996) Neurological complications after allogenic bone marrow transplantation. *Bone Marrow Transplant.* **18**, 1135-1139.

MAGNESIUM, INFLAMMATION AND PAIN

C. DUBRAY[1], Y. RAYSSIGUIER[2]
Centre de Recherche en Nutrition Humaine de Clermont-Jerrand
1 Groupe NPPUA, Laboratoire de Pharmacologie Medicale, Faculte de Medecine, B.P. 38, 63001
Clermont-Ferrand Cedex 1, France
2 Unite Maladies Metaboliques et Nutriments INRA 63122 St-Genes-Champanelle, France.

1. Introduction

The relationships between inflammation and pain were known for a long time, even by antique civilizations. However most of the fundamental knowledge about biological mechanisms explaining some of the links between inflammation and nociception processes have been discovered over the last three decades. Peripheral infection or injury can induce an inflammation and related pain but also some nerves, especially C fibers which are mainly involved in nociception transmission, can generate a so called "neurogenic inflammation". New findings rise some questions about the role of magnesium (Mg) in these two phenomena.

2. Inflammatory response in magnesium deficiency

In 1932, Kruse and coworkers [1] reported peripheral a vasodilatation with hyperemia as a part of symptoms caused by magnesium deficiency in rats. Thereafter, many investigators have recognized the same symptoms in relation with degranulation of mast cells with release of histamine and other inflammatory mediators. The response is analogous to an anaphyactoid reaction [2,3].

2.1 INVOLVEMENT OF CYTOKINES AND NEUROPEPTIDES

Some studies performed in our laboratory contributed to characterize more closely this inflammatory response using weaning male Wistar rats pair-fed a control or Mg deficient diet [4]. Dietary Mg deficiency induces after a few days a characteristic peripheral vasodilatation with hyperemia of the ears. The inflammatory response is associated with a marked increase in the circulating polymorphonuclear cells (neutrophils, basophils and eosinophils). Evidence of the importance of cytokine cascade in the induction and control of the inflammatory reaction is increasing [5]. Elevated levels of neuropeptides (substance P [SP] and calcitonin gene related peptide

T. Theophanides and J. Anastassopoulou (eds.), Magnesium: Current Status and New Developments, 303-311.

[CGRP] followed shortly thereafter by proinflammatory cytokines (IL1, IL6, TNF) have been observed by Weglicki et al [6-8] in Mg deficient animals following 3 weeks of Mg deprivation. Our own studies confirm that Mg deficiency elevates circulating levels of IL6, but the time course of the inflammatory response appears quite different. High levels of IL6 could be detected as early as day 4 on the Mg deficient diet, but there was no concomitant increase in TNF. IL6 is produced by a number of cell types such as macrophages, monocytes and endothelial cells and it induces hepatocytes to form acute phase proteins. The elevation of the IL6 plasma levels after only 4 days on the deficient diet indicated that the release of this inflammatory cytokine is an early pathophysiological event which may be responsible for inducing the hepatic synthesis of acute phase proteins in Mg deficient rats [4].

2.2 CHANGES IN INFLAMMATORY PROTEINS

Changes in plasma levels of positive and negative acute phase proteins have been reported [4, 9-11]. After 1 week on the Mg deficient diet, the concentration of alpha-2 macroglobulin in the plasma was ten times higher than that of control rats. This was accompanied in the liver by an increase in the level of mRNA coding for this protein. Alpha-1 acid glycoprotein concentration in the plasma was also slightly elevated. Moreover, Mg deficient rats showed a significant increase in plasma fibrinogen levels. A second group of plasma proteins including albumin and apo E, showed decreased concentrations. An acute phase response also results in trace elements redistribution. Cytokines may stimulate hepatic metallothionein synthesis leading to sequestration of Zn by the liver, and resulting in a decrease in plasma levels of Zn. Plasma iron is also suqjestered by the liver and/or the reticulo-endothelial system during an acute phase response. Increased iron contents have been reported in tissues of Mg deficient animals. The acute phase response also results in an increase in plasma copper which may be due to the increase in ceruloplasmin level the major copper binding protein [4,9-11].

2.3 THE ROLE OF FREE RADICALS AND EICOSANOID PRODUCTION;

Several studies indicate an increase in free radical production in Mg deficient animals [6,9,12]. Among the radical oxygen species, nitric oxide (NO) is particularly interesting because of its involvement in physiological functions such as the regulation of vascular tone, the cytotoxicity of activated macrophages and the response to noxious stimuli. Recent data demonstrate that stable end products of NO are increased in plasma of Mg deficient rats[13]. This rise results from an activation of the inducible NO synthase rather than in the constitutive form of the enzyme [14,15]. Stimulation of eicosanoid production is also a consequence of Mg deficiency. Arachidonic acid is metabolized by the cyclooxygenase and lipooxygenase pathway with increased production of prostaglandins (prostacyclin, thromboxane, PGE2) and leukotrienes. The liberation may be a response to activation of phospholipase A2 by an increase of intracellular calcium [16].

All these results suggest that Mg deficiency leads to the activation of cellular systems such as monocyte-macrophages, PMN, that synthesize a variety of biological substances, some of which are powerful inducers of inflammatory events (cytokines, free radicals, eicosanoids). However, the precise mechanism by which Mg deficiency can induce inflammatory response remains to be determined. Loss of dietary Mg is associated with release of the neuropeptide substance P (SP) and calcitonin gene related peptide (CGRP) and neurogenic inflammation. Wegkicki et al [6,17,18] postulated that the release of SP may be the earliest event leading to a stimulation of the inflammatory cytokines. We have reported that Mg deficiency produces elevated levels of IgE [19]. Involvement of the complement cascade has also been reported and high plasma levels of anaphylatoxins occurred in Mg deficient animals [20]. Moreover, the ability of Mg to modulate $[Ca^2]$i may provide a way of modifying signal-induced cellular responses [21].

3. Changes in pain thresholds in magnesium deficient animals

The fact that most of the above mentioned mediators play a role in nociception processes [22], led us to look specifically at the effects of Mg deprivation on the response threshold to noxious stimuli by using behavioral tests.

3.1 A MODEL OF INDUCED HYPERALGESIA:

The first studies showed that three days after starting the diet in weaning rats, the animals submitted to the Mg-depleted diet developed a generalized inflammation and in parallel a lowering of the nociceptive threshold. That induced hyperalgesia raised a plateau three to four days later. Surprisingly, we observed in this model different types of responses, depending on the noxious stimulus we applied. The vocalization threshold, in response to mechanical stimulus (paw-pressure test) was significantly lowered in Mg-deprived rats. Conversely, no change in the nociceptive heat (tail immersion in hot water, tail-flick test and plantar test) or chemical (formalin test) thresholds were observed. Several authors depending on the experimental conditions or the analgesic compounds they tested have previously observed such a discrepancy. However, that question has not been answered in a satisfactory fashion, although Haves A.G. et al [23] made an attempt in interpreting these differences in term of number of C or Aδ fibers in peripheral nerves involved in the transmission of nociceptive impulses, according to the pattern of noxious stimulus.
Disregarding the differences due to the tests used, these first experiments raised two questions:

3.1.1 *Do the Mg-depleted rats suffer from hyperalgesia and /or allodynia or both?*

To answer to that question, the animals were submitted to non-noxious stimuli (paw stimulation by Von Frey hairs and tail immersion at 42°C). In these conditions we did not observe allodynia, that is: according to the ISAP Task Force on Taxonomy [24], <<pain due to a stimulus that does not normally provoke pain >>. On the opposite, the Mg-deprived rats showed an hyperalgesia, that is: <<an increased response to a stimulus

which is normally painful>>, especially in response to a mechanical pressure, as previously mentioned.

3.1.2 *Are these changes in nociceptive threshold in relation with the peripheral tissue inflammation or due to a direct effect of low concentration of Mg in the spinal cord tissues?*

That point is much more difficult to clarify due to the complexity of the mechanisms and the number of mediators involved in nociceptive transmission both at peripheral and at spinal or supra-spinal levels

Considering the vasodilatation and inflammatory response observed in Mg deficient rats, the first hypothesis for that induced hyperalgesia is a direct stimulation of nociceptive peripheral neurons by some of the above mentioned mediators. However the inefficacy of nonsteroidal anti-inflammatory drugs (NSAIDs) (piroxican) as well as histamine H1 (astemizole) and NK1 receptor antagonists to reverse the hyperalgesia do not provide any argument in favor of a peripheral dysregulation of the nociceptors, directly or indirectly induced by Mg depletion. Furthermore, the dissociation between mechanical hyperalgesia and unchanged threshold to thermal or chemical noxious stimuli in Mg deprived rats does not fit well with such a peripheral mechanism.

Conversely, two arguments speak for the central (spinal or supraspinal) dysregulation of the nociceptive pathways: 1) the dose depent reversibility of the hyperalgesia by a non-competitive N-Methyl-D-Aspartate (NMDA) receptor antagonist (dizicilpine; MK801) [25]; 2) the dissociation between inflammation and hyperalgesia and the long period of the time required to recover a "normal" nociceptive threshold after reloading the animals with Mg.

3.2 A CRITICAL ROLE OF NMDA RECEPTOR?

There is now good evidence that the excitatory amino-acids, glutamate and aspartate act at a number of receptors throughout the nervous system to produce neuronal excitations. In the spinal cord they appear to play key roles, in concert with peptides, in determining the level of pain transmission and in both the establishment and maintenance of central hypersensitivity. In the dorsal horn, the metabotropic, AMPA and NMDA receptors for the excitatory amino-acids have distinct actions. The AMPA receptor sets the baseline level of activity and when activated, the NMDA receptor then enhances and prolongs transmission and so has been implicated in many states of central hypersensitivity, including hyperalgesia and allodynia. The net result though is that the NMDA receptor-channel complex is not a participant in "normal" synaptic transmission, but when the correct conditions are achieved the complex will suddenly become activate and add a powerful depolarizing or excitatory drive to transmission of pain in the spinal cord which then appears to lead to enhanced synaptic transmission or hypersensitivity [26-28]. Experiments indicate that "pathological pains" where inflammation or neuropathy is present, can be distinguished from simple acute pains on the basis of the sensitivity of only the former to NMDA antagonism. Both the induction and the subsequent maintenance of some of these responses are depended on NMDA processes.

In the specific context of the Mg depletion we have to keep in mind that the NMDA receptor forms a cation-selective channel with a high calcium permeability and sensitivity to channel block by extracellular magnesium [29] There are high levels of NMDA receptors in the spinal cord of a number of species including man [30,31] and they are believed to be important for the induction of mid-term and long-term changes in synaptic strength [32]. These properties of NMDA receptors are in complete agreement with experimental data observed in our model. As a matter of fact, after the Mg deprived rats were given a loading dose of Mg salt and again a normal diet, a slow recovery to a "normal" nociceptive threshold (over a period of 14 to 21 days) occurred. That recovery was clearly dissociated from a much more rapid removal (<48h) of inflammatory symptoms. Such a slow disappearance of hyperalgesia is fully in agreement with the involvement of spinal NMDA receptors, which could directly participate to the neuroplasticity of nociceptive pathways induced by acute Mg deficiency.

If there is good evidence for involvement of NMDA receptors in Mg-depended hyperalgesia in our model, further studies will be necessary to specify the respective role of the peripheral inflammation and a direct effect of Mg depletion on the hyperactivity of NMDA receptors and to confirm these results in other animal species or in "chronic" Mg depletion. Especially, we have to keep in mind that our findings were obtained in young rat (3 weeks) submitted to a fairly acute Mg depletion. Concerning that point, it has to be mentioned that mice fed with a Mg-deficient diet over a period of 42 days [33] did not show any hyperalgesia by using the hot plate test. They even showed a slight increase of the nociceptive response. These discrepancies underline the importance of the experimental conditions for a reliable interpretation of the results, mainly when they are based on behavioral tests. for instance, it has been observed that the rats receiving the Mg-restricted diet show several types neurological abnormalities ranging from hyperexcitability and changes in sleep cycles accompanied by an increase in wakefulness [34] to seizures. To avoid some misinterpretations due to these behavioral changes in Mg-deficient animals, the effects of two reference analgesic compounds: acetaminophen and morphine were tested in control and in Mg-deficient animals. In our experimental conditions, both drugs increased the nociceptive thresholds in a dose-depended manner, thus confirming the hyperalgesia induced by Mg depletion.

4. What about the role of magnesium in painful diseases in man?

4.1. PAINFUL DISEASES AND MG IMBALANCE

Nervous forms of magnesium imbalance represent the best documented of experimental and clinical aspects of magnesium disorders. However, if some variations of intra-and extracellular Mg content have been put forward in several painful diseases in human, such as headache, migraine, colonopathy, diffuse pain syndrope in HLA Bw35+

patients the influence of Mg status in these painful symptoms in human would need to be more deeply investigated.
Circumstantial evidence points to the possible role of magnesium deficiency in the pathogenesis of headaches and has raised questions about the clinical utility of magnesium as a therapeutic regimen in some headaches. Several clinical trials showed the efficacy of intravenous infusion of magnesium sulfate and a relationship between clinical responses and the basal serum ionized magnesium level [35]. However, some other studies concluded that reduced blood magnesium is unlikely to be related to migraine pathophysiology and we can deplore that too many of these trial are not convincing due to the poor methodology of their design (open study, no control group). On the other hand, headaches or migraine are probably not the best model to understand the mechanism(s) of action of Mg in pain relief due to the quite complex interrelationships between neuronal and vascular disturbances in that pathology and the fact that vascular reactivity can be itself modified by Mg imbalance [36]
We have to face a very similar situation concerning the colonopathy. In that often painful pathology, the role of magnesium has sometimes been mentioned. However, most of the studies in that field are quite old and none of them is fully conclusive according to the current methodological criteria for clinical trials. On the other hand, the less advanced understanding about mechanisms involved in the visceral pain (in comparison with somatic pain) is clearly a handicap.

4.2 MAGNESIUM AND PAIN THRESHOLD IN MAN:

If we consider the effect of Mg on pain thresholds, independently of the underlying diseases, we have to distinguish two very different situations:

4.2.1 *Firstly, does a Mg deficiency or a Mg depletion have a clinical incidence on nociceptive threshold in human?*

We have very few arguments, if any, to answer to that question, because as far as we know this point has never been specifically assessed in human. However, one lead is brought indirectly by HLA-Bw35 carriers. As a matter of fact, it has been shown these patients have erythrocyte and plasma magnesium levels lower than those of Bw35 noncarriers [37]. Since several years, it has been described that some of these HLA-Bw35+ patients suffer from a seronegative inflammatory rheumatism with a polyarthralgia predominating in the hands, wrists and knees and/ or a symmetrical acromelic polyarthritis [38]. That rheumatism is often accompanied by non specific and more or less erratic pain which is non far from the symptoms described by the patients suffering from fibromyalgia. It would be interesting to assess the nociceptive threshold in these people by using psychophysical methods to differentiate if pain is related to the rheumatoid symptoms or to a lowered threshold to noxious stimuli. To our knowledge such a study has never been carried out.

4.4.2. *Secondly, do pharmacological doses of Mg have an analgesic effect or do they amplify the efficacy of analgesic compounds?*

The studies which focused on that point are also fairly rare. However, some few data are available, both in animals and in man. In animals, Xiao W.H. & Bennett G.J.[39] showed that systemic Mg^{2+} therapy (magnesium sulfate, 600 mg/kg, s.c.)is able to block the heat-hyperalgesia and mechano-allodynia that are seen in rats with an experimental painful peripheral neuropathy. In these animals, intrathecal (i.t.) injections of Mg^{2+} (185-750 μg) at the level of the lumbar spinal cord significantly reduced heat-hyperalgesia, but perineural application (750 and 7,000 μg) to the site of nerve injury had no effect. Sato E. et al [40] obtained similar antinociceptive effect of intrathecal $MgSO_4$ in Sprague-Dawley rats submitted to the formalin test. In human, a Swiss department of anesthesiology undertook a randomized, double-blind study focused on the role of magnesium sulfate in postoperative analgesia [41]. That clinical trial was carried out in 42 patients undergoing elective abdominal hysterectomy with general anesthesia. The patients received 20% magnesium sulfate or saline (control)15ml intravenously before start of surgery and 2,5 ml/h for the next 20 h. Compared to control subjects, magnesium-treated patients consumed less morphine during the first 48h and experienced less discomfort during the first and second postoperative days. The authors concluded that magnesium could be of interest as an adjuvant to postoperative analgesia.

The background which is put forward for explaining the analgesix effect of pharmacological doses of Mg in both animal and human studies is the role of N-methyl-D-aspartate (NMDA) receptor-gated ion channel as previously described.

4.4.3.*Does the magnesium amplify the effect of some analgesic compounds?*

The last point we would discuss is the potential ability of Mg to increase the magnitude of the effect of some analgesic compounds such as opioids. A lot of physicians reported an improvement of relief obtained in painful patients by combination of morphine and Mg salts used by infusion. These are only cases reports and no specific study has been undertaken to support that point except the previous mentioned Swiss study [41] which reported that cumulative morphine consumption after 48 h was 30% less in magnesium treated patients compared to control patients.

5. Conclusion

To conclude, previous published data showed that Mg-deficiency is able to cause a generalized inflammation and on that Mg injection shows antinociceptive properties in animal models. However, to our knowledge, the hyperalgesia induced by an depletion of Mg has never been described. The variations of intra- and extracellular mg content has often been put forward in several painful diseases in man. However the influence of mg status in these symptoms would need to be more deeply investigated. Some recent clinical results about the role of mg in analgesia are encouraging and we can hope that new clinical data will clarify the potential role of magnesium in that field. Our experimental model using M-depleted rats could provide some help 1) to clarify the relationships between inflammation and pain induced by Mg deficiency. 2) more generally, to better understand the mechanism of action of Mg on nociceptive pathways both in Mg deficiency and in pharmacological use Mg as adjuvant to analgesia.

References

1. Kruse, H.D., Orent, E.R., Mc Collum, E.V. (1932): Studies on magnesium deficiency in animals.I. Symptomatology resulting from magnesium deprivation. *J. Biol. Chem.*, **96**: 519-539.
2. Bois, P., Gascon, A., Beaulnes, A. (1963) Hislamine-liberating effect of magnesium deficiency in the rat. *Nature,* **197:** 501-502.
3. Nishio A., Ishiguro S., Ikegaki I., Matsumoto S., Yoshimitsu F. and Miyazaki A. (1988): Histamine metabolism and pinnal hyperaemia during magnesium deficiency in rats, *Magnes. Res*, **1**: 155-161.
4. Rayssiguier, Y., Malpuech, C., Nowacki, W., Daveau, M., Lebreton, J.P., Rock, E., Mazur, A., Geueux, E. (1994), Evaluation of the inflammatory state during magnesium deficiency in the rat, *Magnes. Res.*, 7, 51
5. Kimball, E.S. (1991): *Cytokines and inflammation*, CRC Press.
6. Weglicki W.B., Phillips T.M. (1992); Pathobiology of magnesium deficiency; a cytokine/neurogenic inflammation hypothesis, *Am J Physiol*, **263**: R734-R737.
7. Weglicki W.B., Phillips T.M., Freedman A.M., Cassidy M.M., Dickens B.F.(1992): Magnesium–deficiency elevates circulating levels of inflammatory cytokines and endothelin, *Mol Cell Biochem,* **110**: 169-173.
8. Weglicki W.B., Stafford R.E., Dickens B.F., Mak I.T., Cassidy M.M., Phillips T.M., (1993):Inhibition of tumor necrosis factor-alpha by thalidomide in magnesium deficiency, *Mol Cell Biochem,* **129**; 195-200.
9. Rayssiguier, Y., Gueux, E., Bussiere, L., Durlach, J., Mazur, A. (1993):Dietary magnesium affects susceptibility of lipoproteins and tissues to peroxidation in rats, *J. Am. Coll. Nutr.* **12**, 133-137.
10. Rayssiguier, Y., Durlach, J., Gueux E, Rock, E. and Mazur, A. (1993), Magnesium and ageing I. Experimental data: importance of oxidative damage, *Magnes. Res.* **6**: 369-378.
11. Durlach, J., Durlach, V., Bac, P., Rayssiguier, Y., Bara, M., Guiet-Bara, A., (1993): Magnesium and ageing II Clinical data: aetiological mechanisms and pathophysiological consequences of magnesium deficit in the elderly, *Magnes. Res.* **6**: 379-394.
12. Weglicki, W.B., Phillips T.M., Mak, I.T., Cassidy M.M., Dickens B.F., Stafford R., Kramer, J.H.(1994): Cytokines, neuropeptides, and reperfusion injury during magnesium deficiency, *Annals.New York Academy Sc.*, **723:** 246-257.
13. Meller, S.T., Gebhart, G.F.(1993): Nitric oxide (NO) and nociceptive processing in the spinal cord. *Pain*, **52**: 127-136.
14. Rock, E., Astier C., Lab, C., Malpuech, C., Nowacki ,W., Gueux, E., Mazur, A., Rayssiguier, Y. (1995): Magnesium deficiency in rats induces a rise in plasma nitric oxide, *Magnes. Res.* **8**; 237-242.
15. Rock E., Astier ,C., Lab ,C., Vignon ,X., Gueux E., Motta, C., Rayssiguier, Y.(1995): Dietary magnesium deficiency in rats enhances free radical production in skeletal muscle, *J. Nutr.*, **125**: 1205-1210.
16. Peplow, P.V.(1992): Modification to dietary intake of sodium, potassium, calcium, magnesium and trace elements can influence arachidonic acid metabolism and eicosanoid production, *Prostaglandins Leuko trienes and Essential Fatty Acids,* **45,** 1-19.
17. Weglicki, W.B., Mak, I.T., Kramer, J.H, Dickens, B.F., Cassidy, M.M., Stafford R.E., Phillips T.M. (1995): Role of free radicals and substance P in magnesium deficiency, *Cardiovasc Res.* **31**: 677-682.
18. Weglicki W.B., Dickens B.F.,Wagner T.L., Chmielinska, J.J., Phillips, T.M.(1996): Immunoregulation by neuropeptides in magnesium deficiency ; ex vivo effect of enhanced substance P production on circulating T Lymphocytes from magnesium-deficient mice, *Magnes. Res.* **9**: 3-11.

19. Prouvost-Danon A., Larvor, P., Rayssiguier, Y et al., (1975): Taux serique d' anticorps reaginiques (IgE) chez la souris en carence magnesique, *Rev. Fr. Allergol.* **I5**: 147-150.
20. Kawanobe, Y. and Sakamoto, M. (1989): Complement systems and phagocytic activities in magnesium deficient rats, *Magnes. Res.* **2**: 81
21. Iseri, L.T. and French, J.H. (1984): Magnesium: Nature's physiologic calcium blocker, *Am. Heart J.* **108**; 188-193.
22. Hill, R.G.(1991); Criteria for the development of analgesic drugs: a personal view from the perspective of the pharmaceutical industry. In ; A.I. Basbaum and J.M. Besson (Eds), *Towards a New Pharmacotherapy of Pain*, John Wiley and Sons, Chichester, pp. 293-329.
23. Haves, A.G., Scadding, W., Skingle, M., Tyers, M.B. (1981): Effects of neonatal administration of capsaicin on nociceptive thresholds in the mouse and rat, *J. Pharm. Pharmacol.* **33**: 183-185.
24. IASP (1994): Classification of Chronic Pain, 2d Edition, IASP Task Force on Taxonomy, H. Merskey and N. Bogduk (eds.), *IASP Press*, Seattle, pp. 209-214.
25. Dubray, C., Alloui, A., Bardin, L., Rock E., Mazur, A., Rayssiguier, Y., Eschalier, A., Lavarenne, J. (1997): Magnesium deficiency induces an hyperalgesia reversed by the NMDA receptor antagonist MK801, *Neuroreport*, **8**: 1383-1386.
26. Woolf, C.J., Thompson, S.W.N.(1991): The induction and maintenance of central sensitization is dependent on N-methyl-D-aspartic acid receptor activation; implications for the treatment of pos-injury hypersensitivity states. *Pain,* **44**: 293-299.
27. Dickenson, A.H., (1994): NMDA receptor antagonsists as analgesics, In: HL Fields, and JC Liebeskind, (eds.) Prog. *Pain research and Management. Seattle:* IASP Press, pp. 173-187.
28. McMahon, SB., Lewin, G.R., Wall, P.D (1993): Central excitability triggered by noxious inputs. *Current Opinion Neurobiol.* **3**: 602-610.
29. Kawajiri, S., Dingledine, R. (1993): Multiple strucrural determinants of voltage-depended magnesium block in recombinant NMDA receptors. *Neuropharmacology,* **32:** 1203-1211.
30. Arancio, O., Yoshimura, M., Murase, K., MacDermott, A.B. (1993): The distribution of excitatory amino acid receptors on acutely dissociated dorsal horn neurons from postnatal rats, *Neuroscience*, **52**; 159-167.
31. Henley, J.M., Jenkins, R., Hunt, S.P. (1993): Localisation of glutamate receptor binding sites and mRNAs to the dorsal horn of the rat spinal cord, *Neuropharmacology*, **32**: 37-41.
32. Burnashev, N., Schoepfer, R., Monyer, H., Ruppersberg, J.P., Gunther, W., Seeburg, P.H., Sakmann, B. (1992): Control by asparagine residues of calcium permeability and magnesium Blockade in the NMDA receptor. *Science.* **257**; 1415-1419.
33. Bac P., Herrenknecht, C., Dupont, Ch.(1996) effects of magnesium acetyltaurinate and morphine hydrochloride on hot plate test in magnesium deficient mice, *XXIVeme Colloque Francais Annuel sur le magnesium, November 1996.*
34. Poenaru, S., Rouhani, S., Durlach, J., Aymard, N., Belkahla, F., Rayssiguier, Y. Iovino, M. (1984): Vigilance states and cerebral monoamine metabolism in experimental magnesium deficiency, *Magnesium*, **3**: 145-151.
35. Mauskop, A., Altura, B.T., Cracco, R.Q. Altura, B.M.(1996): Intravenous magnesium sulfate rapidly allebiates headaches of various types, *Headache*, **36**: 154-160.
36. Luthringer, C., Rayssiguier, Y., Gueux, E., Berthelot, A. (1988): Effect of moderate magnesium deficiency on serum lipids, blood pressure and cardiovascular reactivity in normotensive rats, *Br J Nutr,* **59:** 243-250
37. Henrotte, J.G., Pla, M., Dausset, J. (1990) HLA- and H-2-associated variations of intra-and extracellular magnesium content, *Proc Natl Acad Sci* USA, **87**: 1894-1898.
38. Laroche, M., Ginesty, E., Serrano, E., Cantagrel, A., Bon, E., Moulinier, L., Mazieres, B. (1997): Inflammatory rheumatism and nasal polyposis, *Clin. Rheumatol.,* **16**: 207-211.
39. Xiao, W.H., Bennett, G.J.(1994): Magnesium suppresses neuropathic pain responses in rats via a spinal site of action, *Brain Res.,* **666**: 168-172
40. Sato, E., Takano, Y., Takano, M., Kuno, Y., and Sate, I. (1996): The antinociceptive effect of intrathecally administered magnesium sulfate in rats, *8th World Congress on Pain August 17-22, 1996, Vancouver,* British Columbia, Canada.
41. Tramer, M.R., Schneider, J., Marti, R.A., Rifat, K.(1996): Role of magnesium sulfate in postoperative analgesia, *Anesthesiology,* **84**: 340-347.

INCREASED APOPTOSIS AND FREE RADICAL PRODUCTION IN THYMUS OF MAGNESIUM-DEFICIENT RATS: IMPLICATIONS TO ENHANCED THYMUS INVOLUTION AND IMMUNITY

A. MAZUR[1], C. MALPUECH-BRUGERE[1], W. NOWACKI[1,2], E. GUEUX[1], J. KURYSZKO[2], E. ROCK[1] AND Y. RAYSSIGUIER[1]

[1]Centre de Recherche en Nutrition Humaine d'Auvergne, Laboratoire des Maladies M,taboliques et Micronutriments, INRA, Theix, 63122 St-GenŠs-Champanelle, France and [2]Veterinary Faculty, Academy of Agriculture, 50 375 Wroclaw, Poland

Magnesium plays an important role in the maintenance of host regulatory mechanisms in inflammation and immunity. It is well known that dietary magnesium deficiency in rodents, and especially in rats, causes inflammation (for review[1,2]). This inflammatory reaction is characterized by peripheral vasodilatation, splenomegaly and leukocytosis. Elevated levels of proinflammatory cytokines and acute phase proteins have also been observed during magnesium deficiency [1,2]. On the other hand several observations on the modifications in the immune response during this deficiency have been reported [3]. Since T lymphocytes play a key role in the immune response in the present study we examined modifications occurred in the thymic gland during experimental magnesium deficiency. The thymus presents marked variations in its structure depending on the age and the condition of the organism as a whole. This organ is largest in embryos and gradually and continuously involutes throughout life. However, this process of normal or age involution may be complicated by the rapid changes of "accidental involution" [4]. Since the pioneering work of Selye (1936) it is known that following noxious stimuli, stress results in thymic involution and lymphopenia. Regarding this knowledge we have performed our studies on young rats in which magnesium deficiency was induced by a deficient diet (35 mg/kg) immediately after weanling. Apoptosis plays a crucial role in the thymus involution. This is a cell death mechanism involved in diverse biological processess, which occurs during normal development and homeostasis in response to specific signals [5]. By this process appropriate cells are selectively eliminated during development. Apoptosis is also involved in many diseases. Deregulation of normal cell death can result in profound consequences. Thus, we have focused special attention on the assessment of the extent of the apoptotic process during magnesium deficiency.
Deficient animals after eight days on the experimental diets presented classical signs of inflammation described during this deficiency. Moreover, reduction of thymus weight was recorded in magnesium-deficient rats. After eight days on the experimental diets, the relative thymus weight was lower in deficient rats as compared to controls (0.28±0.01 vs 0.33±0.01 g/100g b.wt; $p<0.01$). Histological modifications were found in thymuses of deficient rats. As compared to controls, thymuses from deficient animals had increased

T. Theophanides and J. Anastassopoulou (eds.), Magnesium: Current Status and New Developments, 313-315.

number of cells per sq mm in the cortex as well as in the medulla. In the medulla there was an increased proportion of macrophages and neutrophils on the second day of deficiency. In cortex an increased proportion of epithelial reticular cells after eight days on the deficient diet was shown. In thymus and blood, lymphocyte populations studied (LCA, Thy1, CD5, CD4, CD8 determined by immunofluorescence with a FACScan cell sorter) were not significantly affected by eight days of deficiency. The detection of histone-associated DNA fragments (Cell Death Detection ELISA, Boehringer), demonstrating the intranucleosomal degradation of genomic DNA, showed significantly greater values in thymuses from magnesium-deficient as compared to controls after two days on the experimental diets (34% increase). Similarly, the tendency to higher values was also observed after eight days of deficiency. In both groups, observed values were greater after two days post-weanling than after eight days. However, despite the significant increase in the level of histone-associated DNA fragments in magnesium-deficient rats, we were unable to detect DNA fragmentation by electrophoresis and ethidium bromide staining. This could be explained by the fact that although in some experimental models apoptotic DNA fragmentation is easly detected, in other cases apoptosis may be found in only a limited number of cells. In the latter case, conventional ethidium bromide staining of DNA may not be sufficiently sensitive to reveal a classical DNA ladder. Since oxidative mechanisms have been implicated in induction of apoptosis in some cell systems, we have evaluated MDA production, assessed on tissue homogenates after induction of oxidation with $FeSO_4$/ascorbate. In both studied periods, greater MDA production was observed in thymus glands of magnesium-deficient rats (about 2-fold after two days of deficiency and 3-fold after eight days) as compared to controls.

From these results it appears that magnesium deficiency enhances thymus involution in relation with an active process of cell death and free radical damage. The precise mechanisms by which magnesium deficiency enhanced thymic apoptosis remain to be determined. A cell will undergo apoptosis as a result of information received from its environment interpreted in the context of internal information, such as its cell type, state of maturity and developmental history [5]. The external information triggering apoptosis can take many forms, e.g. the appearence or disappearence of hormones or cytokines. During magnesium deficiency modifications in cytokine and hormone profile were reported. Calcium entry into cells appears to be a critical early event in apoptosis. The rise in $[Ca^{2+}]_i$ elicited by low extracellular magnesium could be a trigger of endonucleolytic processes leading to apoptosis. Regarding very early modifications in thymus of deficient animals, further studies should be directed to establish if the thymus is one of the first organs affected by low magnesemia.
From our results the question arises: What may be the consequence of enhanced thymus involution and apoptosis during early magnesium deficiency in young animals for the later stage of life? The modified immune response in magnesium-deficient animals has previously been reported [3]. It may be suggested that the modifications in the thymus can participate to harmful consequences of magnesium deficiency for immunity. Other works have shown that long term (2 months) magnesium deficiency resulted in many rats having a reduced thymus weight but there was also in other rats a high occurrence of thymus enlargement and lymphomas (malignant lymphosarcomas) [6]. This supports the

concept that magnesium is important in oncogenesis and malignancy [7]. The intracellular signals involved in induction of apoptosis are often involved in promotion of proliferation or differenciation in other cellular contexts [5]. It has been discussed [8] that although mature lymphocytes proliferate when subjected to immune stimuli, immature lymphocytes instead undergo apoptosis, when stimulated in the same manner.

In summary, this study shows that magnesium deficiency causes accelerated thymus involution associated with ultrastructural modifications, active process of cell death and free radical production. In the light of these observations, it may be hypothesized that disturbances in magnesium status of short duration can have cellular consequences in the immune system on controlling processes such as cell death, differenciation and multiplication which may have various deleterious issues.

References

1. Rayssiguier Y, Malpuech C, Nowacki W, Rock E, Gueux E, Mazur A. (1997). Inflammatory response in magnesium deficiency. In: Smetana R, editor. Advances in Magnesium Research: magnesium in cardiology, London: John Libbey, pp. 415-21.
2. Weglicki WB, Phillips TM, Tong Mak I, Cassidy MM, Dickens BF, Stafford R, Kramer JH. (1994) Cytokines, neuropeptides, and reperfusion injury during magnesium deficiency. *Ann N Y Acad Sci.*;**723**, 246-57.
3. Larvor P. (1980) Magnesium, humoral immunity, and allergy. In: Cantin M, Seelig MS, eds. *Magnesium in health and disease*, New York: Spectrum Publications, pp.201-204.
4. Bloom W, Fawcett DW. (1968). A textbook of histology. Philadelphia: WB Saunders Comp; p 858.
5. Hale AJ, Smith CA, Sutherland LC, Stoneman VEA., Lonhthorne VL, Culhane AC, Williams GT. (1996) Apoptosis:molecular regulation of cell death. *Eur J Biochem.***236**, 1-26.
6. Alcock NW, Shils ME., Lieberman PH, Erlandson RA (1973). Thymic changes in the magnesium-depleted rats. *Cancer Res.*;**33**, 2196-2204.
7. Walker G M. (1986) Magnesium and cell cycle control: an update. *Magnesium*;**5**, 9-23.
8. Khan AA, Soloski MJ, Sharp AH, Schilling G, Sabatini DM, Li S-H, Ross CA, Snyder SH. (1996). Lymphocyte apoptosis: mediation by increased type 3 inositol 1,4,5-triphosphate receptor. *Science*, **273**, 503-507.

MAGNESIUM DEFICIENCY ENHANCES IMMUNE STRESS RESPONSE IN RATS:A CELLULAR APPROACH

E. ROCK, C. MALPUECH , W. NOWACKI*, C. ASTIER , A. MAZUR AND Y. RAYSSIGUIER.

*I.N.R.A.- Centre de Recherches en Nutrition Humaine, Unit, Maladies M,taboliques et Micronutriments, 63122 Saint GenŠs Champanelle, France and *Veterinary Faculty, Academy of Agriculture, 50375, POLAND.*

1. MAGNESIUM DEFICIENCY AND IMMUNE SYSTEM

Pioneer works of Kruse et al.[1] in 1932 have established that magnesium (Mg) deficiency affected the immune system. Peripheral vasodilatation with hyperemia is the most obvious consequence of inflammatory syndrome associated with Mg deficiency. This was later confirmed by greater degranulation of mast cells and release of histamine and inflammatory mediators in Mg-deficient animals [2]. Since then, other studies further showed that Mg-deficient animals were also more susceptible to immune stress. Sensitized guinea-pigs to ovalbumin and fed Mg-deficient diet were found more responsive to anaphylaxis then controls [3]. Similarly, rats fed Mg-deficient diet were shown to be more sensitive to endotoxic shock as evidenced by higher mortality after lipopolysaccharide (LPS) challenge [4,5]. This increased sensitivity to immune stress seems to be specifically related to the decreased Mg status since pharmacological Mg challenge before inducing anaphylactic or endotoxic stress led to decrease of morbidity and mortality of these animals [5,6].

2. MAGNESIUM DEFICIENCY AND IMMUNE STRESSES IN RATS

Experiments were then undertaken to confirm and extend the sensitivity of Mg-deficient animals to immune stress. Weaning rats fed purified diet and containing either 30 or 1000 mg Mg/kg of diet for eight days were used in these studies.
They were challenged with serial dilution of live E. coli and mortality was determined. With $1.9x10^8$ bacteria all 6 Mg-deficient animals dead within 4 hours whereas 5 of 6 control animals dead after 18 hours. With $1.9x10^7$ bacteria, 5 among 6 Mg-deficient rats dead 6 hours after the challenge whereas all control animals were still alive after 48 hours. These data confirm the greater sensitivity to septic shock of Mg-deficient rats previously evidenced by bacterial wall components [4.5]. Similarly, we studied the mortality using platelet activating factor (PAF) as model for anaphylaxis [7]. A lethal

T. Theophanides and J. Anastassopoulou (eds.), Magnesium: Current Status and New Developments, 317-320.

dose of 100 μg PAF per kg b.w. (i.p.) killed the animals after 20 minutes and 40 minutes after challenging in Mg-deficient and control animals respectively. A lower dose of PAF (25 μg PAF per kg b.wt) led to 30% mortality in control animals after 150 minutes following PAF injection whereas during the same time, 70% mortality was observed in Mg-deficient animals. These data confirmed the greater sensitivity to anaphylaxis previously shown in ovalbumin-sensitized guinea-pigs [6]. Together, these results also extend the concept that Mg deficiency greatly enhanced the response of immune system whatever the stimuli used for that activation.

3. MAGNESIUM DEFICIENCY AND IMMUNE CELLS

The reaction of immune system tightly depend on reaction of immune cells including macrophages, polymorphonuclear leukocytes or mast cells. To investigate the effects of Mg deficiency, we studied the functions of peritoneal cells harvested by washing with saline the peritoneal cavity of control and Mg-deficient rats. Previous studies already showed that the number of cells thus recovered were about two fold greater in Mg-deficient animals as compared to controls; however, the proportion of macrophages was 60 percent in both groups [4]. Cytokines release after cell stimulation by LPS and superoxide anion production resulting from respiratory burst induced by phorbol myristate acetate (PMA), had been assessed [4]. The data were in agreement with in vivo studies showing the higher level of tumor necrosis factor (TNFa) found in LPS injected Mg-deficient animals [4]. They also strongly suggest that macrophages could participate to the enhanced reactivity of Mg-deficient animals challenged with LPS through increased level of TNFa production. Indeed a strong correlation has been established between TNFa level in the serum and lethality [9].

4. MAGNESIUM DEFICIENCY AND CELL ACTIVATION

Another point of investigation concerned the mechanisms underlying immune cell activation during Mg deficiency. Enhanced production of superoxide anion by stimulated cells depends on the activity of NADPH oxidase activity. Recent publication from Aoyagi et al.[10] on isolated protein clearly confirmed a previous observation at the cell level [11], that this activity is activated by low level of Mg and inhibited by higher level. On the other hand, this activity also depends on mediators (LPS, cytokines) which lead to activation by signal transduction processes. To further evaluate the respective role of either direct cell modification or its surrounding, we investigated i) the effect of PMA on peritoneal cells isolated from control and Mg deficient animals after incubation in a standard medium and ii) the effect of incubation of normal cells isolated from standard chow-fed rats with serum from control and Mg deficient animals. In the first condition, the data showed that addition of PMA led to 3 to 4 fold increase of superoxide anion production in cells from Mg-deficient animals as compared with controls. This enhanced oxidative burst has been confirmed using three different ways to estimate superoxide anion: luminol-enhanced chemiluminescence, SOD inhibitable ferricytochrome C reduction and spin trapping technique using electron spin resonance spectrometry. These data clearly show that the cells from Mg-deficient animals possess the potential to be more activated by immune stimulator. This enhanced activation looks like the one described during exposure of neutrophils to endotoxin leading to priming the cells for enhanced release of oxygen metabolites during stimulation, likely through modulation of signal transduction components [12]. The second approach confirmed this hypothesis. Indeed, preincubation of standard chow-fed rat peritoneal cells with 8 % serum from Mg-deficient rats leads to significantly higher level of superoxide anions after PMA stimulation. However, this increase is about 25 % of control activity which is much more lower than 300 to 400 % difference observed between the cells from control and Mg-deficient animals. Together, these experiments on cell activity strongly suggest that both primed cells and priming factors present in the serum should be involved in the enhanced susceptibility to immune stress of Mg-deficient animals.

5. MAGNESIUM DEFICIENCY AND SIGNAL TRANSDUCTION

As mentioned above, NADPH oxidase activation is linked to changes in signal transduction components such alteration in membrane potential (Dy), translocation of cytochrome b or of protein kinase C activity from cytosol to plasma membrane and changes in the concentration of intracellular free calcium ($[Ca^{2+}]_i$). Using peritoneal cells from control and Mg-deficient animals, basal level and changes of $[Ca^{2+}]_i$ induced by PAF were investigated after fura2-AM loading. The basal level of $[Ca^{2+}]_i$ does not differ significantly between the two groups. Addition of PAF induced the expected rise of $[Ca^{2+}]_i$; however, as compared to controls, 2.7 fold increase was observed in cells from Mg-deficient animals (435 nM versus 158 nM for controls). Such an increase paralleled NADPH activity indicating that some modifications occurred at the signal transduced by the activator. An important observation is that basal level of calcium is not significantly enhanced in peritoneal cells of Mg-deficient rats despite of priming the cells should result in rising $[Ca^{2+}]_i$ level [12]. To verify this point, we hypothesized that in an earlier phase of magnesium deficiency, circulating factors involved in cell activation should be lower. Experiments were thus undertaken with peritoneal cells obtained from rats fed Mg-deficient diet for 2 days where no clinical signs still appeared. These cells also showed an abnormal rise in $[Ca^{2+}]_i$ induced by PAF (242 nM versus 158 nM for controls) indicating that transduction signal mechanism is affected at very early stage of Mg deficiency. However, the basal level of $[Ca^{2+}]_i$ is 64 nM for both control and Mg-deficient cells confirming that priming could be secondary to other mechanisms involved in the activation process. One possible mechanism could be a modification in the physical state and/or composition of phospholipids surrounding the membrane receptors and proteins which have been shown to be affected in Mg deficient rats [13].

References

1. Kruse,H.D., Orent, E.R. &Mc Collum, E.V. (1932). Studies on magnesium deficiency in animals. I. Symptomatology resulting from magnesium deprivation. *J. Biol. Chem.* **96**, 519-539.

2. Bois, P., Gascon, A. & Beaulnes, A. (1963) Histamine-liberating effect of magnesium deficiency in the rat. *Nature* **197**, 501-502.
3. Ashkenazy, A., Moshonov, S., Fischer, G., Dan Feigel Caspi, A., Kusniec, F., Nen-Ami Sela & Zor, U. (1990) Magnesium-deficient diet aggravates anaphylactic shock and promotes cardiac myolysis in guinea pigs. *Magnesium Trace Elem.* **9**, 283-288.
4. Rayssiguier, Y., Malpuech, C., Nowacki, W., Rock, E., Gueux, E. & Mazur, A. (1997) Inflammatory response in magnesium deficiency. In R. Smetana, Ed., *Advances in magnesium research,* John Libbey, London,. Vol.1, pp. 415-421.
5. Salem, M., Kasinski, N., Munoz, R. & Chernow, B. (1995)□:Progressive magnesium deficiency increases mortality from endotoxin challenge□: protective effects of acute magnesium replacement therapy. *Crit. Care Med.* **23**, 108-118.
6. Kusniec, F., Fischer, G., Sela, B.A., Ashkenazy, Y., Feigel, D., Moshonov, F. & Zor, U. (1994), Magnesium protects against anaphylactic shock and cardiac myolysis in guinea-pigs. *J. Basic Clin. Physiol. Pharmacol.*, **5**, 45-58.
7. Keith, L.H. & Fischer, R.A. (1992): Regulation of hepatic glycogenolysis and vasoconstriction during antigen-induced anaphylaxis. *Am. J. Physiol.*, **262**, G868-G877.
8. Fraker, D.L., Stovroff, M.C., Merino, M.J. & Norton, J.A. (1988)□: Tolerance to tumor necrosis factor in rats and the relationship to endotoxin tolerance and toxicity. J. *Exp. Med.* 167, 95-105.
9. Aoyagi, K., Takeshige, K., Sumimoto, H., Nunoi, H. & Minakami, S. (1992)□: Role of Mg^{2+} in activation of NADPH oxidase of human neutrophils: evidence that Mg^{2+} acts through G-protein. *Biochem. Biophys. Res. Commun.* **186**, 391-397.
10. Korchak, H.M. & Weissman, G. (1978) Changes in membrane potential of human granulocytes antecede the metabolic responses to surface stimulation. *Proc. Natl. Acad. Sci. USA.* **75**, 3818-3822.
11. Forehand, J.R., Pabst, M.J., Phillips, W.A. & Johnston Jr, R.B. (1989)□: Lipopolysaccahride priming of human neutrophils for an enhanced respiratory burst□: role of free calcium. *J. Clin. Invest.*, **83**, 74-83.
12. Heaton, F.W. & Rayssiguier, Y (1987) Magnesium deficiency and membrane properties. In Altura, B.M., Durlach, J. and Seelig, M.S. Eds.,*Magnesium in cellular processes and medicine*. Basel, Karger, S., pp. 121-130.

EFFECT OF MAGNESIUM ADMINISTRATION ON PROLIFERATION AND DIFFERENTIATION OF LYMPHOCYTES

EVA JAHNOVÁ

Institute of Preventive and Clinical Medicine, Department of Clinical Immunology,83301 Bratislava, Limbová 14, Slovak Republic

1. INTRODUCTION

Normal immune responses involve the interaction of several types of cells and soluble products realeased from these cells. Both immunoglobulin production by B-lymphocytes and the immune response system that is mediated by T-lymphocytes require the cooperation of accessory cells producing various soluble cytokines that mediate immune reactions. Previous reviews[1-3] outlined findings from numerous studies, in vitro and in vivo, that showed the detrimental effects of magnesium deficiency environments on lymphocyte proliferation as well as on the immune related functioning of several kinds of T-cells[4,5]. In the studies on humoral immune response in experimental magnesium deficiency were remarkable reductions reported in several classes of circulating immunoglobulins and in antibody activity[6].

The aim of the present study was to investigate whether or not magnesium supplementation induces changes in immunoglobulin production in vitro and in lymphocytes proliferation after administration magnesium to healthy volunteers.

2. MATERIALS AND METODS

Sixteen healthy volunteers (in age ranging from 18 to 41 years) participated in the study. Special dietary regimen and drug administration were elaborated. Blood samples for immunological parameters investigated were collected from probands participating in "Study of relative bioavailability of different magnesium preparations". In the first part of the study at the begining and then after 7 days (saturation of or ganism by Mg) and after 14 days of the study was detected immunoglobulin production. Measurement of IgG, IgA and IgM production in vitro (ex vivo) was performed in a pokeweed mitogen (PWM) stimulated culture by ELISA and proliferation of lymphocytes with [^{3}H]thimidine incorporation into lymphocytes.

3. RESULTS AND DISCUSSION

In this work an increased PWM-stimulated IgG synthesis in vitro ex vivo was found after 7 and 14 days of treatment by magnesium IgM synthesis only after 7 days of

T. Theophanides and J. Anastassopoulou (eds.), Magnesium: Current Status and New Developments, 321-323.

treatment when the organism was saturated by magnesium. The production of IgA was not changed (Tab.1.). An increased DNA synthesis by inference cell proliferation was observed after 7 days of treatmant by magnesium.The increased proliferative response of lymfocytes to mitogenic lektin has been linked only with PHA and ConA mitogen (Tab.2.) Plasma levels and intracellular magnesium were not increased (unpublished data). Some authors found out that magnesium deficient animals exhibit circulating cytokine levels which are indicative of a generalized inflammatory state[7]. These cytokines play an important role in B cell differentiation to produce immunoglobulins. Signals involved in immunoregulation may affect this process. Thus the cytokines that are raised in magnesium deficiency affect cell growth and differentiation to produce immunoglobulins[2,3]

TABLE1. Effect of supplementation by magnesium on PWM-stimulated synthesis IgG, IgA and IgM in vitro by PBMC of healthy subjects

Days of treatment	0	7	14	P
IgG	824±357	3261±602**	3217±1253**	0.001
IgA(ng/ml)	205±47	451±229	159±34	NS
IgM	379±104	997±187**	620±134	0.001

Results are means±SEM (n=16). Statistical significance was calculated using the Wilcoxon test.

Magnesium ions interact to modulate cAMP activity. Changes in the intracellular concentration of cyclic nucleotides have been shown to determine the state of lymphocyte activation in terms of proliferation or differentiation. Experimentally induced increases in intracellular cAMP levels may stimulate or inhibit antibody production in vitro, depending on experimental conditions.

TABLE 2. Effect of supplementation by magnesium on the mitogen induced ^{3}H thymidine incorporation into lymphocytes

Days of treatment				
Mitogenic lectins	Con (mg/ml)	3H thymine incorporation 9dpmx10^{3})		
Control	-	1.2±0.06	1.53±0.157*	1.06±0.069
PHA	0.25	284.18±19.11	320.06±25.97*	260.06±18.96
Con-A	0.01	143.65±7.49	209.65±13.56**	140.78±8.10
Protein A	0.25	89.31±7.15	114.12±9.61	92.81±7.41
Protein A	2.50	143.06±7.34	166.37±9.04	141.68±9.43
PWM	0.0005	19.80±2.95	25.68±4.18	10.88±2.37
PWM	0.005	52.93±4.67	63.37±4.12	47.68±4.66

Results are means± SEM (n=16). Statistical significance was calculated using the Wilcoxon test.

Some evidence suggest that mitogen may modulate free intracellular magnesium[8,9] and following increased free Mg modulate signal transduction[10] . Mg may play an important role in the regulation of levels of second messenger diacylglycerol. Thus, inbalanced production of diacylglycerol in natural B blast may dictate the altered response of these cells[11-13].

In conclusion,magnesium treatment of healthy volunteers may affects of immunoglobulin production and lymphoproliferative response.

References

1. Galland, L.(1988) : Magnesium and Immune Function: An Overview. New York, USA,7: 290-299
2. McCoy, J.H., Kenney, M.A. (1984): Magnesium and immunocompetence. In R.R. Watson: ed, *Nutrition, disease resistance, and immune function*, New York: Marcel Dekker pp. 223-247.
3. McCoy, J.H., Kenney, M.A. (1987): Magnesium and immune function a rewiew. In B.M. Altura, J. Durlach, eds., *Magnesium in cellular processes and medicine*, M.S. Seelig,. Basel: Karger, pp. 196-211
4. Rijkers, G. T., Henriquez, N., Griffioen, A. M.(1993): Intracellular magnesium movements and lymphocyte activation. *Magnes.Res.* **6.** 205-213.
5. Rijkers, G. T., Griffioen, A. W.(1993): Changes in free cytoplasmic magnesium following activation of human lymphocytes. *Biochem. J.* **289**. 373-377
6. Elin,R.J.(1975): The effect of magnesium deficiency in mice on serum immunoglobulin concentrations and antibody plaque forming cells *Proc. Soc. Exp. Biol. Med.* **148**, 620-624.
7. Weglicki, W.B. Phillips, T.M.; Freedman, A.M.; Cassidy, M.M.;Dickens, B.F.(1992): Magmesium-deficiency elevates circulating levels of inflammatory cytokines and endothelin. *Mol.Cell. Biochem.* **110**: 169-173
8. Gershwin, M.E., Lowney, P., Hurley, L.S., Keen, C. L.(1987): Effect of variable magnesium intake on haematological and immunological parameters in the rat. *Magnesium* **6**,153-154. (Abstract.)
9. NG, L. L., Davies J. E.,, Garrido, M.C.(1991): Intracellular free magnesium in human lymphocytes and the response to lectins. *Clinical Science,* **80**, 539-547.
10. Chien, M.M., Cambier, J.C.(1990): Divalent cation regulation of phosphoinositide metabolism. *J. Biol. Chem.* **265**, 9201-9207.
11. Bois, P. (1963): Effect of magnesium deficiency on mast cells and urinary histamine in rats. *Br. J. exp. Path.* **44**: 151-155.
12. Larvor, P.(1980) : Magnesium, humoral immunity, and allergy.In M.Cantin ,M. S. Seelig ,eds, *Magnesium in health and disease*, New York: Spectrum Publications. pp. 201-204

MAGNESIUM PHYSIOPATHOLOGY
The Portuguese Research

MANUEL JUDICE HALPERN
Instituto Superior de Ciencias da Saude 28825 Monte de Caparica-Portugal

PHYSIOLOGY

Several nutritional enquiries done in Western countries showed that there are a subliminal magnesium intake in the studied population, mainly due to the magnesium content in food. In table 1 we give the magnesium contents of some foods

TABLE 1. Magnesium contents of some foods (Mg/100g edible part)

Carrots, Wine, Yogurt	7-9
Brains, Tomatoes, Beer, Milk, Fruits, Potatoes, Eggs	11-13
Kidneys, Meat, Chicken	19-20
Fresh vegetables, Fresh fruits, Pasta, Ham, Liver	22-25
Bread	28-43
Rice	32
Chorizo	36
Shellfish	44
Cheese	47-56
Squid, Calamares, Spinach	100-101
Beans	153

Laires, Rodrigues and Cunha Monteiro [1]

Tap water can have important amounts of magnesium. Studying the magnesium levels in tap water of 247 Portuguese municipalities MASSANO CARDOSO [2] found that 60% of the municipalities had water low in magnesium (less than 6 mg/l). Only water from South showed values between 6 and 30 mg/l.

Great part of Portuguese thermal waters are also poor in magnesium. From 65 different springs R. VALENTIM [3] found 72% with magnesium lower than 6 mg/l, 14% between 6 and 30, 10% between 30 and 60 and 4% superior to 60.

MAGNESIUM DETERMINATION

The most usual means utilized for magnesium assessment is magnesium serum determination. However as magnesium is essentially an intracellular ion, cellular

T. Theophanides and J. Anastassopoulou (eds.), Magnesium: Current Status and New Developments, 325-328.

determination. However as magnesium is essentially an intracellular ion, cellular determinations would be more discriminative. On account of the difficulties for cellular determination, only erythrocytes are generally used for cellular determination. The reference values for Portuguese population are 1,8 to 2,2% for serum magnesium and 4,5 to 5,6% for globular magnesium [1].
Magnesium determination is not discriminative for magnesium carency evaluation. In a study [4] low serum magnesium was observed only in 54% of patients, low globular magnesium in 89,5% and latent tetany symptoms in 100%.
Great part of serum magnesium is complex and protein bound, being the active form only the ionized magnesium. Actually magnesium electrodes afford us a tool to ionized magnesium determinations. The Portuguese reference values are $0{,}5 \pm 0{,}074$ [5]. The reproducibility of the method is 5-9% and coefficient of variation 1,11. Conservation of sera is very food at +4 or –15 [5].

MAGNESIUM IN PREGNANCY AND DELIVERY

In 273 pregnant women has been observed [6] low serum and globular magnesium values as compared to a reference group of non-pregnant women with the same age. Along pregnancy serum magnesium decreases from 16-19th week until 32-35th with the lowest values at 20-23rd and 28-31st. Globular magnesium decreases from 16-19th week to the end of pregnancy with the lowest values at 16-19th and 28-31st weeks.
In delivery there are not changes of magnesium values [7].
In umbilical cord have been observed the same values of latest weeks of pregnancy [8].

MAGNESIUM AND INSULIN

After an oral glucose test we observed a decrease of magnesium in diabetics when insulin reaches the maximum peak, what means that magnesium can be considered a marker of insulin utilization [9]. These results have not been found in non-diabetics.

PHYSICAL EXERCISE

Physical exercise generally decreases magnesium levels. After 30 minutes of bicycle ergometer exercise there are a decrease of globular magnesium [10]. There are also significant decreases of serum and globular magnesium in swimmers and serum magnesium in volleyball players but not in runners [11].
Magnesium supplementation in swimmers and football players increases magnesium levels and improves performance [11,12].

MAGNESIUM CARENCY

It is usual to consider two forms of magnesium carency-*primary* (low magnesium intake) and *secondary* (iatrogenic and non-iatrogenic). In the primary form we must consider deficiency or depletion if it is improved or not by oral magnesium administration [13].

The principal signs of carency are the signs of spasmophylia. The signs we observed in more than 50% of the patients were palpitations, asthenia, cramps, cephalea, precordialgia and ansiety. In a double blind study we observed the suppression or improvement of the signs [13].

MAGNESIUM IN CARDIOVASCULAR DISEASES

In the fourth say after AMI or UA there are an increase of serum magnesium. For globular magnesium the same results are observed in UA but not in AMI [14]. However at 3 months there are 35,7% of AMI and 37,1% of UA with hypomagnesiemia and at 6 months 22,4 and 51,4 what means that at 6 months magnesium decreases significantly ($p<0,05$). Calcium blockers decrease magnesium levels at 3 and 6 months being the results significant in UA at 6 months [15].
In dysrhythmia there are hypomagnesiemia proportional to Killip-Kimball classes [14]. The intravenous administration of magnesium has a therapeutic effect and increases magnesium levels.
In hypertension has been described [16] 17% patients with low serum magnesium and 71 with low globular magnesium.

MAGNESIUM AND DYSLIPEMIA

We have had contradictory results on this subject. In our first study we described low serum magnesium in hypertriglyceridemia [17] but in another study we did not find significative decreases [18]. Magnesium supplementation decreases triglycerides but not significatively [19].

OTHER DESEASES

In *craneoencephalic traumatisms* 66% of patients showed decrease of globular magnesium [20].
In patients with *lung carcinoma* there are increased levels of magnesium on broncho-alveolar fluid [21].
In *chronic alcoholics* we observed a small decrease of serum magnesium in all patients and an increase after 3 weeks of alcohol privation [22].

References

1. Laires, MJ; P. Rodrigues and A.L. Cunha Monteiro (1980) Magnesio em patologia *Rev. Port. Bioq.* **33**: 145-152.
2. Cardoso, S.M. (1996) Magnesium in tap water. In M.J. Halpern and J. Durlach (eds), *Current Research in Magnesium,* John Libbey, London, pp. 141-146.
3. Valentim, R. Personal communication
4. Fernandes, J.S. (1985) Therapeutic effect of a magnesium salt in patients suffering from mitral valve prolapse and latent tetany *Magnesium* **4**: 283-290
5. Halpern, M.J., S.L. Llobet, C. Monteiro, J.A. Carvalho Rodrigues (1996) Ionized magnesium determination: storage conditions, precision studies and reference values. In M.J. Halpern and J.

Durlach (eds) *Current Research in Magnesiu,* John Libbey, London, pp. 5-6.

6. Pedro, E.; A. Sampaio; M.J. Halpern (1985) Contribuicao para o estudo do magnesio na gravidez.*Mom. Obst. Gin.* **3**: 173-176.
7. Pedro, E.; A.T. Sampaio; A.C. Costa; M.J.Halpern (1986) Estudo da magnesiemia e calcemia durante o trabalho de parto. *Mom. Obst. Gin.* **4**: 183-185.
8. Pedro, E.; A.T. Sampaio; J.V. Amaral; O. Aparicio; M.J. Halpern (1986) Estudo da magnesiemia plasmatica e globular do cordao umbilical. *Mom. Obst. Gin.* **4**: 186-188.
9. Pereira, T. (1983) Magnesium as a possible indicator of insulin utilisation by cells. *Magnesium* **2**: 248-249.
10. Laires, M.J. (1992) Magnesium intake and exercise. *Motricidade Humana*8: 57-66
11. Laires, M.J., Monteiro, C.P., Sergio, J., Madeira, F., Saihnas, J. (1994) Content trace elements in blood before and after three standard efforts, In Ph. Collery, L.A.Poirier, N.A. Littlefield, J.-C. Etienne (eds.), *Metal ions in Biology and Medicine*, John Libbey, Paris, Vo.3, pp. 557-563.
12. Laires, M.J., Saihnas, J., Fernandes, J.S. (1992) Indices sanguineos de magnesio. Efeitos de uma suplementacão num exercicio submaximo. *Rev. Port. Med. Desp.* **10**: 203-208.
13. Halpern M.J. (1995) Magnesium physiopathology. In D. Petru and G. Zeno (eds) *Metal elements in environmental and biology Mirton*, Timisoara pp. 133-136.
14. Pereira, J.N.; C. Rabacal; M.J. Laires; T. Pereira; J.S. Fernandes; M.J. Halpern (1988) Serum and red Blood cell Mg levels in acute coronary events. *Magnesium* **7**: 9-15.
15. Franca, A.; J. Nogueira Pereira; Maria J. Laires; Telma Pereira; Carlos Rabacal; Manuel J. Halpern; Jose Simoes Fernandes (1985) Serum magnesium levels in patients after myocardial infarction or unstable angina treated with nitrites and calcium antagonists. *Magnesium* **4**: 291-294.
16. Ruivo, J.; A. Arsenio Personal communication.
17. Laires, M.J.; Ph Chappuis; A.L. Cunhua monteiro; F. Rousselet; M.J. Halpern; A. Legrand. (1979) Etude de la calcemie et de la magnesemie au cours des hyperlipoproteinemies. *Rev. Fr. Endocr. Clin.* **6**: 551-557.
18. Laires, M.J.; T.G. Pereira; M.J. Halpern (1983) Magnesium and hyperlipemias. *Magnesium* **2**: 250-252.
19. Fernandes, J.S.; T.G. Pereira; J. Linder; C. Rabacal; E. Carvalho; J.S. Afonso; J.N. Pereira; F. Mesquita; M.J. Halpern. Magnesium and dyslipemia: a clinical perspective. In M. Fatima Reis, J.M. Pereira Miguel; A.A.S.C.Machado, M. Abdula (eds) *Trace elements in medicine, health and atherosclerosis,* Smith Gordon, Nishimura.
20. Pereira, T.; M.C. Botas; M.J. Laires; M.J. Halpern (1986) Estudo das alteracoes do magnesio e calcio sericos e do magnesio globular nos traumatisados craneo-encefalicos sujeitos a hiperventilacao. *Bul. Clin. Hosp. Civis Lisboa* **43**: 16-20.
21. Laires, M.J.; J. Ruas Silva; A. Teles Araujo; A.L. Cunha Monteiro; M.J. Halpern (1981) Magnesium broncho-alveolaire dans les affections pulmonaires cancereuses ou non. *Magnes. Bull.* **1**: 20-23.
22. Laires M.J.; M. Lucilic de Mello; Orlandina Maia; J. Ruas da Silva; T. Pereira; M.J. Halpern (1985) Hypomagnesemia in chronic alcoholismIn M.J. Halpern; J. Durlach (eds) *Magnesium deficiency in Karger, Basel.*

A COMPARATIVE STUDY OF SERUM Mg^{++}, Ca^{++}, CHOLESTEROL AND TRIGLYCERIDES LEVELS DURING TOTAL INTRAVENOUS ANAESTHESIA WITH PROPOFOL VERSUS THIOPENTONE/N_2O ANAESTHESIA

HARA NASTOU[1], MARIE PALTA[1], GEORGE SARROS[1], ALEXANDER NASTOS[1], SOPHIA NIKOLAIDOU[1], VASSILIKI SARROU[1], NIKOS FOTOPOULOS[2], AND JANE ANASTASSOPOULOU[2]

[1] Ophthalmiatrio of Athens, Anaesthesiology Department 26 Panepistimiou Str., 106 72 Athens, [2]National Technical University of Athens, Chemical Engineering Department, Radiation Chemistry and Biospectroscopy, Zografou Campus, Zografou, 157 80, Athens, Greece.

1. INTRODUCTION

The fact that anaesthetic agents influence blood chemistry has been investigated and proven by many investigators in the past. However, literature referring to the effect of the anaesthetic agents in plasma lipids as well as in serum magnesium levels is still limited in comparison with literature about potassium and calcium. It is well known that many steps in calcium homeostasis are dependent on Mg and hypomagnesemia is also very often accompanied by hypocalcemia. Magnesium is also considered to be a natural calcium antagonist affecting uptake, content, binding, and distribution of calcium in smooth muscle cells [1]. Experimental studies have shown the influence of acute Mg deficiency on lipid metabolism [2]. The investigation of these changes referring to the recent anaesthetic agents is necessary for the understanding and the prevention of side effects during anaesthesia. Propofol, a substituted fenol, is used as a formulated soyabean glycerol, egg-phosphatide emulsion. Prolonged administration of propofol in its current formulation could lead to abnormalities in triglycerides and cholesterol metabolism as has been observed during administration of lipid emuslions for total parenteral nutrition [3].

The purpose of this study was to compare serum Mg, Ca, cholesterol and triglycerides changes during total intravenous anaesthesia with propofol versus thiopentone -N_2O-opioide general anaesthesia.

T. Theophanides and J. Anastassopoulou (eds.), Magnesium: Current Status and New Developments, 329-333.

2. METHODS

Forty five patients aged 40-80 years old, ASA I-II, were studied. They underwent a retinal detachment surgery or vitrectomy under total intravenous anaesthesia with propofol (group A) or general anaesthesia with thiopentone -N_2O which lasted from two to two and a half hours. All patients had received 5 mg i.v. midazolame for premedication, five min. before induction in anaesthesia.

In group A, anaesthesia was induced with 1-2 mg/kg i.v. propofol according to age of the patient and tracheal intubation was performed after administration of 1mg/kg i.v. succinylcholine. Anaesthesia was maintained with continuous infusion using 6mg/kg/h propofol in a syringe pump and 33% oxygen enriched air.

In group B, the induction was done with 5mg/Kg thiopentone while maintenance with N_2O:O_2/4:2. Muscle relaxation as well as analgesia was the same for both groups. All infusions were discontinued 15 min before the end of the operation.

In both groups, neuromuscular block was successful with a 50mg of atracurium in a bolus dose i.v., followed by a 1mg/kg/h continuous infusion while analgesia was done with 1mg alfentanyl in a bolus dose immediately after induction in anaesthesia and continuous infusion of 0.3mcg/kg/h.

Blood samples were taken for the determination of serum magnesium, calcium, cholesterol and triglycerides levels which was done by the method of dry chemistry with a KODAC EKTACHEM DT60 Analyzer. Sampling was performed: before premedication, after induction, after one hour, after two hours, 10 min after recovery and 24 hours after surgery, respectively.

Statistical analyses were done using the multivariate analysis of covariance (MANCOVA) for repeated measures.

3. RESULTS:

Tables 1,2,3 and 4 show the mean and SD of serum Mg, Ca, triglycerides as well as cholesterol respectively.

The multivariate analysis of variance (MANOVA) was used to explore whether the profiles of Mg, Ca, triglycerides and cholesterol were different between the two groups.

<u>Magnesium:</u> There were statistically significant differences in the profile of Mg between the two groups ($F(5,39)=22,64$, $p<.001, \eta^2=.75$). Specifically, it was found that Mg fell significantly at the 2nd time interval in the propofol group ($F(1,43)=17.38$, $p<.001$, $\eta^2=.29$). It remained at about the same level until the 5th time interval and then increased significantly at the 6th one ($F(1,43)=20.20$, $p<.001$, $\eta^2=.32$). On the contrast, there were not significant changes on Mg in the thiopentone group, at any time interval.

<u>Calcium:</u> No statistically significant difference was found between the two groups concerning the profile of Ca ($F(5,39)=2.17$, p>,05).

<u>Triglycerides:</u> The profile of triglycerides was different between the two groups ($F(5,39)=8.16$, $p<.001$, $\eta^2=.51$). Whereas in the thiopentone group there were not

TABLE 1. Mean and SD of serum Mg (mg%) fluctiation in both groups at each time interval

	before premedic.	After induction	1 hour after	2 hours after	after recovery	after 24 hours
	mean (SD)	mean (SD)	mean (SD)	mean (SD)	mean (SD)	mean (SD)
propofol	1.81 (.21)	1.52 (.22)	1.46 (.26)	1.42 (.21)	1.50 (.23)	1.87 (.18)
thiopentone	1.38 (.22)	1.41 (.20)	1.34 (.19)	1.45 (.24)	1.43 (.27)	1.47 (.25)

TABLE 2. Mean and SD of serum Ca (mg%) fluctiation in both groups at each time interval

	before premedic.	After induction	1 hour after	2 hours after	after recovery	after 24 hours
	mean (SD)	mean (SD)	mean (SD)	mean (SD)	mean (SD)	mean (SD)
propofol	8.64 (.77)	8.09 (.57)	7.97 (.69)	8.13 (.77)	8.37 (.79)	8.51 (.43)
thiopentone	8.81 (1.24)	8.53 (.75)	7.89 (1.06)	8.59 (.87)	8.27 (1.30)	8.80 (.81)

TABLE 3. Mean and SD of serum triglycerides (mg%) fluctiation in both groups at each time interval

	before premedic.	After induction	1 hour after	2 hours after	after recovery	after 24 hours
	mean (SD)	mean (SD)	mean (SD)	mean (SD)	mean (SD)	mean (SD)
propofol	158.53 (113.07)	193.37 (123.49)	506.97 (370.16)	469.63 (383.05)	339.25 (61.55)	185.60 (165.42)
thiopentone	144.06 (64.16)	118.27 (50.17)	110.73 (54.96)	116.07 (56.81)	165.42 (67.03)	115.27 (67.04)

TABLE 4. Mean and SD of serum cholesterol (mg%) fluctiation in both groups at each time interval

	before premedic.	After induction	1 hour after	2 hours after	after recovery	24 hours after
	mean (SD)	mean (SD)	mean (SD)	mean (SD)	mean (SD)	mean (SD)
propofol	194.87 (54.55)	174.80 (51.04)	150.93 (51.57)	157.97 (44.84)	165.80 (48.77)	179.90 (37.34)
thiopentone	155.33 (42.58)	148.40 (30.92)	135.40 (39.41)	130.53 (33.83)	1150.53 (33.88)	158.06 (32.07)

significant changes, Triglycerides remained at a high level in the propofol group until the 5th time interval and fell significantly at the 6th one ($F(1,43)=4.31$, $p<.05$, $\eta^2=.09$).
Cholesterol: No statistically significant difference was found between the two groups concerning the profile of cholesterol. ($F(5,39)=2,19$, $p>.05$).
The multivariate analysis of covariance for repeated measures (MANCOVA) was used to investigate the relationship of Mg with Ca, cholesterol and triglycerides between the two groups (propofol-thiopentone).
Mg and Ca: At the multivariate level it was found that magnesium and calcium covaried differently between the two groups ($F(1,214)=8,33$ $p<.01$, $\eta^2=.04$). However, the a posteriori comparisons did not show that Mg and Ca covaried significantly at a specific time interval between the two groups. For this reason the statistically significant difference that was found at the multivariate level is considered to have occurred by chance.
Mg and triglycerides: No statistically significant differences were found between the two groups concerning the covariance of Mg and triglycerides in each time interval ($F(1,214)=0.58$, $p>.05$).
Mg and cholesterol: No statistically significant difference was found between the two groups concerning the covariance of Mg and cholesterol in each time interval ($F(1,214)=.36$, $p>.05$).

4. DISCUSSION

In this study, a reduction in serum magnesium levels was noted in the propofol group during anaesthesia. Specifically: there was a sudden decrease of magnesium after induction in anaesthesia which lasted for the following two hours, until the time of the discontinuation of the drugs which coincided with the 4th sample, when it started to increase, reaching the normal levels the next day. No changes were found in the thiopentone group.

Serum calcium levels, in the propofol group, were decreased in a lesser degree and they reached normal, immediately after discontinuation of the drugs. A small decrease was also noted in the thiopentone group. In a previous study, with urethane and halothane anaesthesia, in rabbits, the decreased plasma levels of Ca were not normalized 2 hours after recovery whereas Mg was increased[4].

It is well known that plasma lipids, as well as, magnesium and calcium cations interact in many metabolic processes. The possibility of lipid emulsions administration adverse effects on serum lipid levels, has been reported in a number of papers describing the use of the anaesthetic for continuous sedation in intensive care, for periods of up to 5 days. Other investigators, in neurosurgical procedures, did not find significant changes in serum lipids after propofol anaesthesia[5]. However, in our study the results showed a statistically significant increase in serum tryglicerides in the propofol group, which was restored to the normal levels by the next day; while no changes were found in the thiopentone group. Serum cholesterol showed a slight decrease in both groups; these changes were not related to magnesium and calcium changes.

None of the patients developed any clinical syndrome related to the sudden changes in the above parameters for the following five days from surgery.

In conclusion, while in thiopentone group no significant changes were found, the transient increase in serum triglycerides, as well as, the decrease in serum magnesium and calcium at the propofol group, seem to be clinically insignificant,.

References:

1. Nastou, H., Sarros, G., Palta, M., Nastos, A., Sarrou, V., Anastassopoulou, J. (1996). Fluctuations in serum magnesium levels during total intravenous anaesthesia with propofol, *Magnesium Res.*, **9**, 47-53 and references therein.
2. Inagaki O., Fugita Y. (1991) Long term effect of magnesium on lipid metabolism in uraemic rats. *Magnesium Research*, **4,** 191 - 195.
3. Schlichtig R, Ayres SM, (1988): *Nutritional support of the Critically ill.* Chicago: Year Book, Medical Publishers Inc., pp. 169-184.
4. Bito L.Z. & Eakins K.E. (1969): The effect of general anesthesia on the chemical composition of blood plasma of normal rabbits. *J. Pharm. Exp.l Ther.* **169,** 277- 286.
5. Magana J., Leal-Zapata, G.F., Igartua-Gascia L. (1991) Cambios en lipidos sericos, colesterol y trigleceridos secundarios a infusion de propofol para induccion y mantenimiento de la anestesia en and isoflurane. *Acta Anaesthesoil. Scand.*, **35**, 302-305.

EFFECTS OF MAGNESIUM ON SPERMIOGENESIS

S. VISKI[1], J. SZÖLLOSI[1], A.S. KISS[2], A. CSIKKEL-SZOLNOKI[3]
[1]*SZOTE Dept of Obstetrics and Gynaecology, Andrology Laboratory,* [2]*Hungarian Magnesium Society,* [3]*JATE Dept. of Inorganic Chemistry and Analitics, Szeged, Hungary*

1. Introduction

Years ago the women had to face society's blame due to their childless marriages. The problem of infertility has increasingly been extended to anomalies in the male partner. Male fertility may also have various causes: reduced volume of ejaculate, decreases number and motility of sperms and increased proportion of abnormal (injured) sperms. It was found that as a result of ATP and cAMP supply in vitro in demembranated models the length of time of the sperms' motility grew thirtyfold (122.0 min) compared to the control (4.2 min) [1,2]. This is made possible by the fact that the energy required for the motion of sperms is released from the ATP by the ATPase [3]. According to Morisava [4] the motility of sperms needs the joint presence of cAMP and Mg ATP. It was found that the formation of ATP, as well as, cAMP is a magnesium intensive process [5,6]. It was also found *in vivo* that magnesium increases the sperm motility, while the sperm production inrcases up to 80% [7].
It was shown that the magnesium deficient diet reduces the spermatogenesis of rats by 40%; damages the axoneme and increases the production of free radicals [8]. This is maybe due to infertility rate of men has increased for the last half century. On the basis of the above we suppose that by per os magnesium therapy both the amount of ejaculate and the number and motility of sperms can increased. In present paper we give account of these investigation as preliminary data.

2. Methods

The examinations were done as is described (see paper in the same volume). The magnesium concentration both in the blood serum and in the seminal plasma, after exposed with nitric acid. qas determined by atom absorption spectrophotometry (AAS). The seminal plasma was obtained by centrifuging the ejaculate for 10 minutes at 5×10 g. The problematic persons who took part in the examinations were treated with magnesium for 3 months (1 g magnesium-citrate/day in tee), while no additional therapy was during that time. After 3 months the above described ejaculate examinations were repeated. The age of patients ranged between 23 and 32 years, the

T. Theophanides and J. Anastassopoulou (eds.), Magnesium: Current Status and New Developments, 335-339.

body weight between 65 and 80 Kg. The couples suffered infertility for 3 and more years. There were no gynecological reasons for infertility in women. There was no sign of magnesium deficiency in the patients. There were no side effects of magnesium therapy during the treatment.

3. Results

The initial conditions and the self-controlled results obtained after the therapy are shown in TAble 2. Our measuring results show that the magnesium therapy applied resulted in favorable, positive changes in both the amount of the ejaculate and the picture of its spermia as well as in the magnesium level, but only with those patients who had regularly taken the magnesium preparation (SD %). Unfortunately, some of the patients neglected taking the "medicine" regularly, so changes (in the magnesium content of the blood serum and ejaculate, and with the spermia, respectively) compared to initial (self control) values were hardly, if at all found. The proportion of abnormal (Injured) sperms was generally reduced by the therapy, and though in some cases it increased, the number of intact sperms was even then higher than in the ejaculate of the initiate control. Such a case occurred first of all with patients of very low sperm number.

4. Discussion

In previous papers [9-11] we made the ejaculate-,spermium-, and magnesium characteristic of normozoo-, oligozoo- and asthenozoospermic men the mumver of deformed, functionally damaged sperms (30-35%) was two- or threefold of the sperm pattern of the nomozoospermic individuals (10-15%) (Table 1). On the bases of the cell

TABLE 1. Summary of human sperm analysis.

	Normozoospermia	oligozoospermia	asthenozoospermia
patient No.	80	30	22
spermium count x million	60±20	11±10	20±10
motility %	70±10	50±10	40±10
abnormality %			
head %	5±5	10±5	10±5
midpiece %	5±5	15±5	15±5
tail %	5±5	8±5	10±5
total abnormality %	15	33	35
magnesium mM (semen plasma)	4.5±0,8	3,3±0,8	2,0±1,2

TABLE 2. Effect of magnesium treatment cn the ejaculatum

Name (Kg)	Treatment	Age year	Body weight Kg	Human Plasma (mM)	Semen	Ejaculatum ml	Spermium Count x milion	Motility%	Abnormality%
Sz.F	Control	23	78	0,86	3,83	6	18	10	30
	Mg			0,96	4,50	7	64	20	30
T.J	Control	27	66	0,80	2,66	5	1	20	50
	Mg			0.92	4.00	3	5	10	30
K.Z	Control	24	65	0.94	3.41	3	10	20	30
	Mg			0,88	4.58	4	20	60	50
M.F.	Control	30	72	0,7	2.41	6	2	0	60
	Mg			0,86	4.16	6	22	60	30
K.G.	Control	28	68	0,98	3.45	8	3	20	30
	Mg			0,98	3.83	6	4	40	45
R.Ch.	Control	31	75	0,85	1.95	4	3	70	25
	Mg			0,95	3.00	5	20	60	30
K.L.	Control	25	77	0,89	3.75	3	5	10	50
	Mg			0,94	4.08	2	11	60	40
H.I.	Control	32	80	0,86	3.29	0.2	0.5	0	100
	Mg			0,98	3.95	3	20	30	50
	Mean value	27	72						
	control			0,86	3.09	4.4	5.3	16	46.9
	Mg			0,93	4.01	4.5	14.5	42.5	38.1
	SD 5%			0,02	0.04	0.49	1.5	10.4	6.1

Correlation coefficient (r) between the human and semsn plasma is r<0.5
Control means before th therapy

biology studies we suggest that the changeswhich were found in the functionally damaged sperms are mainly caused by the free radicals. This statement agrees with the findings of other investigators, that a high amount of active oxygen present in the ejaculate has a damaging effect on the function of the sperm. Peroxidation of multiple-unsaturated fatty acids of the membrane increases its permeability. There was also observed an increase in the calcium ions entering the cell (e. g. in A23187 ionophore or Mg deficiency) induces the formation of active oxygen (free radical) from NADPH through the diaphorase (NADPH-oxydase) enzyme:

$$NADPH + O_2 \longrightarrow NADP^{++} \quad H^{+}+O_2^{-}$$

In magnesium deficiency, in the case of mitochondrial oxidative phosphorylation is uncoopled [5,6] the accumulation of NADPH increases whereby the formation of free oxygen radical becomes more possible. Magnesium deficienvy has same effect.
The flow of glucose in the direction of the so-called hexose-phosphate shunt (pentose-phosphate shunt) increases together with the calcium ion content of cells, wherby the concentration of NADPH grows. The functioning of the entose-phosphate shunt is particularly significant in the testicles.
The activity of ATP-ase, decreases with peroxidation by 20-50%. That is why other investigators found the motility and vitality of sperms to decrease to a considerable extent with an increase in the activity of diaphorase enzyme. Badwey [12] had noticed the increases of concentration of free oxygen radicals produced by the diaphorase enzyme is prevented by the superoxid dismutase (SOD). This is due to the elimination of the free radicals produced by the SOD and catalase (free radicals scavenger) enzymes rather than to the inhibition of free radical formation.
The lipid peroxidation, so membrane injury also increases by the iron ion, whereby both the spermatogenesis and the axonema complexus sustains an injury. The iron accumulation increases by the magnesium deficiency. Matkovics et al [13] found that as a function of an increasing magnesium supply the activity of superoxide dismutase (SOD) and glutathion (GSH) grew by 30-60% in the Drosophilae. In experiments conducted by *Kiss* [5,9] the catalase activity of various tissues increased by 10-50% as a result of magnesium treatment
Our investigations prove that the quantitative and qualitative parameters of the ejaculate can be fabourably influenced by magnesium therapy. Data from the literature and our own results give information on the possible action mechanism.

6. Conclusion

According to the data of our controlled experiment, under the influence of a proper magnesium supply (-therapy) the magnesium concentration increases in all parts of the organism, inluding the generative organs. The pentose-phosphate shunt and so the oxygen free radical formation will thereby be inhibited, and the ATP-ase activity and ATP formation increase, respectively. In response to the magnesium therapy the activity of enzymes eliminating the free oxygen radicals increasing. As a consequence of the above the number, structural soundness and motility of sperms grow, which may mean an increase in fertility due to the magnesium therapy.

References

1. Treetipsatit N., Chulavatnatol M. (1982) Effects of ATP cAMP and pH on initial of flagellar movement in demembraned models of rat epidymal spermatozoa. *Exprtl. Cell. Res.* **142**: 495-499.
2. Mohri Hm, Yanagimachi R. (1980): Characteristics of motor apparatus in testicular epidymal and ejaculated spermatozoa, *Exptl Cell. Res.* **127**: 191-196.
3. Nelson L. (1959) Cytochemical studies with the electron microscope. I. ATP in rat spermatozoa, *Biochim. Biophysic. Acta* **27**. 639 -641.
4. Morisawa M. (1982) Cyclic AMP induced maturation of trout sperm axoneme to initiate motility. *Nature* **295**: 703-704.
5. Kiss A.S. (1983): Magnezziumtragyazas, magnezzium a biologiagan, Mezogazdasagi Kiado Budapest.
6. Balla A., Kiss A.S. (1996): Magnezium a biologiaban, magnezzium a gyermek-gyogyaszatban, Szekelyudvarhely.
7. Kovacsne-Gaal, K. (1994) The importance of magnesium feeding in animal nutrition.In S.Golf (ed.) Magnesium, John Libey, london, pp 115-120
8. Gunther, T. Hollriegl, V. Merker, H.J., Vormann, J., Schumann, K. (1995). Lipid peroxidation and morphology of testis in magnesium deficiency. *Magn. Bull.* **17**, 132
9. Kiss I, Kiss S.S. (1994) Mutagenesis experiments in the presence of Magnesium *Symposium, Balatonszeplak 15.*
10. Kiss A.S., Kiss I. (1995) Effect of magnesium ions on fertility sexx ratio and mutagenesis on Drosophila melanogaster males, *Magnesium Res.* **8**: 243-247.
11. Viski S., Szollosi J, Kiss A.S., Csikkelne-Szolnoki A. (1995): A magnezium szerepe a ferfiak fertilitasaban, Szegedi Akademiai, Bizzottsag "Korneyezzetunk es az Egeszseg Anket" Elondas osazefoglaloi, Szeged.
12. Badwey, J.A., karnovsky, M.L.(1979), Production of superoxide and hydrogen peroxyde by an NADH-oxydase in guina pig polymorphonuclear leucocyte. J. Biol. Chem., 254, 11530-11537
13. Matkovics, B, Kiss, I., Kiss, A.S. (1996), A magneziumkezeles hatasa a Drosophila melanogaster enzimakti-vitasara, 6th Hugarian Magnesium Symposium, Balatonszeplak.

CORRELATION BETWEEN HUMAN SEMEN Mg AND HYDROXYL RADICAL LEVELS AND HUMAN SPERMIOGRAM

A.S. KISS[1], B. MATKOVICS[2], A. CSIKKEL–SZOLNOKI[2], J. SZÖLLÓSI[3], AND S. VISKI[3]
[1]Hungarian Magnesium Society, [2]JATE Dept. of Inorganic Chemistry and Analytics and [3]SZOTE Dept. of Obstetrics and Gynecology, Andrology Laboratory, Szeged, Hungary

1. INTRODUCTION

For some decades the problem of infertility has increasingly been extended to anomalies in the male partner. Male fertility may also have various causes: reduced volume of ejaculate decreased number and motility of sperms and increased proportion of abnormal (injured) sperms. It was found that as a result of ATP and cAMP supply *in vivo* in demembranated models the length of time of the sperms' motility grew thirtyfold (122.0 min) compared to the control (4.2 min) [1,2]. This is made possible by the fact that the energy required for the motion of sperms is released from the ATP by the ATPase [3]. According to Morisava [4] the motility of sperms needs the joint presence of cAMP and MgATP. It was found the formation of both ATP and cAMP to be a magnesium process [5,6]. In our previous experiments with Drosophilae pointed out a close correlation between fertility and the magnesium content of the diet [7,8]; fertility increased compared to the control by 4% with 0.1 M, by 6% with 0.2 M and by 15% with 0.5 M magnesium contained in 1% sugar solution food. On the basis of the above we suppose that by per os magnesium therapy both the amount of ejaculate and the number and motility of sperms can increased.

1. METHODS

The examinations were started with so – called self – control method. For the examinations ejaculates obtained from voluntary childless men married for several years were used. A 72–hour abstaining from sexual life preceding the ejaculate sampling was prescribed. The number and qualitative characteristics (motility, abnormality) of the sperms were determined with the usual microscope technique. The abnormality appears in the membrane injury of the headpiece, midpiece and tail or in two tails and so on. The injuries come probably in consequence of free radicals. The magnesium concentration both in the blood serum and in the seminal plasma, after exposed with nitric acid, was determined by atom absorption spectrophotometry (AAS).

T. Theophanides and J. Anastassopoulou (eds.), Magnesium: Current Status and New Developments, 341-346.

Lipid peroxidation (LPO) was measured by the method of Plancer. Hydroxyl radicals were analyzed by the method of Malliwell and Gutteridge and the protein concentration was determined by Folin.

2. RESULTS

Table 1 shows the effect of Mg treatment on the ejaculate's Mg content quality and quantity (average of 96 measurements). The influence of Mg on sperm count, sperm motility and normal type sperm-count are shown in Figures 1, 2, and 3, respectivelly. Both motility and normal type sperm – count are in correlation with the ejaculate's Mg content according to a maximum curve. Table 2 shows the effect of Mg on protein concentration, LPO and hydroxyl radicals in the case of 10 patients. The results show a change according to a minimum curve for both parameters.

Table 1 Effect of Mg treatment on the ejaculation.

	Treatment	
	Control	Magneium
Mg content mmol/L		
Human plasma	0.86	0.93
	0.7 – 0.92	0.8 – 1.05
Seminal plasma	3.09	4.81
	2.0 – 9.0	4.0 – 12.0
Ejaculatum ml	3.6	4.5
	0.2 – 6.0	2.0 – 8.0
Count x million	5.3	14.5
	0.5 – 10.0	5.0 – 64.0
Spermium		
Motility %	16	42.5
	0 – 20	20 – 60
Abnormality	46.9	38.1
	30 – 100	30 – 50
n = 96	—	—

3. DISCUSSION AND CONCLUSION

In previous papers [9,10] we made the ejaculation, spermium, and magnesium characteristics of normozoo-, oligozoo- and asthenozoospermic individuals. Our results showed that Mg treatment increased both the quality and the quantity of the ejaculate. The fertile capacity of the ejaculate increased even though the count of normal–type sperms remained unchanged or slightly decreased due to the fact that the total number of sperms was higher than before treatment. According to our observations, the optimal Mg concentration of the ejaculate is 5–6 mmol/l. The curve of hydroxyl free radicals shows a minimum under these conditions; the motility and the normal–type sperm

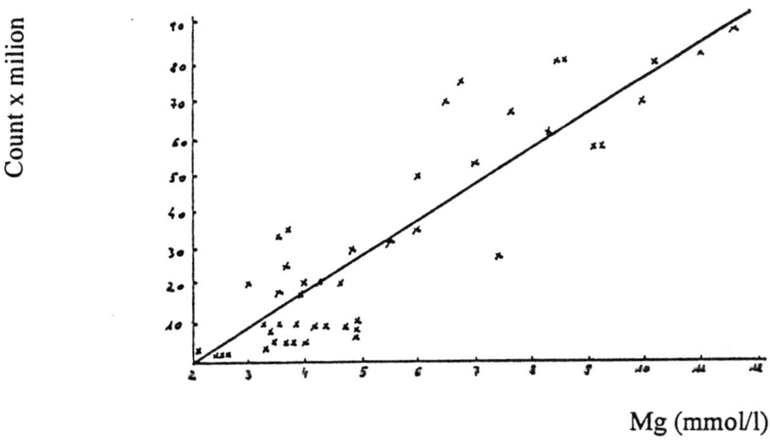

Figure 1. Correlation between the sperm count of human ejaculation and Mg content.

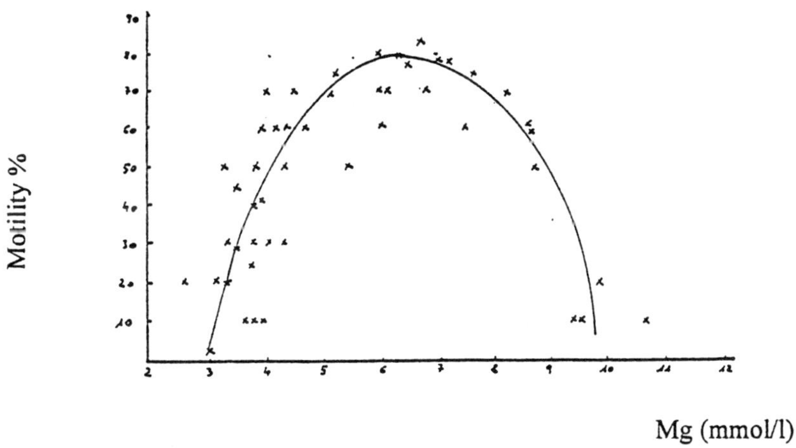

Figure 2. Correlation between the sperm motility of human ejaculation and Mg concentration.

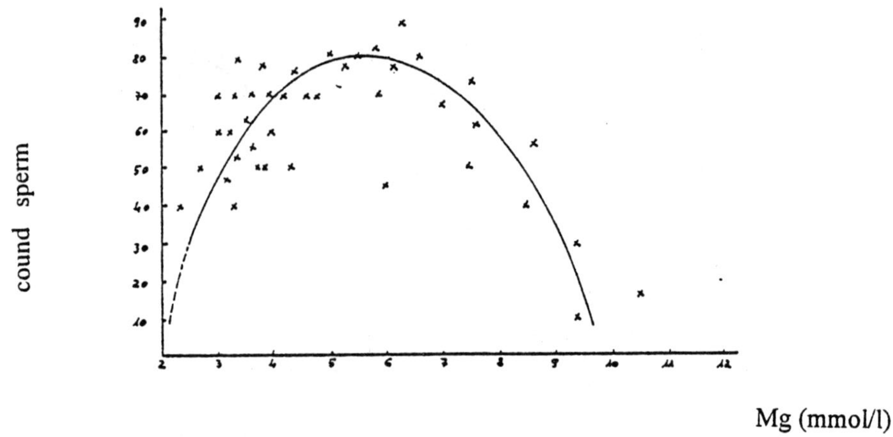

Figure 3. Correlation between the normal type sperm count in the human ejaculation (%) and Mg concentration.

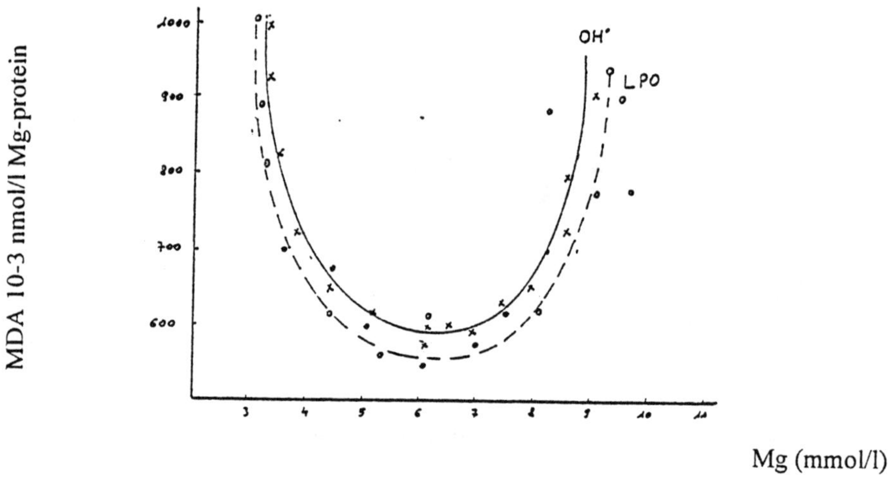

Figure 4. Correlation between hydroxyl radicals in the human ejaculation and LPO in association with Mg concentration.

count is on the maximum. Sperm count of the ejaculate is the most optimal at this Mg concentration. In conclusion we can say that Mg treatment potentates male fertility.

Table 2

Name	Treatment	Protein mg/ml	Mg mmol/l	LPO	$^{\bullet}OH$
				nmol	MDA/mg protein
B.F	Control Mg	34.64	3.00	0.1	0.723
		36.18	3.44	0.827	0.588
B.S	Control	28.00	9.00	0.964	0.795
	Mg	30.02	9.92	0.958	0.637
F.A	Control	33.19	3.56	0.859	1.070
	Mg	35.19	3.86	0.709	0.920
I.A	Control	19.10	10.03	0.999	0.990
	Mg	21.57	12.63	1.090	1.006
K.A	Control	24.66	3.69	1.224	1.109
	Mg	25.98	4.96	1.145	0.826
M.J	Control	25.02	10.91	0.997	0.342
	Mg	26.72	11.17	1.084	0.740
Sz.I	Control	39.10	5.60	0.703	0.500
	Mg	40.90	6.42	0.418	0.267
T.I.	Control	36.00	4.92	0.701	0.542
	Mg	38.48	6.00	0.693	0.412
T.Z.	Control	25.33	4.01	1.000	0.450
	Mg	29.27	4.75	0.884	0.330
V.I.	Control	24.31	2.98	1.103	0.990
	Mg	26.85	3.40	0.809	0.561

4. SUMMARY

Human ejaculated Mg and Oxygen – free radical concentration, sperm count, motility and abnormality were investigated in cases of normozoo -, oligozoo – and asthenozoospermia. The Mg content of human semen was five to ten higher than that of human plasma. Direct correlation was found between sperm count and human semen Mg content. Sperm motility and sperm characteristics (normal/abnormal type of spermia) were changed with the semen Mg level according to a maximum curve. Nevertheless, the concentration of Oxygen – free radicals was changed with the semen Mg level according to a curve. The activity of Oxygen – free radical eliminating enzymes injuring spermia depends on the Mg concentration of ejaculate in a maximum curve. The cause of decreased sperm motility in that ATP concentration was changed with Mg concentration in a maximum curve. The increasing concentration of Oxygen – free radicals was damaged by the ATP synthetase. In conclusion we can say that there is a correlation between Mg level and male fertility.

References

1. Treetipsatit N., Chulavatnatol M. (1982) Effects of ATP cAMP and pH on initial of flagellar movement in demembraned models of rat epidymal spermatozoa. *Exprtl. Cell. Res.* **142**: 495-499.
2. Mohri Hm, Yanagimachi R. (1980): Characteristics of motor apparatus in testicular epidymal and ejaculated spermatozoa, *Exptl Cell. Res.* **127**: 191-196.
3. Nelson L. (1959) Cytochemical studies with the electron microscope. I. ATP in rat spermatozoa, *Biochim. Biophysic. Acta* **27**: 639 -641.
4. Morisawa M. (1982) Cyclic AMP induced maturation of trout sperm axoneme to initiate motility. *Nature* **295**: 703-704.
5. Kiss A.S. (1983): Magnezziumtragyazas, magnezzium a biologiagan, Mezogazdasagi Kiado Budapest.
6. Balla A., Kiss A.S. (1996): Magnezium a biologiaban, magnezzium a gyermek-gyogyaszatban, Szekelyudvarhely.
7. Kiss I, Kiss S.S. (1994) Mutagenesis experiments in the presence of Magnesium *Symposium, Balatonszeplak 15.*
8. Kiss A.S., Kiss I. (1995) Effect of magnesium ions on fertility sexx ratio and mutagenesis on Drosophila melanogaster males, *Magnesium Res.* **8**: 243-247.
9. Kiss A.S., Szollosi J., Viski S. (1995) Connection between concentration of Mg in semen and fertility, *Magnesium Res.* **8:** 107. IV th Hungaro-Romanian Vagnesium Smposium, Szekely-udvarhely, odorheiu Secuiesc.
10. Viski S., Szollosi J, Kiss A.S., Csikkelne-Szolnoki A. (1995): A magnezium szerepe a ferfiak fertilitasaban, Szegedi Akademiai, Bizzottsag "Korneyezzetunk es az Egeszseg Anket" Elondas osazefoglaloi, Szeged.

THE EFFECT OF THE ADDITION OF MAGNESIUM ON THE WEIGHT GAIN OF HEIFERS AND ON FERTILITY PERFORMANCE AND SERVICE PERIOD OF HEIFERS AND COWS

KATALIN KOVACSNE GAAL, LASZLO GULYAS, PETRONELLA STADLER
Pannon University of Agricultural Sciences Faculty of Agricultural Sciences Institute of Animal Breeding
H-9200 Mosonmagyarovar, Var 4

1. Introduction

As for animal breeding, the heavy-producing breeds are very sensitive not only to feeding but also to environmental factors such as management system, housing conditions etc. The animals are subjected to more and more stress factors since the majority of them are not kept under natural circumstances and as a result of this, they have an increased need for minerals and vitamins [1,2]. For decades, the grass tetany of cattle was known to be the only disorder and disease caused by the isufficient supply of magnesium [3-5]. In recent decades lots of researchers have been done with magnesium in animal breeding, the rsults of which do emphasize the importance of an adequate magnesium supply of various species [6-10].

2. Method

A magnesium oxide preparation with a magnesium content of 60.40% (HAMAG LP) was fed to Holstein-Friesian heifers and cows during the experiment. The effect of the preparation on weight gain was determined. The control group consisted of 72 heifers, while another 68 heifers were divided into five test groups eating a diet supplied with 3, 4, 5, 6, 7g Mg/day. Fertility performance of the same heifers was also checked. In addition to this, 20g/head/day magnesium oxide was fed to cows of mixed lactation. Again, the number of fertilizations needed for the impregnation (fertility index) and the interval between calving and the next successful fertilization (service period) were evaluated.

3. Results

The results in Table 1 showing the weight gain of heifers make it clear that the weight gain of heifers was directly proportional to the amount of magnesium fed to them. In

T. Theophanides and J. Anastassopoulou (eds.), Magnesium: Current Status and New Developments, 347-349.

terms of weght gain, groups III and V were superior to the control group at the P<1% level, while group IV showed better results than the control one at the P<0.1% level. The results of the experiments carried out with heifers eating diets supplemented with magnesium are summarized in Figure 1. The fertility oerformance of the control group was 86.11% whereas that of the test group was 97.15%. Besides the improved pregnancy results, the fertility index was also lower (i.e. better) in each month-1.53 in the magnesium-fed group and 1.68 in the control one.

TABLE 1 The weight gain of heifers in the test and control groups

	Control	I.	II.	III.	IV.	V.	Total
Amount of Mg (g/day)	-	69,3	92,4	115,5	128,7	150,1	-
Group individuals (head)	72	14	14	14	13	13	68
Weight gain (g/day) 28th Feb 1996-31st May 1996	623	641	656	792*	911*	815*	760

*P<1% **P<0.1%

The effects of a three-month-long addition of magnesium on cows are shown in figure 2. The fertility performance of the test group during the first, second and tird month was 14.3%, 19.0% and 22.4 % respectively superior to that of the control group. Studying the interval between calving and the next successful fertilization of the same group according to lactation and in total, it is seen that the magnesium-fed group was superior to the control one (Table 2). Because of culling reasons, the initial number of individuals decreased to some extent.

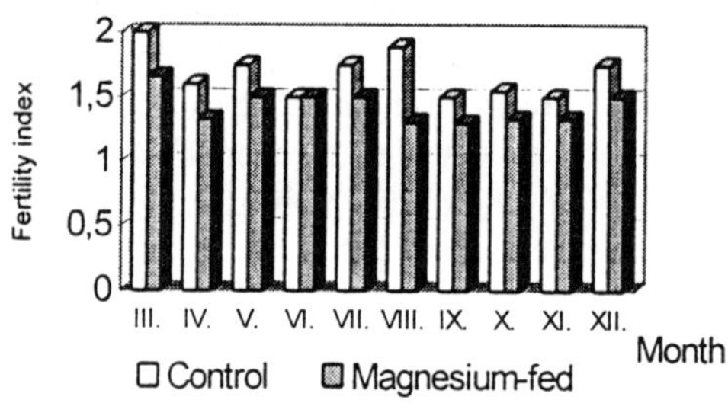

Figure 1 Fertility performance of control and test heifers (1996)

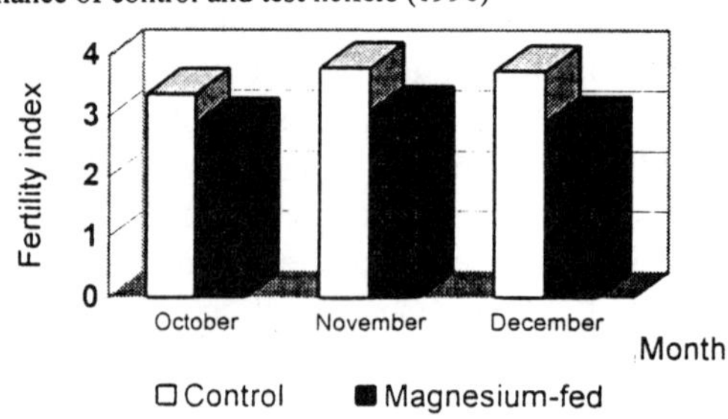

Figure 2 Fertility performance of control and test cows

TABLE 2 Service period of cows according to lactation

Lactation	n	Control			n	Test		
		average	s.d.	CV%		average	s.d.	CV%
1-2	29	108,86	39,52	36,30	24	91,46	25,64	28,03
3-4	19	129,58	32,65	25,20	25	90,20	26,54	29,42
5<	14	130,93	27,12	20,71	16	86,69	22,97	26,50
Total	62	120,19	36,10	30,03	65	89,80	25,05	27,90

References

1. Meyer, H. (1981) Mg-Bedarf und Mg-Versorgung bei Haustieren, *Magnesium Bulletin* **1**, 13
2. Lough, D.S. Beede, D.K. Hirchert, E.M. (1988) Lactional performance of mid-lactation dairy cows fed varying dietary concentrations of magnesium supplemented by organic magnesium oxide, *J. Dairy Sci.* **71** 1, 155
3. Samson, B.F. Manston, R. Vagg, M.J. Mallinson, C.B. Conteras, P.A. (1982) *Berichte vom 12 Weltkongress fuer Rinderkrankheiten* Bd. 1 Amsterdam, 574
4. Kolb, E. (1985) Neure biochemische Erkentnisse zum Mg-Stoffwechsel der Wiederkauer sowie zur Entstehung und Behandlung der Hypomagnesaemie unter besonderer Berucksichtgung des Praeparates Tetamag, *Mh. Vet-Med.* **40**, 615.
5. Gonye, S. (1987) *Az asvanyianyag-forgalom es zavarai. In : A szarvasmarha anyagforgalmi betegsegei es mergezese,* Mezogazdasagi Kiado, Budapest.
6. Kovacsne Gaal, K. Molnar, G. (1980) A magneziumadagolas hatasa a tej magneziumtartalmara, *Mosonmagyarovari Mezogazdasagtudomanyi Kar Kozlemenyei* **4**, 143-156.
7. Kovacsne Gaal, K. (1981) Amagnesiumadagolas hatasa a tojas nehany jellemzo tulajdonsagara, *ATE Mezogazdasagtud Kar Kozl. (Mosonmagyarovar)* **23**, 181.
8. Kovacsne Gaal, K. (1986) Magneziumkiegeszites hatasa a pecsenyecsirkek hizlalasi eredmenyiere, *Baromfitenyesztes es feldolgozas* **2**, 87.
9. Kovacsne Gaal, K. (1989) Effect of magnesium supplement on digestibility and biological value of feed in pigs, *Magnes. Res.* **2**, 228.
10. Kovacsne Gaal, K. Szerdahelyi, A. (1987) A magneziumkiegeszites hatasa a kocasuldok szaporodasi teljesitmenyere, *Allattenyesztes es Takarmanyozas* **26**, 441.

A NEW METHOD DEVELOPMENT FOR THE MASS MEASUREMENT OF ^{26,29}Mg AND THE EXCITED STATES OBSERVATION VIA THE ^{12}C(^{16}O, 2p) ^{26}Mg AND ^{13}C(^{18}O, 2p) ^{29}Mg NUCLEAR HEAVY-ION REACTIONS

EVANGELOS N. GAZIS[1] AND APOSTOLOS D. PANAGIOTOU[2]
[1]National Technical University, Physics Department, Zografou GR-157 80, Greece, [2]Nuclear Physics Laboratory, University of Athens, Ilisia, Athens, Greece

1. INTRODUCTION

Over the last few years there has been a growing interest in measuring the masses of 'exotic' nuclei, lying far from the valley of stability [1-4]. A possible way to produce very neutron-rich nuclei, such as ^{29}Mg, ^{34}Si, is to use reactions initiated by heavy ions, in which both projectile and target are already neutron-rich. Then, with the emission of protons from the compound nucleus, one may hope to reach these exotic nuclei. Reactions such as ^{18}O(^{19}F, 2p)^{35}P and ^{7}Li(^{7}Li, 2p)^{12}Be have been reported previously [2]. The particle stability of large number of light (sd-shell) nuclei, with very large neutron excess, has been observed [5,6]. The mass excess of a number of these nuclei, with $T_z = 5/2$ and 3, has been measured in the past years [7,8]. Previous measurements of the mass excess of the ground state of ^{29}Mg have been made by β-decay [9] and direct heavy ion reactions [4,10]. Scott et al.[4], have also indicated four excited states, while Detraz et al.[11], have obtained γ-rays from unidentified transitions within ^{29}Mg. We report here the measurement of the mass excess of ^{26}Mg and ^{29}Mg(T_z = 5/2) and the observation of four excited states of this nucleus, using the ^{13}C(^{18}O,2p)^{29}Mg reaction.

2. EXPERIMENTAL METHOD

2.1 ^{26}MG INVESTIGATION AND NOVEL METHOD DEVELOPMENT

We have set out to investigate and develop a new method for detecting the two protons in space and time coincidence in a single counter-telescope. The experimental merit of this method is that by using a very simple system and geometry with large solid angle we have been able to obtain 2p energy spectra with good resolution and low background. To develop the method, we used the (^{16}O,2p) reaction on targets of ^{12}C, ^{16}O and ^{27}Al. These reactions are expected to proceed through the formation of compound nucleus and the subsequent evaporation of two protons.

T. Theophanides and J. Anastassopoulou (eds.), Magnesium: Current Status and New Developments, 351-357.

The emission of two protons from the compound system may take place via : **(i)** emission from opposite sides of the compound nucleus, with $E_r = E_{2p}^{CM}$ and $\Phi_r = 180°$, or **(ii)** emission from the same position (and in the same direction) as a 'diproton', with $\Phi_r = 0°$ and $E_r = 0$ and **(iii)** emission with $0° < \Phi_r < 180°$ and $0 < E_r < E_{2p}^{CM}$, where E_r, Φ_r are the relative energy and angle of emission of the two protons in the centre-of-mass system, the compound nucleus is at rest, and E_{2p}^{CM} is the available energy of the two protons in this system. In the first and third cases the protons cease to interact between themselves almost immediately outside the nucleus. This gives rise to a typical three-body problem. In the second case of 'diproton' emission (with $E_1^{CM} \approx E_2^{CM}$), the relative distribution of the two protons in the laboratory depends on their final-state interaction. The two protons are considered to be in the $^1S_0$, T = 1 unbound state. This state is a broad distribution [12], for which we have assumed an 'average ' break-up energy of about 380 keV.
The two protons emitted simultaneously and in the same direction from the compound nucleus, were detected by a single telescope consisting of ΔE(150μm), E(2000μm) and A/C(300μm), Si surface barrier detectors and collimated so as to subtend an angle 8°-10°. Because the ^{2}He system is unbound, the two protons move out within a cone whose maximum half-angle is determined by E_{2p}^{lab} and the assumed break-up energy.
Considering the simultaneous passage of two protons through the ΔE, E detectors and assuming Bethe's relation for the energy loss, we arrive at a particle identification function (PI) :

$$PI \propto mz^2[E_1 + E_2)^2] / (E_1E_2)$$

where E_1, E_2 are the individual energies of the two protons in the laboratory and m, z are their mass and charge. This PI function assumes continuous values between '4' for $E_1 \approx E_2$ and '5.3' for $E_1 = (1/3)E_2$, for all practical purposes. With our experimental conditions we expect then to see a broad distribution in the mass spectrum, corresponding to the 2p detection and identification, just after the triton mass. The particle identification was performed on-line using the semi-empirical relation $\Delta E(E + K\Delta E + K_0)$, with properly adjusted constants. To minimize pile-up and chance ΔE-E coincidences we used fast pile-up rejection and fast timing with a time-window of about 6ns. We bombarded targets of ^{12}C and Al_2O_3 with 33 and 35 MeV ^{16}O beams and 2p spectra were obtained at angles between 0° and 70° in the laboratory frame. At very forward angles the ^{16}O scattered beam was stopped by a 41 mg cm^{-2} Au backing. Figure 1 shows a typical mass spectrum obtained with a ^{12}C target. The '2p mass' group identified as '^{4}H', is clearly seen. From the width of the '^{4}H' distribution we calculate that the two protons have very similar energies ($E_1 - E_2 \leq 5$ MeV).

2.2 ^{29}Mg INVESTIGATION

The experiment was carried out at the MP-Tandem Accelerator Laboratory of the Institut de Physique Nucleaire, Orsay. Beam of $^{18}O^{+5}$,with 32 MeV energy and about

100 nanoampere intensity, bombarded a 50 μg/cm^2 ^{13}C target, enriched to 99.5%. The ^{13}C was deposited into 19.3 mg/cm^2 Au backing, used to stop the beam and allow very forward measurements with large solid angle of about 20 msr.

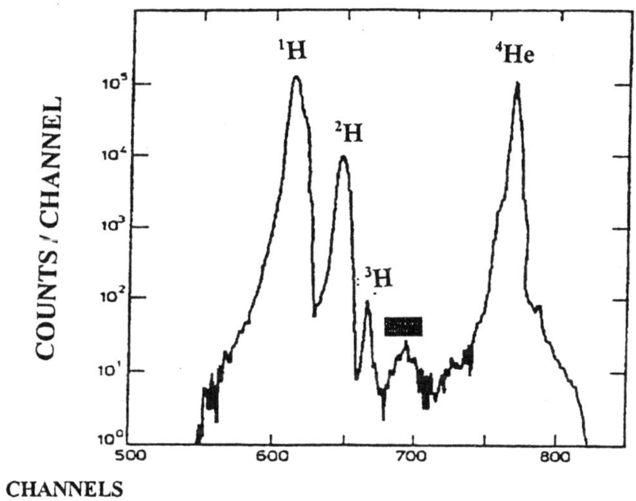

Figure 1. Mass spectrum obtained with 35 MeV ^{16}O beam and ^{12}C target

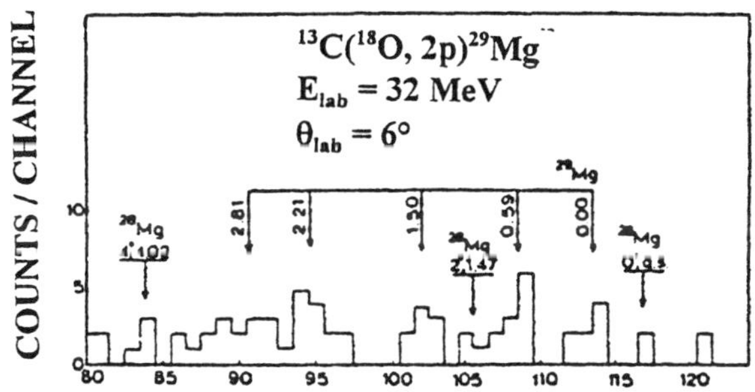

Figure 2. 2p-energy spectra, obtained in the reaction ^{13}C(^{18}O,2p) ^{29}Mg

The emitted two protons from the compound nucleus ^{31}Si were detected by a detector telescope, consisting of ΔE_1(150 μm), ΔE_2(100 μm), E(3000 μm) and A/C(500 μm) silicon surface barrier detectors. For the identification of the 2p's we have employed the method described obove and obtained twice the particle identification $PI_1 = \Delta E_1, (\Delta E_2 + E)$ and $PI_2 = (\Delta E_1 + \Delta E_2), E$.

Two integrating, two dimensional evens events were obtained : ΔE_1 versus $(\Delta E_1 + \Delta E_2 + E)$. Figure 2, the projection of the 2p-particle group onto the Etot$(\Delta E_1 + \Delta E_2 + E)$ axis, shows the $6°_{Lab}$ 2p-energy spectrum after having subtracted the background. In the spectrum we determine the ground state of ^{29}Mg and four excited levels at 0.59 ± 0.08 MeV, 1.50 ± 0.08 MeV, 2.21 ± 0.09 MeV and 2.81 ± 0.09 MeV. We also indicate the position of the ground, first and second excited states of ^{28}Mg, populated by the (^{18}O,2p) reaction on the ^{12}C contaminant of the target. The location of these states was determined in a separate run of the $^{12}C(^{18}O,2p)\,^{28}Mg$ reaction at precisely the same energy (32 MeV) and angle the ground and the four excited states of ^{29}Mg observed do not coincide with any of the low-lying ^{28}Mg states.

The very weak population of ^{28}Mg as compared to the high isotopic purity of the target. this is important in view of the approximately 100 times larger total cross section for production of ^{28}Mg [13]. The cross section for the ground state of ^{29}Mg was determined from this experiment to be of the order of 1 nb/sr, taking into account a 2% efficiency was calculated assuming a Watson-Migdal distribution for the unbound 2p-system.

The energy calibration was obtained from the d, t and α-particle energy spectra, which were recorded simultaneously and produced clearly identified low lying states. We have also used in this calibration the first two states of ^{26}Mg and ^{28}Mg obtained in the $^{12}C(^{16}O, 2p)^{26}Mg$ and $^{12}C(^{18}O,2p)\,^{28}Mg$ reactions, at 30 MeV and 32 MeV beam energy, respectively.

3. RESULTS AND DISCUSSION

3.1 ^{26}Mg RESULTS

Figure 3 shows the 2p energy spectrum obtained in the $^{12}C(^{16}O,2p)^{26}Mg$ reaction at 35 MeV with a 40 μgcm^{-2} natural ^{12}C target. We clearly identify the first few levels of ^{26}Mg and notice a relatively strong population of the 0^+ ground state, with a cross section at 10° of about 80 μbsr^{-1}.

The energy resolution in this spectrum is about 500 keV, due to the large effective acceptance plane (±5°) of the telescope and the unequal energy loss of the two protons transversing the 41 mg cm^{-2} gold foil. The differential cross sections at 10° for the

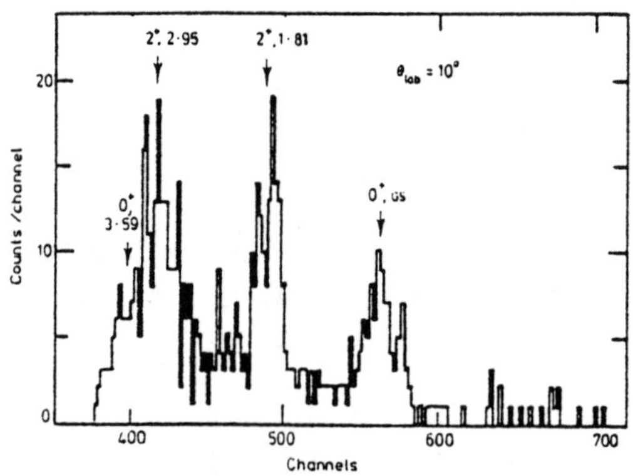

Figure 3 2p - energy spectrum obtained with 35 MeV ^{16}O in the reaction $^{12}C(^{16}O, 2p)^{26}Mg$

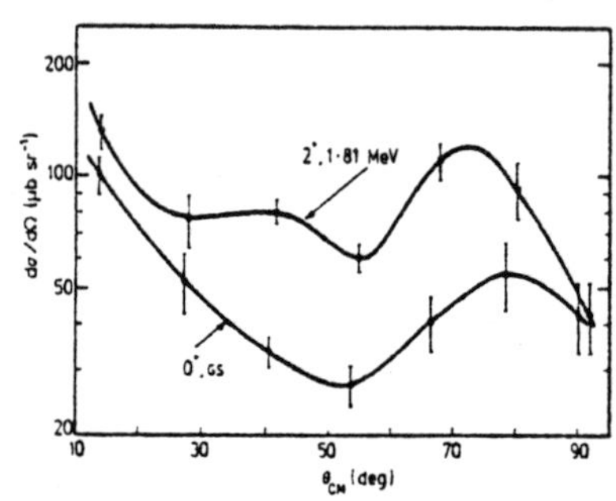

Figure 4 Angular distribution of the 2p's from the reaction $^{12}C(^{16}O, 2p)^{26}Mg$, leading to the 0^+ (g.s.) and 2^+ (1.81 MeV)states

ground states of ^{30}Si and ^{41}K, produced with ^{16}O and ^{27}Al targets at 35 MeV beam energy, are about 8 mbsr $^{-1}$ and 3 mbsr $^{-1}$ (±40%) respectively. The relatively weak population of ^{30}Si and ^{41}K may be partly due to the higher Coulomb barrier in the entrance channel and the poorer matching of the total spin in the incident and exit channels.We have also obtained an angular distribution of the 2p's leading to the 0^+ ground state and the 2^+, 1.81 MeV state of ^{26}Mg, between 14° and 90° in the centre-of-mass frame (figure 4). The absolute cross sections quoted are correct to about ±40%. This large uncertainty is due mainly to the estimated 'average' break-up energy of the ^{2}He unbound system, which strongly determines the 2p detection efficiency.

3.2 ^{29}Mg RESULTS

From this experiment we deduced the mass excess of ^{29}Mg to be -10.78 ± 0.05 MeV, which compares excellently with the previous measurement [4],as well as with the prediction of modified shell model scheme [16]. Table 1 compares the experimental and calculated values for the mass excess of ^{29}Mg.

TABLE 1. ^{29}Mg Mass Excess (MeV)

Experimental		Theoretical	
[9]	-10.59 ± 0.40	[15]	-10.66
[4]	-10.75 ± 0.05	[16](Mod.)	-0.75
[10]	-10.65 ± 0.055	[16](trns)	-10.70
work	-0.78 ± 0.05	[17]	-11.58

The two excited states at 1.50MeV and 2.21MeV are in fair agreement with two states of [4] at 1.4MeV and 2.34 MeV, Table 2. The first excited state we observe at 0.59MeV has not been seen by Scott [4], probably due to its very much reduced cross-section in heavy ion transfer reaction at high energies. The small spin of this level is not well matched by high transferred angular momentum in that reaction. This state may correspond to the weak 603keV γ-ray, observed by Detraz [11]. The reason for the weak transition may be that this level is not fed strongly by a γ-cascade, not directly by the β-decay of ^{29}Na. A similar situation exist in the 750 keV γ-ray transition from the first excited to the ground state of ^{31}Si.

Table 2 shows the energy levels of ^{29}Mg obtained in this work and in [4], as the theoretical predictions of Cole et al.[18]. For comparison we also show the known energy levels of the related nucleus ^{31}Si.

TABLE 2. ^{29}Mg and ^{31}Si Energy Levels (MeV)

^{29}Mg		^{31}Si	
This Work		Experimental	
[9]	0.00	0.00	$3/2^+$
[10]	0.59 ± 0.08	0.75	$1/2^+$
[11]	1.50 ± 0.08	1.70	$5/2^+$
[12]	2.21 ± 0.09	2.32	$3/2^+$
[13]	2.81 ± 0.09	3.13	$7/2^-$
[14]	--	3.53	$3/2^-$

Our observed level scheme for ^{29}Mg is similar to that ^{31}Si .This can be understood if we assume that the three neutrons, coupled to ^{26}Mg and ^{28}Si occupy the same orbitals, namely the $\{(s_{1/2})_2 (d_{3/2})_1\}$ or $(d_{3/2})_3$, independently of the proton orbitals.
The shell model calculation, using the Predom-Wildenthal (PW) interaction, produces a ground state doublet $3/2^+$, $1/2^+$, separated by only 20 keV, Fig.5.

4. CONCLUSIONS

We notice oscillations in the angular differential (fig. 4) cross section and a decrease as we move away from very forward angles. The overall shape of the distribution is not, however, reminiscent of direct transfer but rather of compound system formation.
We have also estimated from the ^{12}C (^{16}O, 2p)^{26}Mg (GS) reaction the total experimental error in the mass of ^{26}Mg. The most important sources of error are the uncertainties in

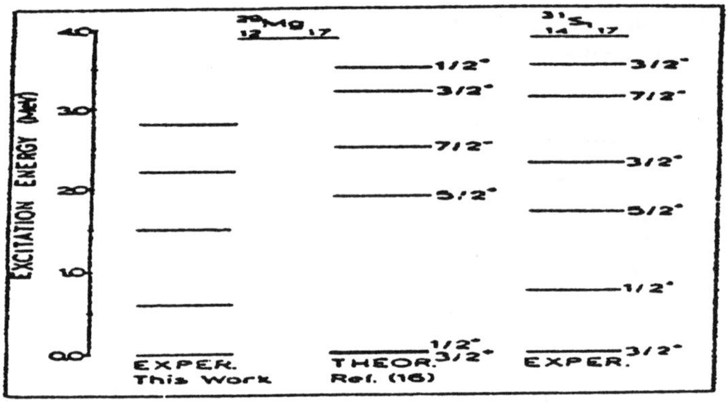

Figure 5 Energy level diagram of ^{29}Mg, obtained in this work. The theoretical predictions of Cole et al. As well as the lebel diagram of ^{31}Si are also shown.

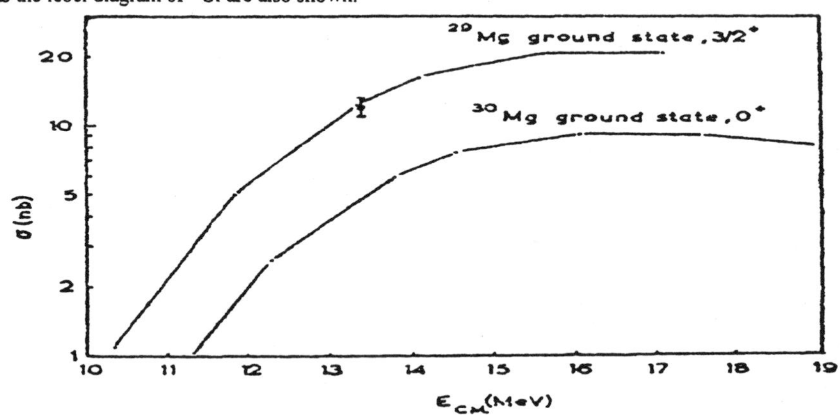

Figure 6 Calculated total cross sections for the ground state of ^{29}Mg and ^{30}Mg at 13.4 MeV centre of mass energy is also shown.

the target thickness, the detection angle, the peak centroid and the beam energy. With our present conditions an overall error of about ±120 keV was obtained. Having sufficiently developed this new and simple method for detecting the two protons in a single counter-telescope, we proceed below undertaking mass measurements of light exotic nuclei like the ^{29}Mg.

The predicted energy of the first excited $1/2^+$ state in ^{29}Mg is much lower than the known position of the same state in the related nuclei ^{31}Si(0.75 MeV)and ^{33}S(0.84 MeV). However, this prediction is highly sensitive to small changes in the interaction and should be considered with caution. Our candidate for the first $1/2^+$ state at 0.59 MeV is in good agreement with the corresponding states in the related nuclei ^{31}Si and ^{33}S.

The cross section for production of the ground state of ^{29}Mg was calculated using the code STAPRE [14], which incorporates the Hauser- Feshbach scheme.

The absolute value of the calculated cross section agrees surpisingly well with the experimentally determined one, at 13.4 MeV centre of mass energy, Fig. 6. The calculation was also performed for ^{30}Mg, using the ^{14}C(^{18}O, 2p) ^{30}Mg reaction.

References

1. Robertson R G H et al, (1974) Highly proton rich Tz -2 nuclides: ^{8}C and ^{20}Mg, *Phys. Rev. Lett* **32** 1207
2. Goosman D R and Alburger D E, (1972)Mass and decay of a highly neutron-rich isotope of phosphorous: 48-sec ^{35}P, *Phys. Rev.* **C6** 820
3. Thibault C et al, (1975) *Atomic Masses and Fundamental Constants* 5 eds J H Sander and A H Wapstra (New York : Plenum) p 205
4. Scott D k et al (1974) Spectroscopy of exotic nuclei using heavy-ion transfer reactions, *Phys. Rev. Lett* **33** 1343
5. Bowman , J. D. et al, (1974), Detection of neutron-excess isotopes of low -Z elements produced in high-energy nuclear reactions, *Phys. Rev.* **C9**, 836
6. Westfall , G. D. et al, (1979), Production of neutron-rich nuclides by fragmentation of 212-MeV/ amu ^{48}Ca, *Phys. Rev. Lett.* **43**,1859
7. Goosman , D.R., Alburger, D.E. (1973),Mass and β decay of ^{25}Ne, a Tz=+5/2 nuclide, *Phys. Rev.* **C7**, 1133.
8. Nathan, A. M., Alburger, D.E. (1977) Mass and β decay of ^{34}Si, *Phys. Rev.* **C15**, 1448,
9. Goosman, D.R., Davis, C.N., Alburger, D.E. (1973) Mass and β decay of the new isotope ^{29}Mg: Systematics of masses of Tz=5/2 nuclides in the 2s-1d shell, *Phys. Rev.* **C8**, 1331
10. Paschopoulos, I., et al (1978) Energy levels of 27Na and masses of ^{29,30}Mg odserved via ^{18}O-induced transfer reactions on ^{26}Mg *Phys. Rev.* **C18**, 1277
11. Detraz, C., et al (1979) Beta decay of ^{27}Na-^{32}Na and their descendants, *Phys. Rev.* **C19** ,164
12. Rutkowski R W and Gross E E The p-p final-state interaction in the ^{3}He(d,t)2p reaction, (1971) *Phys. Lett,* **35B**, 151
13. Blann, M. and Plasil, F, (eds.). Alice Overlaid, a Nucleon -Evaporation code.
14. Uhl., M. Acta Phys. Austriaca 31, 245 (1970)
15. Bassichis, W. H., Ali S.S. (1974) Garvey-Kelson mass formula in the s-d shell, *Phys. Lett.* **52B**, 282
16. Jelley, N.A., et al (1975), Predictions of the masses of highly neutron rich light nuclei, *Phys. Rev.***C11**, 2049
17. Thibault. C,.Klapisch, *R.* (1974),New mass-table parameters for nuclei Z≤16,. *Phys. Rev.* **C9**, 793
18. Cole, B. J. et al (1974), Theoretical prediction of ^{29}Mg excited energy states, *Nucl. Phys.* **A7**,1399

NEW DATA ON THE SCAVENGING ROLE OF PREPARATION TIOMAG IN ANTIOXIDANT THERAPY

J. SZÁNTAY, J. KENDERESI-SZÁNTAI, MINERVA SZÁNTAY
IIIrd Medical Clinic, Univ. of Medical and Pharmacy, Cluj-Napoca, 3400, Romania

1. INTRODUCTION

It is generally admitted that thiol groups and thiol-group-donor substances are the most important scavengers of free radicals [1]. The scavenging activity of N-acetyl-cysteine (NAC) and other thiol-group donors, such as methionine, was demonstrated in respiratory diseases, toxicology mutations and induction of neoplasms, dysfunctions of the immune system, vascular injuries and cardiac diseases. The shortcoming of NAC and methionine administration in antioxidant therapy is based exclusively on the presence of Mg in the daily diet only, which is not sufficient. The Mg requirement (6mg/kg/day) is rarely met. Our diet is marginal in Mg, and we live with a chronic Mg deficiency. It is well-known that thiol groups can exhibit their scavenging activity only in presence of Mg ion. This fact emphasises the imperative necessity of Mg for assuring the scavenging activity of thiol groups, offered by cytoprotective preparation TIOMAG, which simultaneously ensures partly the inactivation of catecholamines excess via O-methylation and on the other hand it is an important donor of thiol groups itself.

2. RESULTS AND DISCUSSION

The mechanisms by which oxidants cause injury are complex. Because of their avidity for accepting electrons from target molecules, oxidants can modify the structure and/or the function of the target. In this regard oxidants can injure a cell by interacting with its structural components [2]. There is a number of areas in clinical medicine where oxidant injury is believed to be important: respiratory diseases, toxicology, mutations and induction of neoplasms, dysfunctions of the immune system, vascular injuries, cardiac diseases [2].

T. Theophanides and J. Anastassopoulou (eds.), Magnesium: Current Status and New Developments, 359-362.

It is generally admitted that thiol groups and thiol-group-donor substances are between the most important scavengers of oxygen free radicals (OFR). But it is also well-known that thiol groups can exhibit their scavenging activity only in the presence of Mg^{2+} ion [3].

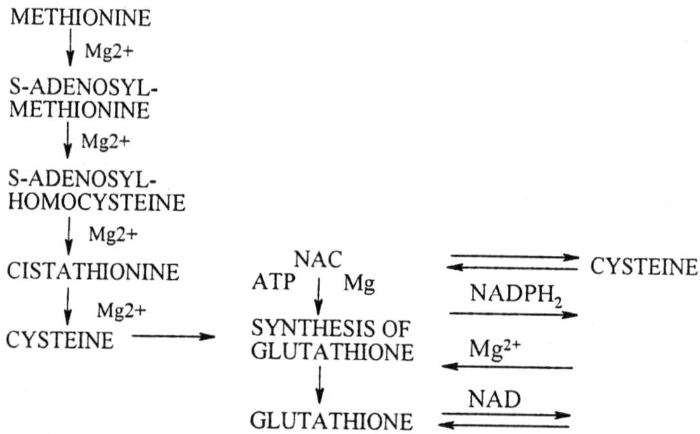

Figure 1. The scavenging mechanism of methionine and NAC in presence of Mg through neutralisation of oxidants and OFR.

There are evidences that the antioxidant therapy with methionine and N-acetyl-cysteine (NAC) benefits only the Mg brought by food, which is not sufficient [2]. But for what reason?

The objective answer to this question is as it follows: first of all, according to J. Durlach, the lack of Mg in the diet is a fundamental factor at the origins of Mg deficit. Mg requirement is more than 6 mg/kg/day. However, this is rarely met. Our diet is marginal in Mg and we live with a chronic Mg deficiency [4]. Secondly, it is well-known that Mg takes part at the non-specific adaptation of the organism to: psychical and physical stress, physical over-exercise, sports, biological stress represented by diseases of various aetiology, toxic stress such as aggression of environmental toxic agents, labour in toxic environment, or alcohol abuse. The global result is an increase of Mg deficit.

In conclusion, these two facts emphasise once again, that Mg substitution therapy is absolutely necessary for assuring the scavenging activity of NAC, methionine and its metabolites. The composition of cytoprotective preparation TIOMAG is: methionine as source of thiol groups and Mg gluconate as source of Mg. Substitution therapy regimen with TIOMAG consists of Mg - 300-600 mg, methionine - 1.5-2.0 g, vit. B1 - 60-100 mg, vit. B6 - 250-750 mg, Ca - 185 mg. Ca administration was 185 mg/day, performed when serum Ca level was under 4.5 mEq/l.

The performance of cytoprotective preparation TIOMAG can be summarised as follows:

1. Correction of Mg deficiency with normalisation of the intracellular rate of Mg/Ca;
2. Inactivation of cathecolamine excess via O-methylation and thus realising a sparing effect in thiol groups;

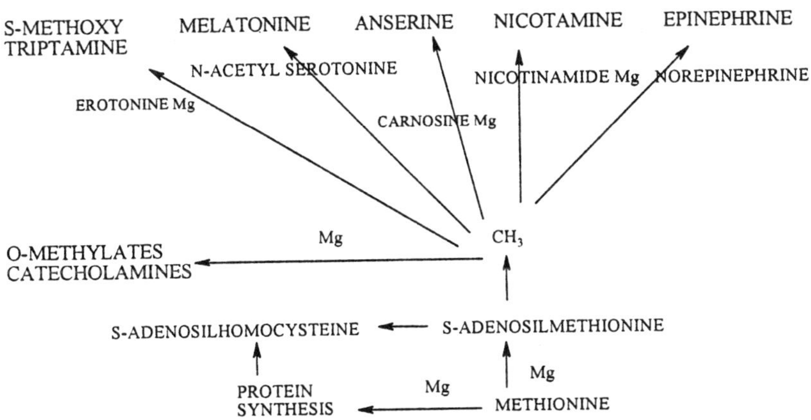

Figure 2. The patway of catecholamine inactivation with O-methylation

3. Control of intracellular Ca increase through Mg ions via Mg/Ca antagonism, that contributes to normalisation of ATP synthesis;

4. neutralisation of OFR through the glutathione - glutathione S-S and NADH/NAD systems;

5. Stabilisation of the control system of Mg intracellular homeostasis.

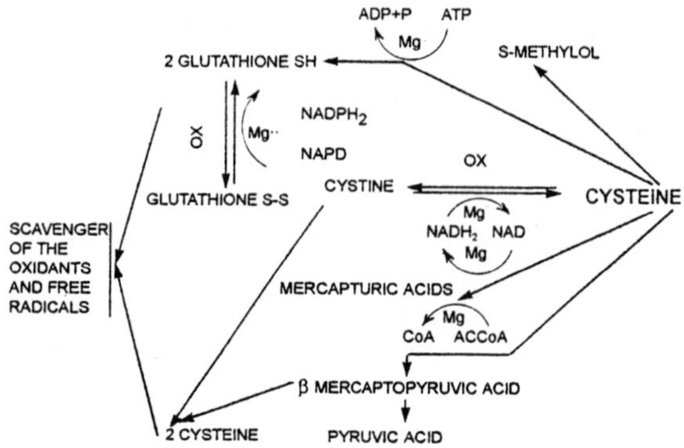

Fiure 3. Thiol groups scavenger activity against oxidants and free radicals

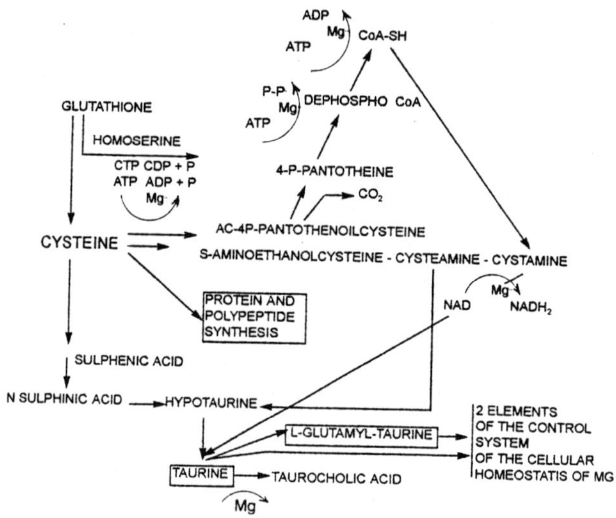

Figure 4. The control system of the cellular homeostasis of Mg^{2+} ions

References

1. Anastassopoulou, J. (1994), Free radicals in Biological Systems, in N. Russo, J. Anastassopoulou and N. Barone, eds., in *Properties and Chemistry of biomolecular systems*, Kluwer Academic Publishers, The Netherlands, pp. 23-30.
2. Crystal, R. G. (1991), *J. Am. Med.* , **91, Suppl. 3C**, 1 S-145.,
3. Szántay, J. (1992) In Abstracts Volume of *The European Congress on Magnesium*. Intern. Society for the Development of Research on Magnesium (S.D.R.M.) Giessen, , p. 47.
4. Durlach J. (1989), Magnesium In Clinical Practice. John Libbey Eurotext, London - Paris, p. 74.

EFFECT OF Mg THERAPY ON THE OUTCOME OF PREGNANCY IN CERTAIN OBSTETRIC DISEASES

L. LUKACSI, A. SEBESTYEN, F. BANHIDY,A. LASZLO AND F. PAULIN
2nd Department of Ob/Gyn. Semmelweis University of Medicine, Budapest, H-1082, Hungary

OBSECTIVE

Preeclampsia (PE) and intrauterine retardation (IUGR) are known to significantly alter perinatal morbidity, mortality and the outcome of pregnancy. The smooth muscle-relaxant effect of the intravenously applied $MgSO_4$ was first described in 1905 [1]. Rissman was the first to use Mg in order to prevent eclampsia in 1916 [2]. The method came into general use first of all in the USA [3,4], followed by the wide-spread use of it both in the prevention and treatment of eclampsia in Europe, too [5,6]. The chemical and biochemical basis of the therapeutic effect was summarized by Gunther et al. [7]. In our study the effect of Mg infusion treatment on pregnancy outcome in patients with PE, IUGR, imminent abortion and imminent premature labor was examined.

METHOD

For infusion treatment, $MgSO_4$ (CORMAGNESIN 200) was used in the following patient groups: PE(n=13), IUGR(n=30), Imminent abortion(n=21), and premature labor (n=28). Further parameters, like mother's age, obstetric history, duration of hospitalization (Table 1), mother's serum Mg level before and after treatment (Tables 3-7), way of the parturition, weight of the fetus were examined (Table 2). PE was defined as mild or severe. Atomic absorption spectrophotometry Varian Spectra A-30 on 285,2 nm was used in order too measure mother's serum Mg level.

RESULTS

In Tables 2-7 we demonstrate the results achieved with application of Cormagnesin infusion. In Table 2 we accentuate from among the obstetrical events of the different patient groups:

1. The $MgSO_4$ infusion applied in premature labor resulted in the prolongation of pregnancy.

T. Theophanides and J. Anastassopoulou (eds.), Magnesium: Current Status and New Developments, 363-366.

TABLE 1. Characteristics of different patient groups ($\bar{x}$±SD)

	n	Age (year)	Parity	Previous abortions	Admittance week of gestation	Duration of hospitalization (day)
Imminent abortion	21	27.29±6.18	0.52±0.68	0.43±0.75	21.24±5.02	27.71±25.96
Imminent premature labor	29	27.14±5.13	0.90±0.94	0.24±0.64	32.45±2.53	17.86±11.71
Retardation (IUGR)	30	27.83±6.45	0.53±0.78	0.83±1.21	34.83±2.00	21.30±9.24
Preeclampsia	13	28.92±5.39	1.00±1.47	0.15±0.38	34.77±4.02	15.77±10.96

TABLE 2.Characteristics of obstetrical experience after cormagnesin treatment

	Spontaneous abortion	Premature labor	Immature labor	Term labor	Weight percentil between 5-10 (g)	Eclampsia	Frequency of Cesarean section (%)
Imminent abortion (n)	1	2	3	9	1	0.00	23.81
Imminent preamature labor (n)	0	1	4	23	9	0.00	37.93
Retardation (IUGR) (n)	0	0	14	16	15	0.00	50.00
Preeclampsia (n)	0	1	6	6			
-mild n=7 (n)		1	1	3	2	0.00	38.46
-severe n=6 (n)		0	3	2	1	0.00	38.46

2. In IUGR, despite the treatment hypotrophy frequency was 50% (between 5-10 weight percentile).
3. During the period of examination no eclampsia was observed.
4. In cases with PE and IUGR we are more often compelled to accomplish the parturation with Cesarean section.

In Tables 3-7 we demonstrate the serum Mg levels during the infusion treatment. According to our results we ascertain that the serum Mg content of the mother's venous blood essentially increases, in contradiction to our observations in PE.

CONCLUSION

The Cormagnesin infusion containing $MgSO_4$ proved to be effective in the prolongation of pregnancy, in the prevention of eclampsia, and improved perinatal morbidity and mortality. During its application no side effects were detected. During hospitalization of these kind of patients we recommend $MgSO_4$ infusion as a first choice of treatment. Effects of longer Mg treatmnent with higher doses in patients with PE and IUGR on metabolism of the pregnant uterus and the maternal-fetal-placental unit are to be further examined.

TABLE 3. Mg Level of maternal serum during cormagnesin in application ($\bar{x}$±SD)

	n	Duration of treatment (day)	Before treatment mmol/l		Ater treatment mmol/l
2g/day	78	6.36±3.79	0.74±0.12	p<0.022	0.78±0.12
4g/day	20	4.85±3.77	0.75±0.15	p<0.068	0.82±0.12

Table 4. Mg Level of maternal serum during cormagnesin in infusion in IUGR ($\bar{x}$±SD)

	n	Duration of treatment (day)	Before treatment mmol/l		Ater treatment mmol/l
2g/day	26	8.19±3.80	0.74±0.10	p<0.11	0.78±0.11
4g/day	5	5.80±5.35	0.79±0.11	p<0.067	0.89±0.08

TABLE 5. Mg Level of maternal serum during cormagnesin in imminent abostion ($\bar{x}$±SD)

	n	Duration of treatment (day)	Before treatment mmol/l		Ater treatment mmol/l
2g/day	16	5.37±2.90	00.73±0.103	p<0.20	0.76±0.11
4g/day	6	5.16±4.53	0.73±0.19	p<0.33	0.78±0.11

TABLE 6. Mg Level of maternal serum during cormagnesin in imminent premature labor ($\bar{x}$ ±SD)

	n	Duration of treatment (day)	Before treatment mmol/l		Ater treatment mmol/l
2g/day	27	5.18±2.84	0.74±0.11	p<0.04	0.79±0.12
4g/day	5	2.60±1.34	0.80±0.19	p<0.45	0.81±0.17

TABLE 7. Mg Level of maternal serum during cormagnesin in preeclampsia ($\bar{x}$ ±SD)

	n	Duration of treatment (day)	Before treatment mmol/l		Ater treatment mmol/l
2g/day	9	6.33±5.78	0.76±0.16	p<0.49	0.76±0.17
4g/day	4	6.00±1.82	0.68±0.01	p<0.07	0.80±0.12

References

1. Meltzer, S.I., Auer, J. (1905), Physiological and pharmacological studies of magnesium salts. I. General anestesia by subcuteneous injection. *Am. J. Physiol.* **14**, 366.
2. Rissman, P. (1916) Neue Wegw der Eklampsibehandlung. *Zeitschr. Gebursth. Gynaek.* **78**, 447.
3. Lazard, E.M. (1925) A preliminary report on the intravenous use of magnesium sulfate in puerpal eclampsia. *Am. J. Obstet. Gynecol.* **9**, 178.
4. Dorsett, L. (1926) Intramusclar injection of magnesium sulfate control of convulsion in eclampsia. *Am. J. Obstet. Gynecol*, **11**, 227.
5. Lippert, T.H. (1983) Uber die Anwendung von Magnesiumsulfat als Antikonvulsium bei schwerer Schwangeschaftsgestose und Eklampsie. *Geburtsh. U. Frauenheilk.* **43**, 589.
6. Conradt, A., Weidinger, H., Algayer, H. (1984), reduzierte Frequenz von Gestosen bei betamimetika behandelten Riskoschwangerschaften mit Magnesium-Zusatztherapie. *Geburtsh. U. Frauenheilk*, **44**, 118.
7. Gunther, T. (1991), Biochemical bases of the therapeutic actions of magnesium. *Mag. Bull.* **13**, 46

PROTECTIVE EFFECTS OF MAGNESIUM ON INNER EAR BLOOD FLOW IN THE ACOUSTIC TRAUMA. ANIMAL STUDIES

F. SCHEIBE, H. HAUPT
Department of Otorhinolaryngology, Charité Hospital, Humboldt University, Berlin, Germany

1. INTRODUCTION

We have recently demonstrated in the guinea pig [Scheibe et al., 1997] that preventive oral magnesium (Mg) administration significantly reduces the rate of the acoustic trauma caused by impulse noise exposure (simulated small-weapon noise). Animals with an optimal Mg status revealed a permanent hearing loss (ABR) which was 15-20 dB lower (frequency range, 3.75-30 kHz) than that in animals with a suboptimal status. The exact mechanism underlying this protective effect of Mg is unknown at this time, but it may involve a direct vascular effect. To test this hypothesis, we conducted this study to investigate the effect of Mg supplementation on cochlear blood flow (CBF) in the acoustic trauma using laser Doppler flowmetry.

2. MATERIALS AND METHODS

The experiments were performed on 25 pigmented anesthetized (ketamine-xylazine) guinea pigs (260-420 g) with either a physiologically high or low Mg status produced by different diets. The animals were fed an experimental diet with a Mg content of 0.01% and received tap water with either an additive of 39 mmol Mg (high-Mg group) or without Mg admixture (low-Mg group). Body temperature was maintained at 38°C. Cochlear temperature, heart rate (HR) and mean arterial blood pressure (MABP) were monitored. The animals were exposed to the same impulse series (L_{peak} 167 dB, 1/s, 38 min), which was used in our previous studies mentioned above. Using a MBF3D dual channel laser Doppler monitor (Moor, U.K.), CBF *(Flux)* as well as concentration *(Conc)* of red blood cells (RBCs) and their mean velocity *(Speed)* were recorded continuously in the basal and upper parts of the cochlea.

3. RESULTS

3.1. MAGNESIUM STATUS.

The oral preventive administration of the two diets (for ≥ two weeks) resulted in siginificant differences in the total Mg concentrations of perilymph, cerebrospinal fluid and plasma between the two groups (Table 1), as measured in a separate series (using

T. Theophanides and J. Anastassopoulou (eds.), Magnesium: Current Status and New Developments, 367-370.

Table 1. Total Mg concentration (mmol/l) in the perilymph, cerebrospinal fluid and blood plasma of the high-Mg and the low-Mg groups. All values are expressed as mean ± SD for the number of animals (*n*) tested.

Fluid	High-Mg group	Low-Mg group
Perilymph	0.76±0.10[11]	0.53±0.11[11]
Cerebrospinal	0.88±0.09[9]	0.77±0.10[11]
Blood plasma	1.23±0.18[11]	0.71±0.27[8]

*/** P < 0.05/0.001 vs. Low-Mg group

atomic absorption spectrometry) in order to test the animals' Mg status. The fluid Mg levels of the two groups were at the upper and lower limits of the physiological range.

3.2. CIRCULATION.

Initial mean HR, MABP and cochlear temperature were 191 ± 15 min^{-1} (n = 25), 41 ± 9 mmHg (n =19) and 34.4°C (n = 15), respectively, and did not show any significant changes at the end of the experiment. There were no significant difference in the MAPB between the groups.

3.3. COCHLEAR BLOOD FLOW.

Significant differences in all laser Doppler parameters tested *(Flux*, *Speed* and *Conc)* were found between the two groups, as measured during and up to one hour after the exposure (Fig. 1 and 2), though the signals revealed considerable interindividual variations. Furthermore, there were also significant differences in the laser Doppler signals as recorded from the basal and apical parts of the cochlea. The basal CBF *(Flux)* showed significant decreases (relative to baseline), as measured in the low-Mg group during the exposure, which recovered in the post-exposure period (Fig. 1), while no significant change was found to occur in the high-Mg group. The apical *Flux* signals measured in the low-Mg group revealed only a tendency towards a noise-induced decrease, while they showed significant increases in the high-Mg group during the exposure, which returned to baseline in the post-exposure period. The basal and apical *Flux* differences between the two groups were found to amount up to about 15% and 25% , respectively.

The *Speed* and *Conc* signals revealed an inverse Mg-related response to the noise exposure in both parts of the cochlea. Fig. 2 shows the basal data of both signals. There were significant *Speed* decreases in the low-Mg group after the exposure, while in the high-Mg group a clear tendency towards a *Speed* increase was observed. In contrast, the *Conc* signals found in the low-Mg group showed an increase which was significant 20 min after the exposure, while in the high-Mg group significant decreases were found during and following the exposure.

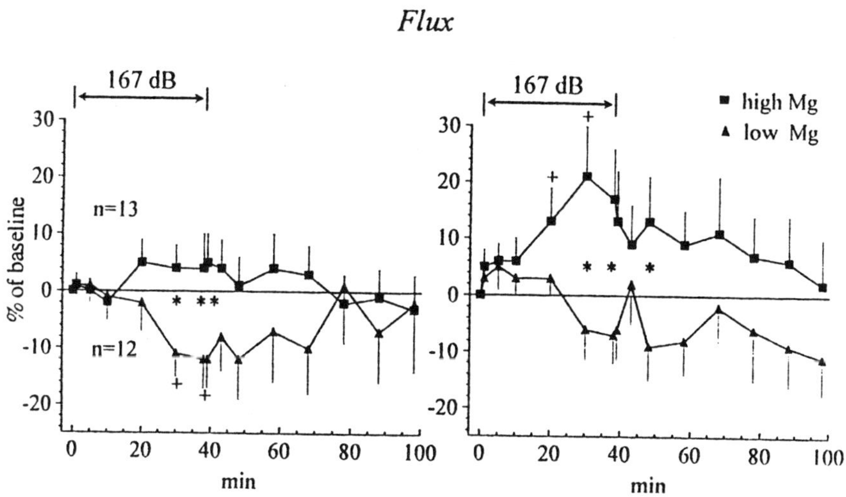

Figure. 1. Time courses of the basal (left) and apical (right) *Flux* responses (mean ± SEM) to the impulse noise series (L_{peak} 167 dB, 1/s, 38 min) for the high-Mg group and the low-Mg group. (*/+ $P < 0.05$ between the two groups / vs. baseline).

4. DISCUSSION

Although there were considerable interindividual variations in the laser Doppler responses to the impulse noise exposure, this study clearly demonstrates that an optimal Mg status, produced by prophylactic Mg supplementation, has a protective effect on the CBF in the acoustic trauma. The differences in the noise-induced *Flux* responses which were found to occur between the high-Mg and the low-Mg groups as well as between the basal and apical parts of the cochlea (Fig. 1) are in line with the functional findings (ABR) observed in our previous studies. The higher *Flux* values (up to about 25%), which were found in the high-Mg group in both parts of the cochlea, correspond to the lower hearing loss (TTS) measured in our functional studies and vice versa. Furthermore, the *Flux* signals which were found to be lower in the basal part of the cochlea than in the apical part, are in line with the frequency-related impairment of the cochlear function which was more pronounced at the high frequencies. When comparing the basal and apical *Flux* difference, it has to be considered, however, that the laser probe may measure different vascular beds in the two parts of the cochlea.

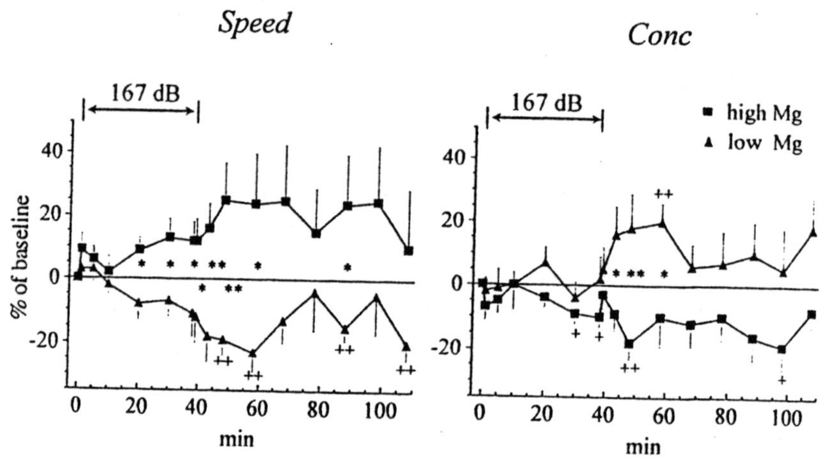

Figure. 2. Time courses of the basal *Speed* and *Conc* responses (mean ± SEM) to the impulse noise series (L_{peak} 167 dB, 1/s, 38 min) for the high-Mg group and the low-Mg group. (*/** $P < 0.05/0.01$ between the two groups; $^{+}/^{++}P < 0.05/0.01$ vs. baseline).

The decrease in the *Speed* signals associated with an increase in the *Conc* signals, which were observed in the low-Mg group (Fig. 2), indicates a vasodilation of the cochlear blood vessels and/or a RBC agglomeration following the noise exposure. The inverse noise-induced *Speed* and *Conc* responses found in the high-Mg group, suggest an improvement of the hemorheology, which may compensate the noise-induced vascular impairment. In addition to a possible effect of Mg on the vascular tonus, a rheological improvement of the CBF may contribute to the protective effect of Mg in the acoustic trauma. First results of a parallel study to test the effect of Mg on blood viscosity support this assumption.

Acknowledgement. This work was supported by the German Defence Ministry *(InSan I 0593-V-2694)*

References

1. Scheibe, H. Haupt, E. Rebentisch, M. Wenzel (1997) Prophylactic and therapeutic effects of magnesium on acoustic trauma: Animal studies. *Proceedings of the International Congress on Noise Control Engeneering, inter · noise 97*, Budapest, Hungary

INVESTIGATIONS ON THE MAGNESIUM STATUS MARKER IN HUMANS. THE ENVIRONMENTAL VARIABILITY OF THE MAGNESIUM LEVELS IN THE HAIR, PLASMA AND ERYTHROCYTES OF SCHOOL CHILDREN FROM POLAND

Z.ZACHWIEJA[1], H.BARTOŃ[1], M.SCHLEGEL-ZAWADZKA[1], M.FOŁTA[1], J.CHŁOPICKA[1], M.KROŚNIAK[1], J.PIETRZYK[2], A.HUZIOR-BAŁAJEWICZ[2], E.KAMIŃSKA[3], M.KOMOROWSKA[3]

[1]Department of Food Chemistry and Nutrition, Jagiellonian University, College of Medicine, 30-084 Kraków, Podchorążych str. 1, Poland

[2]First Department of Pediatrics, Polish-American Children's Hospital, Wielicka str. 265, 30-663 Kraków, Poland

[3]Department of Clinical Biochemistry, Polish-American Children's Hospital, Wielicka str. 265, 30-663 Kraków, Poland

1. Introduction

A deficiency in magnesium status is observed in different age groups of the Polish population [1]. It is believed there is a connection with inappropriate diet and with environmental pollution. The aim of the present study was to investigate the influence of some environmental factors and gender on children's magnesium status.

2. Materials and Methods

A population of around 300 randomly selected school children (T) consisting of an approximately equal number of boys (M) and girls (F) was included in the study. The children were 11 (± 0.5) years old and lived in southern Polish towns of various levels of industrialisation and environment contamination. The investigated population was analysed according to the groups of children (number of groups investigated, code) associated with the particular schools (7, C), classes (13, K) or type of environment (3, S: rural, urban, industrial).

The hair were washed according to the IAEA recommendation for environmental studies (1985), then digested in HNO_3 using the microwave system (MDS2000). The quantitative analyses of magnesium in the hair, erythrocytes and plasma were carried out by means of an AAS flame method (5100 ZL Perkin-Elmer). The results were analysed with either nonparametric or parametric tests for logarithmic values. The statistical tests of significance of the group's differentiation were performed separately for the (S-, C- or K-) groups of children of the same sex. The geometric means of Mg content for the particular groups were normalized to equal statistical weights of both sexes (the normalized geometric mean - GMC - was defined as the square root of the product of the geometric means for the male and female subgroups).

T. Theophanides and J. Anastassopoulou (eds.), Magnesium: Current Status and New Developments, 371-372.

3. Results and Conclusions

1. The variability of magnesium level for the whole population under investigation (standard deviation - SD compared vs. GM) is the lowest for plasma (ca. 9%), but higher for erythrocytes (28%) and hair (92%).

Table 1. The magnesium content mean values for the plasma, erythrocytes and hair of children for the total (T) and for ranges within the investigated enviromental groups*

Material	GM (T)	SD (T)	GMC (C)	GMM (C,M)	GMF (C,F)	GMM/GMF
Plasma	0.79	0.07	0.74 - 0.87	0.73 - 0.86	0.77-0.87	0.93 - 1.04
Erythrocyte	1.48	0.41	0.98 - 1.87	0.94 - 2.01	1.03-1.85	0.79 - 1.17
Hair	20.0	18.3	11.6 - 37.6	9.3 - 25.9	12.0-55.6	0.45 - 0.84

*the data are associated with the subgroup type given in parentheses, other symbols are defined in the text

2. The differentiation of Mg levels for the environmental groups is the highest for hair. This can be exemplified for the S-groups of females where the Kruskal-Walis and Median tests gave the following results:

Material	Kruskal-Walis test		Median test		
Plasma	H (2, N= 150) = 8.3	p =0.016	$\chi 2 = 6.7$	df = 2	p = 0.035
Erythrocyte	H (2, N= 113) = 24.7	p =0.000	$\chi 2 = 12.4$	df = 2	p = 0.002
Hair	H (2, N= 148) = 66.9	p =0.000	$\chi 2 = 49.1$	df = 2	p = 0.000

The significances became even less for finer fractions respective to the age (e.g. for girls of 11-years old - p= 0.22 for the median test for plasma).

3. The gender effect is statistically insignificant for plasma and erythrocytes, but became significant for hair ($p < 0.05$, at least). The girls had ca. 50% higher Mg levels in hair than boys (see GMM/GMF ratio in the table 1).

4. The geometric means of Mg content in the hair of boys and girls from a given group show both a positive logarithmic and linear correlation. {for K-groups: r-Spearman=0.83, r-Pearson=0.81, N=13, log Mg_GMM=0.28+0.68 x log Mg_GMF, p<0.001; Mg_GMF = (1.60 ± 0.18)xMg_GMM (p<0.0001)}

5. A positive, highly significant correlation of geometric means of Mg in the hair of boys and girls shows that factors which affect the Mg level in their hair are common for both sexes and linked to the environment.

6. Recognising such a high variability of the Mg levels even within precisely selected subgroups of children we believe that, currently, there are no bases for establishing the norms for magnesium levels in hair.

7. On the other hand, the high variability of the Mg content in the hair of children, as opposed to plasma and erythrocytes, observed both in connection with environment and with gender, leads us to hope that hair analysis will prove suitable for determining deficiencies of magnesium in the human population.

4. Reference.

Zdrojkowska, B., Rutkowska U., Szponar L. (1996) *Magnesium in the health profilaxis* Żyw.Człow.Met. **23**,169-178.

CONTENTS OF CALCIUM, MAGNESIUM, ZINC, COPPER AND IRON IN HAIR OF CHILDREN AND THE YOUNG

TADEUSZ KOZIELEC, PIOTR MICHOÑ
Departament of Family Medicine, Pomerian Medical Academy, Szczecin, Poland ul. Podgórna 22/23

Magnesium as well as a number of other bioelements like calcium, zinc, copper and iron play an important role in a proper work of the whole „chain" of intracellular metabolism.
Demand for trace elements in the developmental age of a man is increased and their supply in food often too small.
One of the methods used to determine the level of elements in a human organism is the biochemical analysis of hair. Diagnostic usability of that method is proved by the works of many authors, who revealed a correlation between elements in hair and their intracellular contents in an organism, both in the case of physiological balance and in pathological states.
The aim of this job was determination of current contents of some bioelements (magnesium, calcium, zinc, copper and iron) in hair of developmental age population i.e. from 1st to 20th year of life and their comparison with normal ranges published by other authors.

1. MATERIAL AND METHOD

The Examination was performed in the group of 695 people from the area of Szczecin in the years 1993-1996. The selection of children and the young for the examination was random. The examined were divided into age groups for each year or for a few years: 1st, 2-4th, 5-9th, 10-14th, 15-18th, 19-20th year of life. All the examined persons were healthy and nourished themselves with a standart diet.
Hair used for the determination had not been subjected to any procedure, which might have had an influence on the contents of the determined bioelemnts. Hair sections of the lenght of 3-4 cm were collected from 6-8 points of a head. The total weight of the hair egualled to 0.3-0.4 g.
Contents of elements was determined using the method of atomic absorption spectroscopy.

T. Theophanides and J. Anastassopoulou (eds.), Magnesium: Current Status and New Developments, 373-375.

The values of obtained concentration of the determined bioelements have been drawn up according to the applied division into age groups: infants, 2-4, 5-9, 10-14, 15-18 and 19-20 year old children. The applied division onto above age groups results from different development rate and demand for macro and microelements in children and the young in different age. The above clasiffication of age groups was based especially on the criterion connected with their habitation in changing environmental circumstances: infants nursery, nursery school, primary and secondary school, college. The obtained results were subjected to statistical analysis.

2. RESULTS AND DISCUSSION

The average values of the determined bioelements for the whole population (from 1st to 20th year of life) have been presented in micrograms and they egualled to: 386,20 μg/g d.m. for calcium, 28,37 μg/g d.m. for magnesium, 154,36 μg/g d.m. for zinc, 12,02 μg/g d.m. for copper and 20,13 μg/g d.m. for iron.

The results of obtained bioelements concentration in the six applied age groups were subjected to detailed analysis. Differences between average values of calcium concentration in the stated groups were statistically significant except the group of 2-4 and 5-9 year old children. The lowest average values have been obtained in the group of 2-4 year old children.

The average contents of magnesium in particular age groups differed in the way statistically significant and was lowest in the group of 2-4 year old and highest in the group of 15-18 year old children.

In case of zinc differences statistically significant occured between all selected age groups. The lowest values occured in the group of 2-4 and the highest in the group of 19-20 year old children.

The lowest average values for copper occured in the group of 2-4, and the highest in the group of 10-14 year old children. Differences in copper concntration between particular age groups were statistically significant.

The lowest average values for iron on the basis of biochemical analysis of hair have been obtained in the group of 10-14, and the highest in the group of 1 year old children. As a result of statistical calculation it was found that the differences in concentration of this element between the groups (similarly as in case of magnesium, zinc and copper) infants to 2-4 year old children., 5-9 to 10-14, 10-14 to 15-18, 15-18 to 19-20 year old children were statistcally sigificant.

The obtained results of average concentration of determined bioelements have been compared with the results from other authors (Table 1). Norms guoted in table 1 comprise the examined in various ranges of age and usually refer to adults. We did not manage to find in the available literature any tests concerning concentration of determined bioelements in the population

TABLE 1.Confrontation of own arithmetical means study bioelements in hair to other authors (µg/g s.m.).

Author	Calcium	Magnesium	Zinc	Cuprum	Ferrum
	$\bar{X}$	$\bar{X}$	$\bar{X}$	$\bar{X}$	$\bar{X}$
Own studies	386.20	29.37	154.36	12.02	20.13
Aalbers Th.G. de Goeij J.J.	440	38	165	11	6
Aleksandrowicz J. i wsp.	335	31	186.00	14.00	19.00
Baku S.B. Parr R.M.	522	60.2	174.1	10.23	23.70
Matsubara J. Chomori S.	410	38	150	9.0	10
Schrauzer G.	400	60	-	-	-

$\bar{X}$ - arithmetical meansfrom 1st to 20th year of life

Therefore it was difficult to perform any comparative analysis of our results with other normal ranges published in the literature. The obtained average values of determined bioelements in children and the young approximate the norms published by Matsubar J. and by Aleksandrowicz et al.

The concentration values of dicussed bioelements presented in our material are the results gathered in the last few years and concern the population of developmental age in Poland. They can be useful in determining deficiency and updating the values recognised as referential in the area of trace elements on the basis of biochemical analysis of hair.

SERUM MAGNESIUM EVOLUTION AS RELATED TO TREATMENT IN PATIENTS WITH LUNG CANCER

L. ALOUANE, Ch. GHAWAR, F. CHERIF, D. BELHABIB, Ch. BEN RAYANA, M. MAALEJ and K. NAGATI
Ecole superieure des Sciences et Technipues de la Sante de Tunis.
Institut National de Nutrition de Tunis. Hopital de Pneumo-phtisiologie, service ibn Nefis. Ariana

1. Introduction

In a previous study we did found that patients with lung cancer had serum magnesium levels significantly lower than in normal controls[1]. We then suggested the hypothesis stipulating an increased magnesium uptake by the tumour cells. Malignant cells have an increased requirement for substances involved in cell multiplication, and are able to take Mg from extracellular pool[2].

In our present work we tried to study the effect of treatment on serum magnesium evolution in patients with lung cancer.

2. Patients and methods

Serum Mg was determined in 55 patients affected by lung cancer. Treatment types and other characteristics of the different groups are reported in Table 1. All subjects gave their informed consent to the study.

TABLE 1. Treatment types and characteristics of the patients

Treatment	n	Histologic type	n	age years	Tobacco pack year
symptomatic	41	Epidermoid	13	64±9	52±40
		Adenocarcinoma	23	60±15	61±23
		Mycrocytoma	5	73±13	48±32
Chemotherapy	4	Microcytoma	4	43±6	37±48
Radiotherapy	6	Adenocarcinoma	6	5 55±11	34±15
Surgery	4	Adenocarcinoma	4	57±3	57±16

Serum Mg was assayed by Atomic Absorption Spectroscopy.
The results are expressed as means $\pm$ SD. Statistical analysis was performed using the Student's t-test.

T. Theophanides and J. Anastassopoulou (eds.), Magnesium: Current Status and New Developments, 377-380.

3. Results

For the forty one patients who received a symptomatic treatment, serum Mg was evaluated twice, 1 and 2 months after diagnosis. By the end of the second moth, a significant decrease in serum magnesium was observed in all patients (p<0,002) Table 2

TABLE 2. Mg evolution in patients with symptomatic treatment.

Hystologic type	Initial value Mg mmol/l	After 1 month Mg mmol/l	After 2 months Mg mmol/l
Epidermoid	0.8±0.19	0.73±0.13	0.56±0.11*
Microcytoma	0.81±0.15	0.60±0.16	0.59±0.08**
Adenocarcinoma	0.83±0.16	0.64±0.11	0.62±0.16*

*p<0.002; **p<0.02

Four patients underwent chemotherapy for six cures. Serum magnesium was determined just before the first treatment administration and two weeks after every cure. Serum Mg showed a slight progressive decrease which reached statistical significance after the second cure (p<0.01) Table 3.

Six patients hace had radiotherapy treatment in fice session. After the first session of radiotherapy there is a significant decrease in serum magnesium levels when compared to initial levels (p<0.02). The Mg values started to progressively increase to reach normal concentration at the end of the fifth session (Table 4).

Four patients underwent surgery. Serum Mg showed a significant increase on 1,3 and 7 days after intervention (p<0.001) and returned to initial lebels on day 52. Serum magnesium reached normal levels by the end of the forth month and remained at steady level nine months following the surgery (Table 5).

TABLE 3. Evolution of magnesium in chemotherapy patients

cure	initial value	First	Second	Third	Fourth	Fifth	Sixth
Mg mmol/l	0.79±0.06	0.67±0.07	0.63±0.02*	0.74±0.11	0.75±0.06	0.69±0.05	0.61±0.08*
	0.72-0.89	0.58-0.77	0.59-0.66	0.56-0.82	0.72-0.86	0.59-0.74	0.48-0.67

*p<0.01 vs initial value.

TABLE 4. Radiotherapy effect on serum magnesium

Session	Initial value	First	Second	Third	Fourth	Fifth	Sixth
Mg mmol/l	0,79±0,06	0,67±0,07	0,63±0,02*	0,74±0,11	0,75±0,06	0,69±0,05	0,61±0,08*
	0,72-0,89	0,58-0,77	0,59-0,66	0,56-0,82	0,72-0,86	0,59-0,74	0,48-0,67

[1]p<0,02, [2]p<0.03, [3]p<0.007 vs initial value

TABLE 5. Effect of surgery on serum magnesium of patients

Days after surgery	Initial value	Just after operation	3	7	52	112	285
Mg mmol/l	0.75±0.04 0.71-0.79	1.14±0.04• 1.08-1.18	1.04±0.05 0.95-1.10	0.92±0.04 0.88-0.98	0.80±0.06 0.75-0.88	1.06±0.07• 1.10-1,17	1.08±0.04• 1.04-1.15

•p<0.001

4. Discussion:

Mg is believed to play an important role in the growth of cancer and in oncogenesis[3]. There is an absolute requirement for Mg in all enzyme reactions catalyzed by ATP. It is required for FNA synthesis, ribosomal integrity and protein synthesis[4].

The study of serum magnesium evolution in our patients with lung cancer revealed that serum magnesium was significantly decreased 2 months after diagnosis when patients were only having a symptomatic treatment. As no regression in the tumor growth is observed, malignant cells continue to have an enhanced uptake of Mg from the extracellular circulating pool. This explains the lowered concentration in serum Mg.

Radiotherapy did not improve magnesium status in our patients at the begining of treatment. Nevertheless after five sessions of treatment the Mg levels reached normal values. This is may be due to a reduction of tumor cell proliferation.

The decrease of serum Mg after chemotherapy administration could be explained by an enhanced cellular Mg uptake from extracellular pool, in order to repair cellular, subcellular and nuclear structure damage due to chemotherapy drugs.

In the postoperative period magnesium rose on days 1.3 and 7, and returned to preoperative levels on day 52. Our results are in agreement with those of Cordova[5] who believe that these variations are due to a release of magnesium from muscle caused by a high catabolic rate during the convalescence period or to a release of magnesium from damage tissues. Other researchers[6] have described how surgical stress can provoke an early period of hypermagnesaemia; this may occur as a result of a reduction in magnesium excretion.

After four months the magnesium levels reached normal values with an amelioration of the patients health status. Tumor resection ameliorated the serum magnesium level. These results support our previous findings stipulating that an increased consumption of magnesium by tumor cells is basically behind a lowered serum level in patients with.lung cancer.

References

1. Alouane L., Alguemi Ch., Cherif F., Hedhili A., Maalej M.. & Nagati K. (1996): Serum magnesium, zinc and copper in lung cancer. *Cur. Res. in Magnes.* Ed. By MJ Halpern & J. Durlach, Chapt. **22**: 87-89.
2. Sartori S., Nielson I., Tassinari D., Mazzotta D., Vecchiatti G., Sero A. & Abbasciano V. (1992): Serum and erythrocyte magnesium concentrations in solid tumors: relationship with stage of malignancy. *Magnes. Res.* **5**: 189-192.

3. Durlach J., Bara M., Guiet-Bara A. & Collery p. (1986): Relationship between magnesium, cancer and carcinogenic or anticancer metals. *Anticancer Res.* **6**: 1353-1362.
4. Tazawa T., Itoh T., Kobayashi M., Saito K. (1990): Changes in blood magnesium and calcium concentrations in patients with stomach cancer. . *Magnes. Res.* **4** , 275-281.
5. Cordova A. (1995): Variations of serum magnesium and zinc after surgery, and postoperative fatigue. *Magnes. Res.* **8**: 367-372.
6. Hessov I., Hasselblad C., Fasth S. & Hulten L. (1983): Magnesium deficiency after ileal resections for Crohn's Disease. *Scand. J. Gastroenterol.* **18**: 643-649.

LATE PAPERS

The following papers were submitted
too late to be included in the appropriate section.
Their content is not indexed.

MAGNESIUM AND ATHEROGENESIS

B.M. ALTURA AND B.T. ALTURA
Departments of Physiology and Medicine and The Center for Cardiovascular and Muscle Research, The State University of New York, Health Science Center at Brooklyn, Brooklyn, N.Y. 11203 USA

Summary: Dietary Mg-deficiency in rabbits and rats induces a hyperlipemic state, oxidation of lipoproteins, vasospasm, ischemia, increased vascular reactivity, enhanced Ca^{2+} uptake in arterial walls (smooth muscle and endothelial cells) often followed by atherosclerosis or a pro-atherogenic state. Experimentally, Mg-deficiency can lead to rapid changes in vascular and cardiac muscle membrane fatty acids and phospholipids with concomitant alterations in cellular sphingolipid second messengers. Mg-deficiency can also lead to alterations in vascular, cardiac and brain pH and cellular bioenergetics. Clinically, a large segment of the cardiovascular high risk population (e.g., ischemic heart disease, hypertension and stroke) demonstrate a high incidence of abnormally lowered levels of serum ionized Mg^{2+}, but not serum total Mg levels. Such lowered levels of ionized Mg^{2+} (e.g., 0.3-0.48 mM) when mimicked in-vitro with primary cultured cerebral, aortic and coronary vascular smooth muscle cells result in spasm-contraction, intracellular Ca^{2+} ($[Ca^{2+}]_i$) overload and a resistance to normalization of $[Ca^{2+}]_i$ when normal or high $[Mg^{2+}]_o$ is reintroduced. Collectively, the available data suggest that lowered serum and tissue Mg^{2+} levels results in a pro-atherogenic state which eventuates in hypertension, ischemic heart disease, acute myocardial infarction and/or stroke. Extracellular Mg^{2+} appears to be a true regulatory cation. Lowered serum ionized magnesium should be added to the list of other well-known cardiovascular risk factors.

1. Introduction

Hypercholesterolemia, hypertension, diabetes mellitus, cigarette smoking, immune injury, end-stage renal disease, renal dialysis, and prolonged stress all are widely accepted as risk factors for atherosclerosis. No common link has been identified which forms a rational underlying basis to these disorders and atherogenesis. Moreover, it is not clear how lipoproteins, Ca^{2+} and macrophages gain access to the normally impermeable arterial walls or what allows plyable, physiologic vascular smooth muscle (VSM) cells to change their state (phenotype) from a contractile cell to a non-contractile

T. Theophanides and J. Anastassopoulou (eds.), Magnesium: Current Status and New Developments, 383-395.

synthetic (or modulatory) cell [1, 2]. Nor is it clear how these diverse risk factors lead to ischemic heart disease (IHD) and stroke.

During the past two decades, information and data has accumulated to suggest that magnesium (Mg) deficiency caused either by poor diet and/or errors in body Mg metabolism may be a missing link between the diverse cardiovascular risk factors and atherosclerosis [3-16]. New evidence from our laboratories points to potential intracellular signals linking extracellular Mg^{2+} deficiency to etiology of atherogenesis and vascular disease. Numerous clinical studies performed in conjunction with several colleagues indicates the utility of Mg^{2+} ion selective electrodes (ISE's) in the diagnosis and treatment of several of the the above vascular disorders.

Calcium ions are believed to be the most important divalent cations which can act as second messengers. Intracellular levels of free Ca^{2+} is usually below 100 nM in most eukaryocytes and prokaryocytes, and these levels of Ca^{2+} are increased with various types of agonists, hormones, etc., by way of receptors which in turn trigger various signal transduction cascades. Mg^{2+} ions on the other hand are not thought, generally, to act as second messengers. Recent studies on animal cells and tissues as well as human blood (and serum) and red blood cells, reviewed below, indicate that deficits in ionized Mg^{2+} can result in activation of numerous vascular and cardiac cellular and molecular events. Such events, we believe, are suggestive of the probability that Mg^{2+} ions can act as extracellular signals in the cardiovascular system. These data also suggest rather strongly that low serum ionized Mg^{2+} levels should be added to the above major cardiovascular risk factors.

2. Magnesium Modulates Blood Lipids and Atherogenesis

Evidence from both animals and humans suggests that dietary and blood levels of Mg^{2+} may modulate serum levels of lipids and lipoproteins [10, 16-23]. Using rabbits, our laboratory has reported that the dietary level of Mg modulates the serum level of cholesterol and triglycerides in normal animals; the lower the Mg intake, the higher the serum lipid levels [10, 16, 19].

In rabbits fed a high-cholesterol diet, Mg deficiency, comparable with the reduced dietary intake seen today (i.e. 30-40% normal) in the adult population, exacerbates atherogenesis and lipid deposition in arterial muscle and it stimulates macrophages. Pretreatment with orally administered Mg aspartate attenuated the atherosclerotic state and lowered serum cholesterol and triglycerides [10, 16, 19]. In these studies, the extent of atherogenic lesions was poorly correlated with the level of serum cholesterol and highly dependent on the level of dietary Mg and on the Ca:Mg ratio.

High blood levels of lipids and lipoproteins can mask a "true" Mg-deficient state [16]. A similar hidden state of Mg deficiency has been observed by our group in renal transplant patients as well as in end-stage renal disease (ESRD) [14, 23, 24]. These people are characterized by extensive, unexplained atherogenesis. In all such patients, we found a marked lowering of free but not total Mg, decreased urinary output of Mg, hyperinsulinemia, and elevated total serum cholesterol or triglycerides or both.

ICa^{2+}:IMg^{2+} ratios, signs of elevated vascular tone, increased vascular reactivity, and risk of atherogenesis were also significantly elevated in our stable renal transplant recipients and subjects with ESRD. TMg in these patients is usually normal or elevated, respectively, suggesting that in advanced atherogenesis and states of hyperlipidemia, a normal TMg level would often be seen, despite the fact that the body and, specifically, the vascular walls would be deficient in free Mg^{2+}.

3. Magnesium Exists in Three States in Blood

Blood normally contains magnesium ions (Mg^{2+}) in three states: bound to proteins; complexed to small anion ligands such as bicarbonate, phosphate, citrate, lactate, acetate, etc; and free. In tissues, most of the Mg^{2+} is bound to ATP. Most clinical laboratories measure total Mg (TMg) levels by colorimetry or atomic absorption spectroscopy. However, it is the free ionized form of Mg that is physiologically active in blood and body fluids. Usual estimates of free Mg^{2+} concentration have relied upon TMg measurements in protein-free ultrafiltrates which, of course, exclude the protein-bound Mg.

Since the levels of anions can vary significantly in pathologic states, and in view of the role apparently played by Mg in cellular homeostasis, it is desirable to directly measure free Mg^{2+} in blood and other body fluids. With this in mind, Mg^{2+}-sensitive ion-selective electrodes (ISEs) have been designed to obtain these measurements in the presence of potential cationic interferences and other interferents [25].

4. Deficits in Ionized Mg^{2+} in Ischemic Heart Disease, Acute Myocardial Infarction and Stroke

If deficiency in cardiac and vascular tissue Mg^{2+} does play a role in etiology of atherogenesis and vascular disease, as the above suggests, then one would anticipate that a large number of patients with ischemic heart disease (IHD), acute myocardial infarction (AMI) and stroke should demonstrate blood, cardiac and vascular tissue deficiency of Mg. Although some evidence for decreased content of blood TMg, and cardiac and coronary VSM TMg has been seen in some patients with IHD, AMI and stroke, the data are not consistent; most of the consistent data has been noted in cardiac tissue (i.e., mean 20% loss of TMg in hearts) of patients who died of sudden death ischemic heart disease [3, 5, 18, 26-33]. We believe a great deal of these inconsistencies, up to recently, have been due to the inability to measure serum IMg^{2+} and$[Mg^{2+}]_i$.

Data obtained recently, by our group, clearly shows, for the first time, consistent deficits in serum IMg^{2+} in many IHD patients (NY Heart Assn. classes 2-4), patients early after AMI, and patients early after stroke (e.g., cerebral infarction; subarachnoid hemorrhage; cerebral hemorrhage) [10, 13, 14, 19, 34-39]. Measurements of $[Mg^{2+}]_i$ in cardiac and vascular tissue are not as yet available. Since we have recently demonstrated a close correspondence between extracellular serum IMg^{2+} measurements (but not TMg) and RBC $[Mg^{2+}]_i$ measurements in untreated hypertension [5, 37], NIDDM [40], gestational diabetes in pregnant women [41], pregnant women during labor [42] and ESRD patients [22, 29], we believe that the deficits in IMg^{2+} noted in the IHD, AMI and stroke patients are most likely reflecting cardiovascular cell and tissue loss of $[Mg^{2+}]_i$.

5. Low $[Mg^{2+}]_o$ Produces Vasospasm, Increased Vascular Tone, and Elevation in $[Ca^{2+}]_i$

Approximately 25 years ago, we demonstrated, for the first time, that reduction in $[Mg^{2+}]_o$ resulted in sustained vasospasm of most types of small peripheral and cerebral blood vessels; the lower the $[Mg^{2+}]_o$, the greater the developed tension and decreased arteriolar lumen sizes [15, 17, 29, 43-45]. Reduction of $[Mg^{2+}]_o$ also resulted in potentiation of most vasoconstrictor and pressor agents, and decreased responses to dilator agents [29, 44, 45]. The end result of lowered serum IMg^{2+} and $[Mg^{2+}]_i$ would be vasospasm, increased vascular tone, increased vascular reactivity, reduction in peripheral blood flow (hypoxia), and thus local vascular injury, a potent stimulus for hypertension and atherogenesis [1].

Using a combination of radiolabeled ^{45}Ca, digital image analysis with fluorescent probes, and confocal laser scanning microscopy with fluorescent probes, we have found that Mg^{2+} gates special membrane Ca^{2+} channels in both VSM and endothelial cells as well as controlling the release of intracellular free cytosolic Ca^{2+} ($[Ca^{2+}]_i$) from these cells [17, 44-55]. The entry and release of $[Ca^{2+}]_i$, and loss of membrane-Mg^{2+}, most likely initiates a number of intracellular signals (i.,e., growth factors, adhesion factors, cytokines, sphingolipids, and phospholipids; see below) which could result in VSM cell profiferation (and cell transformation), alteration in endothelial cell wall permeability, macrophage attraction, adhesion and cell wall entry, as well as increased vascular resistance (hypertension), coronary vasospasm and occlusion (IHD, AMI), and cerebral vasospasm (cerebral infarction, cerebral hemorrhage). Low $[Mg^{2+}]_o$-induced vasospasm could accelerate uptake of oxidized lipids and/or stimulate generation of oxygen radicals (see below).

6. Low $[Mg^{2+}]_o$ Results in Formation of Pro-inflammatory Agents

Feeding rats and hamsters diets deficient in Mg, for periods of from 14-21 days, results in cardiomyopathic lesions, cytokine formation and release (e.g., interleukins [IL]-1, 2, and 6; tumor necrosis factor [TNF]-α), and endothelin formation [56-58]. Although other growth factors (e.g., transforming growth factor β; fibroblast growth factor), have not as yet been identified in Mg deficient states, it is likely that Mg^{2+}-deficiency will be found to cause production of many of the well-known growth factors implicated in atherogenesis.

7. Low $[Mg^{2+}]_o$ Results in Lipid Peroxidation and Free Radicals

Until recently it was not known how Mg deficiency could promote cardiovascular damage. During the past several years, reports from several laboratories, including ours, have suggested that Mg deficiency can lead to formation of ferrylmyoglobin and lipid peroxidation products.

We suggest that the link between Ca^{2+} overload induced by Mg deficiency and cardiac myocyte membrane damage is formation and action of the ferrylmyoglobin radical [59]. Using rat perfused working hearts, we have prevented such injury by using one-electron reductants such as ascorbate [59].

Other lipid peroxidation products appear 7-21 days after dietary Mg deficiency is initiated in rats and hamsters [13, 56-58]. Several different antioxidants have been reported to afford complete protection against myocardial necrosis in these animal models [56, 57]; and against cardiac failure in our perfused hearts [59], strengthening the idea that free radicals are important in vascular injury induced by Mg deprivation. It is also possible that these antioxidants prevent loss of cellular Mg^{2+} or promote enhanced transport of Mg^{2+} into the vascular, endothelial, and cardiac cells [13].

8. Low $[Mg^{2+}]_o$ Enhances Platelet Aggregation and Thrombotic Tendencies

Several studies dating back to the late 1950's indicate that low $[Mg^{2+}]_o$ promotes blood coagulation [32, 61]. More recently, several investigators have demonstrated that platelet agregation is enhanced in low $[Mg^{2+}]_o$ environments [62] and that megakaryocyte numbers and white cell counts are increased in Mg-deficient hamsters [63]. Such pathways would obviously tend to produce thrombotic tendencies in atherogenesis and might play important roles in the "cracking and bleeding" of atherosclerotic plaques known to be seen in IHD and AMI patients. These pathways could also be set into motion by Ca^{2+} overload, ischemia and stasis, and elaboration of growth factors induced by Mg deficiency.

9. Low $[Mg^{2+}]_o$ Results in Deficits in Coronary Flow, Reduced Cardiac Bioenergetic Reserves, and Cardiac Failure

Although Prinzmetal angina, SDIHD and AMI all can lead to arrhythmias and cardiac failure, until recently it has not been possible to explain these events as due to a Mg deficient state. Using intact perfused working rat hearts, exposed to low, normal or high $[Mg^{2+}]_o$, we have been able to establish a rational basis for how Mg^{2+}-deficiency might provoke cardiac failure.

In-vitro ^{31}P-NMR spectroscopy studies performed in our laboratories, on the intact perfused rat heart, indicates that acute elevation of extracellular free Mg^{2+} concentrations can increase phosphocreatine (PCr) levels, intracellular pH, and intracellular free Mg^{2+} levels [64]. Intracellular inorganic phosphate (Pi) levels decreased, and the cytosolic phosphorylation potential and free energy of ATP hydrolysis increased accordingly [64]. Elevated levels of $[Mg^{2+}]_o$ also resulted in increases in coronary flow, cardiac output, and stroke volume, and there was a concomitant fall in oxygen consumption [64, 65]. Many of these beneficial hemodynamic effects were also observed more than 12 years ago in intact dog hearts in-vivo [66].

Acute reductions in $[Mg^{2+}]_o$ result in opposite effects on cardiac performance, high-energy phosphates, intracellular pH, $[Mg^{2+}]_i$, cytosolic phosphorylation potential, and free energy of ATP hydrolysis [67]. Low $[Mg^{2+}]_o$ also resulted in decreased coronary flow, stroke volume, and cardiac output, leading to cardiac failure. These observations suggest that Mg^{2+} regulates cardiac performance, oxygenation, and substrate delivery in the myocardium. Elevated intracellular free Mg^{2+} and pH could be expected to enhance creatine kinase activity as well as that of other myocardial cellular enzymes, increasing cardiac efficiency, stroke volume, and coronary flow, whereas reductions in $[Mg^{2+}]_i$ and pH (cytosolic acidosis) would do the opposite.

10. Low $[Mg^{2+}]_o$ Results in Oxidation of LDL

Oxidized modification of lipoproteins is a well-recognized step in atherogenesis [1, 68] and Mg deficiency is known to result in alterations in plasma lipoproteins and triglycerides (see above) [16, 21, 22]. A recent study, from France, has now clearly demonstrated that Mg levels affect the susceptibility of lipoproteins to peroxidation [69]. Oxidized lipoproteins (e.g., oxidized very-low density lipoproteins [VLDL] and LDL) are clearly cytotoxic to cells in culture [1, 68]. This Mg^{2+}-mediated pathway may be an important step whereby oxidized LDL stimulates monocyte adhesion, as has been shown recently for LDL treatment of endothelial cells [70].

11. Magnesium-Dependent Changes in Fatty Acid Saturation, Double Bond Content and Double Bond Oxidation of VSM: Possible Relation to Intracellular Signaling Pathways

In atherogenesis and vascular disease, it has been demonstrated that membrane phospholipids are altered. The failure to focus on the latter may be one of the chief reasons that neither the frequency of restenosis after percutaneous transluminal coronary angioplasty (PTCA) has improved nor have the varied recommended treatments met with success. Approximately 15 years ago, our laboratory reported that dietary Mg-deficiency in rats results in alterations of membrane phospholipids and fatty acid saturation in intact cardiac tissue [71]. Our newer work has strengthened and extended these findings [72]. This early finding may be particularly important in light of recent evidence that indicates that phosphoinositide-derived second messengers are important in regulation of Ca^{2+} in VSM [73].

With such a background, we recently investigated whether Mg^{2+} can modulate fatty acid saturation, double bond oxidation and plasmalogen (α, β-unsaturated ether) content in rat aortic and canine cerebral VSM membranes [74, 75]. In our preliminary observations, using ion exchange chromatography to separate lipids and ^{1}H-NMR spectroscopy, we noted that exposure of primary cultured aortic and cerebrovascular VSM cells to low $[Mg^{2+}]_o$ for 18 hr increased the fatty acid content of aortic VSM about three-fold. Fatty acid chain length and bouble bond (DB) content clearly was increased two-fold when aortic VSM cells were exposed to a medium containing 0.17 mM or 0.3 $[Mg^{2+}]_o$ vs. one containing 1.2 mM $[Mg^{2+}]_o$; the lower the $[Mg^{2+}]_o$, the lower the fatty acid chain length and DB content. These data show that the level of fatty acid unsaturation is quite sensitive to $[Mg^{2+}]_o$ concentrations over the human pathophysiological range (0.3 to 0.6 mM) found by our ISEs in patients with hypertension, ESRD, IHD, AMI, NIDDM, and stroke (see above). A concomitant decrease in fatty acid chain length and DB indicates oxidation of DB resulting in truncation of fatty acids.

We also found changes in the *α, β*-unsaturated ether protons of extracts of canine cerebrovascular smooth muscle cells incubated for 18 hrs in different $[Mg^{2+}]_o$ [74, 75]. In contrast to the classical double bonds, discussed above, the *α-β*-unsaturated ether content of the N-containing phospholipid fraction increases in VSM cells as the $[Mg^{2+}]_o$ decreases. Collectively, such data, when viewed in light of our earlier data on cardiac muscle [71, 72] suggests that extracellular Mg^{2+} can modulate phospholipid moieties and content in VSM cell membranes. Such attributes of extracellular Mg^{2+} could be pivotal in generating and controlling intracellular signal pathways.

Using primary rat aortic and canine cerebral VSM cell cultures, fura-2, indole-1,and fluo-3, digital image microscopy and laser scanning confocal microscopy, we have found that a reduction of $[Mg^{2+}]_o$ by 35-40% can result in an approximate 35-45fold rise in nuclear $[Ca^{2+}]_i$, but only a 6- and 10-fold rise, respectively, in perinuclear and cytosolic cell $[Ca^{2+}]_i$ [5, 45]. Since large pre-RNA and tRNA transcription could be activated by such rises in nuclear $[Ca^{2+}]_i$, it is possible that this rapid elevation in $[Ca^{2+}]_i$ is pivotal in setting many subcellular signaling and molecular events into motion which are causal in transforming VSM cells and in their proliferation.

12. Conclusions

By demonstrating that reduction in $[Mg^{2+}]_o$ can induce an entire array of pathophysiological phenomena (e.g., vasospasm, increased vascular reactivity, elevation in $[Ca^{2+}]_i$, formation of pro-inflammatory agents, oxygen radicals, platelet aggregation, reduction in cardiac bioenergetics, cardiac failure, oxidation of lipoproteins, changes in membrane fatty acid saturation, plasmalogen content and N-phospholipids [suggesting changes in intracellular signals]), we have provided a rationale for how extracellular Mg-deficiency could result in hypertensive vascular disease, atherogenesis, IHD, AMI and stroke. Inadequate dietary intake of Mg or errors in Mg metabolism could result in dyslipidemias, atherosclerosis, vascular thrombosis, stroke, and sudden cardiac death.

13. Acknowledgments

We are indebted to The National Institutes of Health (AA-08674) who has supported many of our basic research and clinical investigations summarized herein. We are also very much indebted to several long-term collaborators: Asefa Gebrewold, Raj K. Gupta, Sara M. Handwerker, Adele B. Kostellow, Mariana S. Markell, Gene A. Morrill, and Aimin Zhang.

14. References

1. Ross, R. (1993): Atherosclerosis: a defense mechanism gone awry. Am. J. Pathol. 143:987-1001.
2. Schwartz, S.M.; deBois, D. and O'Brien, E.R.M. (1995): The intima. Soil for atherosclerosis and restenosis. Circ. Res. 77:445-465.
3. Altura, B.M. and Altura, B.T. (1990): Role of magnesium in pathogenesis of high blood pressure: Relationship to its actions on cardiac and vascular smooth muscle. In: Hypertension: Pathophysiology, Diagnosis, and Management. Laragh, J.H.; Brenner, B.M., eds., pp. 1003-1025, New York: Raven Press.
4. Altura, B.M. and Altura, B.T. (1994): Role of magnesium and calcium in alcohol-induced hypertension and strokes as probed by in-vivo television microscopy, digital image microscropy, optical spectroscopy ^{31}P-NMR, spectroscopy and a unique magnesium ion-sensitive electrode. Alcoholism: Clin. Exp. Res. 18:1057-1068.
5. Altura, B.M. and Altura, B.T. (1995): Role of magnesium in the pathogenesis of hypertension updated: Relationship to its actions on cardiac, vascular smooth muscle, and endothelial cells. In: Hypertension: Pathophysiology, Diagnosis, and Management, 2nd Edition. Laragh, J.H.; Brenner, B.M., eds., pp. 1213-1242, New York: Raven Press.
6. Altura, B.M.; Altura, B.T.; Gebrewold, A.; Ising, H. and Gunther, T. (1984): Magnesium deficiency and hypertension: Correlation between magnesium-deficient diets and microcirculatory changes in situ. Science 223:1315-1317.
7. Altura, B.M.; Altura, B.T.; Gebrewold, A.; Gunther, T. and Ising, H (1992): Noise-induced hypertension and magnesium: Relationship to microcirculation and Ca. J. Appl. Physiol. 72:194-202.
9. Altura, B.M. and Altura, B.T. (1983): Influence of magnesium on vascular smooth muscle and serum biochemical parameters from diabetic and hypertensive rats. Magnesium 2:253-266.
10. Altura, B.M. and Altura, B.T. (1996): Role of magnesium in pathophysiological processes and the clinical utility of magnesium ion selective electrodes. Scand. J. Clin. Lab. invest., 56(suppl. 224):211-234.
11. Ng, L.L.; Davies, J.E. and Ameen, M. (1992): Intracellular free-magnesium levels in vascular

smooth muscle and striated muscle cells of the spontaneously hypertensive rats. Metabolism 41:722-727.

12. Nadler, J.L.; Goodson, S. and Rude, P. (1987): Evidence that prostacyclin mediates the vascular action of magnesium in humans. Hyperten. 9:379-383.

13. Altura, B.M. and Altura, B.T. (1995): Magnesium in cardiovascular biology. Scientific Am. Science & Med. 2:28-37.

14. Altura, B.T. and Altura, B.M. (1997): Ionized magnesium measurements in serum, plasma and whole blood in health and disease. In: Advances in Magnesium Research, Vol. 1, ed. R. Smetana, pp. 538-546, London: John Libbey, in press.

15. Altura, B.M. and Altura, B.T. (1974): Magnesium and contraction of arterial smooth muscle. Microvascular Res. 7:145-155.

16. Altura, B.T.; Brust,M.; Barbour, R.L.; Bloom, S.; Stempak, J. and Altura, B.M. (1990): Magnesium dietary intake modulates blood lipid levels and atherogenesis. Proc. Nat. Acad. Sci. USA 87:1840-1844.

17. Altura, B.M. (1982): Magnesium and regulation of contractility of vascular smooth muscle. Adv. in Microcirculation 11:77-113.

18. Altura, B.M. (1988): Ischemic heart disease and magnesium. Magnesium 7:57-67.

19. Altura, B.M. and Altura, B.T. (1995): Magnesium metabolism and cardiovascular pathobiology. In: Magnesium and Physical Activity, ed. L. Vecchiet, pp. 37-70. Lancashire: Parthenon.

20. Davis, W.H.; Leary, W.P.; Reyes, A.J. and Olhaberry, J.V. (1984): Monotherapy with magnesium increases abnormally low high density lipoprotein cholesterol: a clinical assay. Curr. Therap. Res. 36: 341-346.

21. Rayssiguier, Y. (1986): Magnesium, lipids and vascular disease. Magnesium 5:182-190.

22. Rasmussen; Autrup, P.; Golstein, K.; Mcnair, P.; Mortensen,P.B.; Larsen, O.G. and Lawaetz, H. (1989): Influence of magnesium substitution therapy on blood lipid composition in patients with ischemic heart diseases. Arch. Intern. Med. 149:1050-1053.

23. Markell, M.S.; Altura, B.T.; Barbour, R.L. and Altura, B.M. (1993): Ionized and total magnesium levels in cyclosporin-treated renal transplant recipients: relationship with cholesterol and cyclosporin levels. Clin. Sci. 85:315-318.

24. Markell, M.S.; Altura, B.T.; Sarn, Y.; Delano, B.G.; Ifudo, O.; Friedman, E.A. and Altura, B.M. (1993): Deficiency of serum ionized magnesium in patients receiving hemodialysis or peritoneal dialysis. ASAIO J. 39:M801-M804.

25. Altura, B.M. and Lewenstam, A. (1994): Unique magnesium ion selective electrodes. Scand. J. Clin. Lab. Invest. 54 (Suppl. 217): 1-100.

26. Crawford, T. and Crawford, M.D. (1967): Prevalence and pathological changes of ischaemic heart disease in a hard-water and in a soft-water area. Lancet 1:229-232. 1967.

27. Chipperfield, B. and Chipperfield, J.R. (1973): Heart muscle magnesium, potassium, and zinc concentrations after sudden death from heart disease. Lancet 2:293-296.

28. Chipperfield, B. and Chipperfield, J.R. (1978): Differences in metal content of the heart muscle in death from ischemic heart disease. Am. Heart J. 95:732-737.

29. Turlapaty, P.D.M.V. and Altura, B.M. (1980): Magnesium deficiency produces spasms of coronary arteries: Relationship to etiology of sudden death ischemic heart disease. Science 208:198-200.

30. Rasmussen, H.S. (1993): Justification for magnesium therapy in acute ischaemic heart disease. Clinical and experimental studies. Danish Med. Bull. 40:84-99.

31. Altura, B.M. (1979): Sudden-death ischemic heart disease and dietary magnesium intake: Is the

target site coronary vascular smooth muscle? Medical Hypoth. 5:843-849.

32. Seelig, M.S. (1980): Magnesium deficiency in the pathogenesis of disease. New York: Plenum Press.

33. Eisenberg, M.J. (1991): Magnesium deficiency and sudden death. Am. Heart J. 10:269-280.

34. Altura, B.M. and Altura, B.T. (1995): Magnesium and cardiovascular biology: An important link between cardiovascular risk factors and atherogenesis. Cell Molec. Biol. Res. 41:347-360.

35. Altura, B.T. and Altura, B.M. (1992): Measurement of ionized magnesium in whole blood, plasma and serum with a new novel ion-selective electrode in healthy and diseased human subjects. Magnesium Trace Elem. 10:90-98.

36. Altura, B.T., Shirey, T.L., Young, C.C., Dell'Orfano, K., Hiti, J., Welsh, R., Yeh, W., Barbour, R.L. and Altura, B.M. (1994): Characterization of a new ion selective electrode for ionized magnesium in whole blood, plasma, serum, and aqueous samples. Scand. J. Clin. Lab. invest. 54 (Suppl. 217):21-36.

37. Altura, B.T.; Burack, J.; Cracco, R.Q.; Galland, L.; Handwerker, S.M.; Markell, M.S.; Mauskop, A.; Memon, Z.S.; Resnick, L.M.; Zisbrod, Z. and Altura, B.M. (1994): Clinical studies with the NOVA ISE for Mg^{2+}. Scand. J. Clin. Lab. Invest. 54(Suppl. 217):53-68.

38. Altura, B.T., Bertschat, F., Jeremias, A., Ising, H. and Altura, B.M. (1994): Comparative findings on serum IMg^{2+} of normal and diseased human subjects with the NOVA and KONE ISE's for Mg^{2+}. Scand. J. Clin. Lab. Invest. 54 (Suppl. 217): 77-82.

39. Altura, B.T.; Memon, Z.I.; Zhang, A.; Cheng, T.P.-O.; Silverman, R.; Cracco, R.Q. and Altura B.M. (1997): Low levels of serum ionized magnesium are found in patients early after stroke which result in rapid elevation in cytosolic free calcium and spasm in cerebral vascular muscle cells. Neuroscience Lett. 230: 37-40.

40. Resnick, L.M.; Altura, B.T.; Gupta, R.K.; Alderman, M.H. and Altura, B.M. (1993): Intracellular and extracellular magnesium depletion in type 2 (non-insulin-dependent) diabetes mellitus. Diabetologia 36:767-770.

41. Bardicef, M.; Bardicef, O.; Sorokin, Y.; Altura, B.M.; Altura, B.T.; Cotton, D.B. and Resnick, L.M. (1996): Extracellular and intracellular magnesium depletion in pregnancy and gestational diabetes. Am. J. Obstet. Gynecol. 172:1009-1013.

42. Bardicef, M.; Bardicef, O.; Sorokin, Y,; Altura, B.M.; Altura, B.T.; and Resnick, L.M. (1996): Perinatal cellular ion metabolism: ^{31}P-nuclear magnetic resonance spectroscopic analysis of intracellular free mangesium and pH in maternal and cord blood erythrocytes. J. Soc. Gynecol. Invest. 3: 66-70.

43. Altura, B.T. and Altura, B.M. (1980): Withdrawal of magnesium causes vasospasm while elevated magnesium produces relaxation of tone in cerebral arteries. Neuroscience Lett. 20:323-327.

44. Altura, B.M. and Altura, B.T. (1981): Magnesium ions and contraction of vascular smooth muscles: Relationship to some vascular diseases. Federation Proc. 40:2672-2679.

45. Altura, B.M.; Zhang, A. and Altura, B.T. (1993): Magnesium, hypertensive vascular diseases, atherogenesis, subcellular compartmentation of Ca^{2+} and Mg^{2+} and vascular contractility. Mineral Electrolyte Metab. 19:323-336.

46. Turlapaty, P.D.M.V. and Altura, B.M. (1978): Extracellular magnesium ions control calcium exchange and content of vascular smooth muscle. Eur. J. Pharmacol. 52:421-423.

47. Altura, B.M. and Altura, B.T. (1981): Magnesium modulates calcium entry and contractility in vascular smooth muscle. In: Ohnishi, T.; Endo, M., ed. The Mechanism of Gated Calcium Transport Across Biological Membranes, pp. 137-145. New York: Academic Press.

48. Altura, B.T. and Altura, B.M. (1982): The role of magnesium in etiology of strokes and cerebrovasospasm. Magnesium 1:277-291.

49. Altura, B.M.; Altura, B.T.; Carella, A. and Turlapaty, P.D.M.V. (1982): Ca^{2+} coupling in vascular smooth muscle: Mg^{2+} and buffer effects on contractility and membrane Ca^{2+} movements. Canad. J. Physiol. Pharmacol. 60:459-482.

50. Altura, B.M. and Altura, B.T. (1984): Magnesium, electrolyte transport and coronary vascular tone. Drugs 28(Suppl. 1):120-142.

51. Altura, B.M.; Altura, B.T.; Carella, A.; Gebrewold, A.; Murakawa, T. and Nishio, A. (1987): Mg^{2+}-Ca^{2+} interaction in contractility of vascular smooth muscle: Mg^{2+} versus organic calcium channel blockers on myogenic tone and agonist-induced responsiveness of blood vessels. Canad. J. Physiol. Pharmacol. 65:729-745.

52. Zhang, A.; Carella, A.; Altura, B.T. and Altura, B.M. (1991): Interactions of magnesium and chloride ions on tone and contractility of vascular muscle. Eur. J. Pharmacol. 203:223-235.

53. Zhang, A.; Cheng, T.P.-O. and Altura, B.M. (1992): Magnesium regulates intracellular free ionized calcium concentration and cell geometry in vascular smooth muscle cells. Biochim. Biophys. Acta Molec. Cell Res. 1134:25-29.

54. Zhang, A.; Cheng, T.P.-O.; Altura, B.T. and Altura, B.M. (1993): Mg^{2+} and caffeine-induced intracellular Ca^{2+} release in human vascular endothelial cells. Brit.J. Pharmacol. 109:291-292.

55. Delpiano, M.A. and Altura, B.M. (1996): Modulatory effect of extracellular magnesium ions on K^+ and Ca^{2+} currents of capillary endothelial cells from rat brain. FEBS Lett. 394: 335-339.

56. Freedman, A.M.; Atrakchi, A.H.; Cassidy, M.M. and Weglicki, W.B. (1990): Magnesium deficiency-induced cardiomyopathy: protection by vitamin E. Biochem. Biophys. Res. Comm. 170:1102-1106.

57. Freedman, A.M.; Cassidy, M.M. and Weglicki, W.B. (1991): Captopril protects against myocardial injury by magnesium deficiency. Hypertension 18:142-147.

58. Weglicki, W.B.; Phillips, T.M.; Freedman, A.M.; Cassidy, M.M. and Dickens, B.F. (1992): Magnesium-deficiency elevates circulating levels of inflammatory cytokines and endothelin. Molec. Cell. Biochem. 118:105-111.

59. Wu, F.; Altura, B.T.; Gao, J.; Barbour, R.L. and Altura, B.M. (1994): Ferrylmyoglobin formation induced by magnesium deficiency in perfused rat heart causes cardiac failure. Biochim. Biophys. Acta Molec. Dis. 1225:158-164.

60. Dickens, B.F.; Weglicki, W.B.; Li, S. and Mak, I.T. (1992): Magnesium deficiency in vitro enhances free-radical intracellular oxidation and cytotoxicity in endothelial cells. FEBS Lett. 311:187-191.

61. Weaver, K. (1980): A possible anticoagulant effect of magnesium in pre-eclampsia. In; Cantin, M.; Seelig, M.S., eds. Magnesium in Health and Disease, pp. 833-838. Jamaica: Spectrum Publ. Inc.

62. Nadler, J.L.; Buchanan, T.; Natarajan, R.; Antonipillai, I.; Bergman, R and Rude, R. (1993): Magnesium deficiency produces insulin resistance and increased thromboxane synthesis. Hypertens. 21(part 2):1024-1029.

63. Rishi, M.; Ahmad, A.; Makheja, A.; Karcher, D. and Bloom, S. (1990): Effects of reduced dietary magnesium on platelet production and function in hamsters. Lab. Invest. 63:717-721.

64. Barbour, R.L.; Altura, B.M.; Reiner, S.D.; Dowd, J.L.; Wu, F.; Altura, B.T. and Gupta, R.K. (1992): Influence of Mg^{2+} on cardiac performance, intracellular free Mg^{2+} and pH in perfused hearts as assessed with ^{31}P-nuclear magnetic resonance spectroscopy. Magnes. Trace Elem. 10:99-116.

65. Wu, F.; Zou, L.Y.; Altura, B.T.; Barbour, R.L. and Altura, B.M. (1992): Low extracellular magnesium results in cardiac failure in isolated perfused hearts. Magnesium Trace Elem. 10:364-373.

66. Friedman, H.S.; Nguyen, T.N.; Mokroaui, A.M.; Barbour, R.L.; Murakawa, T. and Altura, B.M.

(1987): Effects of magnesium chloride on cardiovascular dynamics in the neurally intact dog. J. Pharmacol. Exp. Ther. 243:126-130.

67. Altura, B.M., Barbour, R.L.; Dowd, T.L.; Wu,F.; Altura, B.T. and Gupta, R.K. (1993): Low extracellular magnesium induces intracellular free Mg deficits, ischemia, depletion of high-energy phosphates and cardiac failure in intact working rat hearts: A ^{31}P-NMR study. Biochim. Biophys. Acta 1181:328-332.

68. Morel, D.W. and Chisolm, G.M. (1989): Antioxidant treatment of diabetic rat inhibits lipoprotein oxidation and toxicity. J. Lipid Res. 30:1827-1834; 1989.

69. Rayssiguier, Y.; Gueux, E.; Bussiere, L.; Durlach, J. and Mazur, A. (1993): Dietary magnesium affects susceptibility of lipoproteins and tissues to peroxidation in rats. J. Am. Coll. Nutr. 12:133-137.

70. Berliner, J.A.; Territo, M.C.; Sevanian, A.; Ramin, S.; Kim, J.A.; Bamshad, B.; Esterson, M. and Fogelman, A.M. (1990): Minimally modified low density lipoprotein stimulates monocyte endothelial interactions. J. Clin. Invest. 85:1260-1266.

71. Brautbar, N. and Altura, B.M. (1987): Hypophosphatemia and hypomagnesemia result in cardiovascular dysfunction: Theoretical basis for alcohol-induced cellular injury. Alcoholism: Clin. Exp. Res. 11:118-126, 1987.

72. Altura B.M.; Gebrewold, A.; Altura, B.T. and Brautbar, N. (1996): Magnesium depletion impairs myocardial carbohydrate and lipid metabolism and cardiac bioenergetics and raises myocardial calcium content in-vivo: Relationship to etiology of cardiac diseases. Biochem. Molec. Biol. Intl. 40: 1183-1190.

73. Nahorski, S.R.; Wilcox, R.A.; Mackrill, J.J. and Challiss, R.A.J. (1994): Phosphoinositide-derived second messengers and the regulation of Ca^{2+} in vascular smooth muscle. J. Hypertens. 12(Suppl. 10):S133-S143.

74. Kostellow, A.B.; Morrill, G.A.; Ma,. G.-Y.; Zhang, A.; Altura, B.T.; Altura, B.M. and Gupta, R.K. (1995): Mg^{2+}-dependent changes in fatty acid unsaturation and plasmalogen content of vascular smooth muscle cells: A ^{1}H-NMR study. Proc. intl. NMR Meeting, Nice, August 1995, Abstracts.

75. Morrill, G.A.; Gupta, R.K.; Kostellow, A.B.; Ma, G-Y.; Zhang, A.; Altura, B.T. and Altura, B.M. (1997): Mg^{2+} modulates membrane lipids in vascular smooth muscle: a link to atherogenesis. FEBS Lett. 408: 191-194.

VASCULAR DISEASES AND IONIZED MAGNESIUM

B.T. ALTURA, A. ZHANG, AND B.M. ALTURA
Departments of Physiology and Medicine and The Center for Cardiovascular and Muscle Research, State University of New york, Health Science Center at Brooklyn, Brooklyn, N.Y. 11203 USA

Summary: Use of ion-selective electrodes (ISE's) for IMg^{2+} on plasma, whole blood and sera from patients with a variety of cardiovascular disease syndromes (e.g., hypertension, myocardial infarction, diabetes, long-term renal transplants, liver transplants, during and before cardiac surgery, ischemic heart disease [IHD], non-insulin dependent diabetes (NIDDM), end-stage renal disease [ESRD], peritoneal [HEM] and continuous ambulatory dialysis (CAPD), and stroke syndromes), has revealed interesting data. The results indicate that these vascular disease syndromes are associated with significant depression in serum IMg^{2+} but not TMg. 60-65% of patients early after acute myocardial infarction exhibit significant depression in serum IMg^{2+} levels associated with concomitant rises in CPK and CKMB enzyme levels; most patients with mild infarctions fail to exhibit significant depression in IMg^{2+}. Use of ^{31}P-NMR spectroscopy on red blood cells, from several of these disease states, to assess free intracellular Mg ($[Mg^{2+}]_i$), demonstrates a high correlation ($r = 0.5$-0.8) between IMg^{2+} and $[Mg^{2+}]_i$. Correlations between the clinical course of several of the above disease syndromes and the fall in IMg^{2+} and $[Mg^{2+}]_i$ were found. The ICa^{2+}/IMg^{2+} ratio appears, from our data, to be an important guide for signs of peripheral vasoconstriction, ischemia or spasm and possibly atherogenesis. Overall, our data point to important uses for ISE's for IMg^{2+} in the diagnosis and treatment of vascular disease states.

1. Introduction

Estimation of the ionized Mg^{2+} (IMg^{2+}) level in serum or plasma by analysis of ultrafiltrates (complexed Mg + IMg^{2+}) is somewhat unsatisfactory, as the methods employed do not distinguish the truly ionized form from Mg^{2+} bound to organic and inorganic anions. Moreover, whole blood can not be used for ultrafiltration. Because the levels of these ligands can vary significantly in numerous pathological states, it is desirable to directly measure the levels of IMg^{2+} in complex matrices such as whole blood, plasma and serum. Using novel ion selective electrodes (ISE's), we have found that there is virtually no difference in IMg^{2+}, irrespective of whether one samples whole blood, plasma or serum [1-4]. These data demonstrate that the mean concentration of IMg^{2+} in blood is about 600 μmoles/liter(0.54-0.65 mM, 95% CI); 65-72% of total Mg

T. Theophanides and J. Anastassopoulou (eds.), Magnesium: Current Status and New Developments, 397-405.

being free or biologically-active Mg^{2+}.

Over the past 10 years, data has been brought forth to suggest that the level of extracellular ionized Mg^{2+} ($[Mg^{2+}]_o$) may play an imprtant, crtical role in the control of vascular tone, vascular reactivity, blood flow and cardiac dynamics [for review, see 5]. In view of the importance of $[Mg^{2+}]_o$, we have been attempting to discern whether human cardiovascular disease (CD) syndromes are associated with deficits in serum IMg^{2+} and whether correlations of IMg^{2+} to other important CD parameters and/or risk factors exist. Below, we present a brief summary of our findings so far.

2. Total Serum Mg Does Not Reflect Serum Ionized Mg

Using 237 samples obtained from 31 patients undergoing cardiopulmonary bypass surgery, who received Mg cardioplegia, we found dramatic differences between IMg^{2+} and TMg [2, 4]. The scatter about the regression line in these reports illustrates the uncertainty that would exist in predicting ionized Mg concentration based only on total Mg determinations measured by atomic absorption spectrophotometry.

3. Frequency of Abnormal IMg^{2+} High in Hospitalized Patients and in Patients After AMI

In collaboration with Hartmut Ising in Berlin, we determined in a double-blind study, using the NOVA Stat Profile 8 Analyzer (NOVA ISE's) and KONE Mg^{2+} electrodes, that randomized hospitalized patients clearly exhibit a high incidence of low serum ionized Mg levels. Fifteen out of 20 patients (75%) exhibited IMg^{2+} below our lower limit of IMg^{2+}, i.e.., 0.54 mM/L [6].

Using similar double-blind studies in Berlin and New York, we found that 83% of patients admitted for coronary heart disease and acute myocardial infarction exhibited low serum IMg^{2+} but normal total serum Mg, resulting in significantly lowered % IMg^{2+} levels [6]. On the basis of such data, we speculated that even borderline hypertensive subjects might exhibit depressed serum IMg^{2+}.

3. Ionized Mg Low in Moderate and Severe Untreated Hypertension

Our data indicates that even mild hypertensive patients, that is patients exhibiting arterial blood pressures of 140/90 to 150/95 mmHg (systolic/diastolic), usually exhibit significantly depressed serum ionized, but not total Mg (TMg) levels. The serum ICa^{2+}/IMg^{2+} levels also were elevated [7]. Essential, untreated hypertensive subjects, diagnosed on the basis of elevated blood pressure readings found repeatedly to be in excess of 150/95 mmHg, in the absence of secondary hypertension, exhibited significantly lowered serum free Mg^{2+} and elevated Ca:Mg ratios [5-7].

5. Serum IMg^{2+} Reflects Red Blood Cell Ionized Mg^{2+} in Hypertension

Do serum IMg^{2+} levels reflect intracellular free Mg^{2+} in tissues and cells of hypertensive subjects? In collaboration with Lawrence M. Resnick in Detroit, and using ^{31}P-NMR spectroscopy to measure red blood cell intracellular free Mg, we have recently found a high correlation of RBC free Mg to extracellular serum ionized Mg [6, 7].

6. Correlation of IMg^{2+} Levels with CPK and CKMB Enzyme Levels In Patients Early After Acute Myocardial Infarction

Examination of 85 patients (42-90 years old), randomly admitted to the emergency room for possible acute myocardial infaction (AMI), indicated that 52 of these subjects on hospitalization were diagnosed with AMI. Of these, 39 exhibited early after AMI, 15-50% depression in the level of serum IMg^{2+} with the mean value (0.467 ± 0.008 mM) exhibiting a 22% deficit in serum IMg^{2+}; but no significant deficit in serum TMg. CPK and CKMB enzyme levels were elevated by more than 100- and 15-fold, respectively. The 13 patients diagnosed with milder AMI's exhibited no significant difference in either mean IMg^{2+} (i.e., 0.61 ± 0.01 mM) or TMg levels; CPK and CKMB levels were elevated approximately 25- and 10-fold, respectively. These new data suggest that there is a spectrum or mosaic of AMI patients. It does, however, appear that the more severe forms of the syndrome present with profound deficits in IMg^{2+}. These patients may turn out to be the best candidates for early IV Mg^{2+} therapy as contrasted with those subjects who do not show early deficits in serum IMg^{2+}.

7. Low Levels of Serum Ionized Magnesium Are Found in Patients Early After Stroke

Ninety-eight patients admitted to the emergency rooms of three of our urban hospitals with a diagnosis of either ischemic stroke or hemorrhagic stroke exhibited early and significant deficits in serum IMg^{2+}, but not TMg. Twenty-five percent of these stroke patients exhibited >65% reductions in the mean serum IMg^{2+} found in normal healthy human volunteers or patients admitted for minor bruises, cuts or deep lacerations. The stroke patients also exhibited significant elevation in the serum Ca^{2+}

$(ICa^{2+})/IMg^{2+}$ ratio, a sign of increased vascular tone and cerebrovasospasm [8]. These new results suggest that changes in Mg^{2+} metabolism play important roles in stroke syndromes and in the etiology of cerebrovasospasm associated with hemorrhage.

8. IMg^{2+} in Renal Transplant Recipients

Cyclosporine (CSA) A-treated renal transplant recipients develop atherosclerotic disease and hypertension at an accelerated rate [9]. It seemed to us that such patients would represent a test of the hypothesis that such a classic syndrome should be associated with deficits in serum IMg^{2+}.

The leading cause of death in long-term renal transplant recipients is cardiovascular disease [10]. Hypertension is prevalent in more than 50% of these patients.

We obtained data on 39 renal transplant subjects [11, 12]. The time post-operatively after transplant was between 6 months and 3 years, averaging 26 months. Our lab tests revealed that these transplant patients had elevated serum creatinine, total cholesterol, ionized calcium, and elevated ICa^{2+}/IMg^{2+} ratios.

Renal transplant patients, treated with cyclosporine A, clearly exhibit deficits in IMg^{2+} and % IMg^{2+} but not total Mg [9].

Our data reveals that the most profound drop in serum IMg^{2+} levels in these patients corresponds to the peak blood level of cyclosporine A [9]. The ICa^{2+}/IMg^{2+} ratios, signs of elevated vascular tone, increased vascular reactivity, and risk of atherosclerosis [5], were also elevated [11] and related to the cyclosporin A level. Deficits in IMg^{2+} may be observed as late as 3 years post-transplant in CSA-treated renal transplant patients. It is clear that hypomagnesemia may not be apparent from measurement of TMg alone. Deficiency in IMg^{2+} is not related to dose of CSA or diuretic, or body weight, Ca values, phosphate levels, renal function or time following transplant [11, 12]. Our data indicate that IMg^{2+} values show a strong inverse correlation with the CSA level. IMg^{2+} values fall further as CSA levels rise following drug ingestion, suggesting an acute effect of CSA on redistribution of Mg. In addition, IMg^{2+} values show a strong positive correlation with cholesteral values [11, 12]. We also noted that low IMg^{2+} values are associated with fasting hyperinsulinemia, which has been associated with accelerated atherosclerosis in other populations [13]. Low IMg^{2+} values probably reflect total body Mg depletion as evidenced by decreased urinary excretion in hypomagnesemic patients when compared to patients remaining normomagnesemic [11, 12].

9. IMg^{2+} is Abnormal in End-Stage Renal Disease

End-stage renal disease (ESRD) patients have a high risk for sudden cardiac death, which is unexplained [14]. The relationship to dialysate Mg concentrations remains unclear [15]. Studies of Mg in hemodialysis patients are controversial, showing either normal or increased TMg. But, since hypomagnesemia is associated with cardiovascular disease, carbohydrate intolerance and hypertension, all of which ESRD patients have, we wondered if ionized hypomagnesemia could be important in end-stage renal disease and sudden cardiac death seen in these patients, particularly since low serum Mg^{2+} has been associated with a high frequency of arrhythmias and coronary vasospasm [5]. Studies were therefore performed in 26 randomly selected hemodialysis (HEMO) patients, 10 continuous ambulatory peritoneal dialysis (CAPD) patients, and 66 age-matched control subjects. IMg^{2+} and ICa^{2+} were analyzed utilizing NOVA Mg^{2+} and Ca^{2+} ISE's and a NOVA Biomedical Stat Profile 8 Analyzer. Hematocrit, BUN, creatinine, cholesterol, phosphorous, TCa, and parathyroid hormone were routinely measured.

Hemodialysis patients were dialyzed for 3-4 hours twice weekly. CAPD patients performed 4-5 exchanges with 2 liters of peritoneal fluid. Dialysate Mg was 0.375 mM/L for the hemodialysis patients and 0.25 mM/L for the CAPD patients [16].

The ages for the two groups were similar. Creatinine and BUN levels were elevated in both groups with the CAPD patients showing the higher levels. In both groups of subjects, the serum phosphorous, PTH, and albumin levels were clearly abnormal [16].

Our data indicate that the serum IMg^{2+} values are significantly lowered in both HEMO- and CAPD patients, with CAPD patients showing the greatest deficits. TMg levels are elevated in HEMO but <u>not</u> CAPD patients, whereas there is no change in serum ICa^{2+}. Although both HEMO and CAPD patients exhibit major reductions in % IMg^{2+}, HEMO- patients clearly exhibit the more profound deficit in serum IMg^{2+} [16].

In conclusion, both hemodialysis and CAPD patients had higher ICa^{2+}/IMg^{2+} ratios than control patients. Both hemodialysis and CAPD patients had decreased % IMg^{2+} compared with controls. It is, thus, clear that TMg may be normal or increased in these patients resulting in decreased % IMg^{2+}. Deficits in IMg^{2+} did not correlate simply with indices of nutrition, dialysis efficiency or length of time on dialysis [16]. TCa and ICa^{2+} are normal, resulting in increased ICa^{2+}/IMg^{2+} ratios.

10. Non-Insulin Dependent Diabetes and Ionized Mg in Serum and RBCs

Since 30-40% of patients on dialysis are diabetics, who have atherosclerosis and hypertension, we wondered whether there is a relationship to IMg^{2+} levels. Are diabetes, hypertension, dyslipidemia and accelerated vascular disease linked to insulin resistance [17]? If so, are these in part consequences of Mg deficiency and/or errors in Mg metabolism?

Using the NOVA Stat Profile 8 Analyzer, and the NOVA Mg^{2+} ISE, to measure fasting subjects with and without type II diabetes, we, in collaboration with Lawrence M. Resnick at Cornell University School of Medicine, found that IMg^{2+} levels were significantly reduced in diabetics compared with nondiabetics [18].

Using ^{31}P-nuclear magnetic resonance (NMR) spectoscopy to get an idea of free intracellular Mg^{2+}, we found that there is a correspondence between the deficit in IMg^{2+} and intracellular free Mg^{2+} in Type II diabetics [18].

11. IMg^{2+} in Insulin-Resistant versus Insulin-Sensitive Subjects with NIDDM

In view of these findings, we wanted to test the hypothesis that Mg^{2+} ions play an important role in the pathogenesis of Type-2, non-insulin-dependent diabetes by leading to insulin resistance. In order to do so, and to evaluate the Mg status of two well-characterized black groups of patients with NIDDM characterized by Harold Lebovitz et al [19, 20], we evaluated the two groups, one insulin-resistant, and one insulin-sensitive, in euglycemic remission, after euglycemic clamping. In collaboration with Michelle Barron and Harold Lebovitz at our institution, we desired to determine whether a correlation between IMg^{2+} levels, blood pressure and lipid levels exists in the insulin-resistant NIDDM subjects.

Although not shown, the insulin levels were clearly higher in the insulin-resistant group (14.89 ± 3.42 vs. 7.77 ± 2.4 mU), with a tendency to higher blood triglyceride levels (87.6 ± 9.53 vs. 68.8 ± 9.0 mg/dl).

Examination of our preliminary data on TMg, IMg, % IMg, ICa and the ICa/IMg ratios in the two groups indicate that the IMg^{2+}, $\%IMg^{2+}$ and ICa/IMg^{2+}, but not TMg, in the insulin-resistant group, as predicited from our hypothesis, are all different from the insulin-senitive group, with the IMg and % IMg indicative of deficits.

However, when we examined the insulin-resistant patients more carefully, we found that they could be stratified into two groups if we utilize a cut-off point of our lower normal level of IMg^{2+}, i.e., <0.54 mM/L. Here, it was seen that such stratified low IMg^{2+} patients also exhibit elevated arterial blood pressure, elevated diastolic blood pressure, elevated blood triglycerides and very elevated ICa/IMg ratios [21].

We therefore conclude that 60% of our black Type-2 diabetic subjects with insulin resistance have serum IMg^{2+} levels equal to or below our lower limit of serum IMg^{2+}. None of our type-2 diabetic subjects with insulin-sensitivity exhibit abnormalities in serum IMg^{2+}, TMg or ICa/IMg^{2+} levels. Our new data could thus be used to strengthen our hypothesis.

12. Conclusions

The data gathered, so far, by our group of collaborators in a wide variety of cardiovascular diseases, are consistent with the hypothesis that Mg^{2+} ions play important regulatory roles in cardiovascular cellular dynamics. A fairly large number of diseases which pose high risk for atherogenesis and hypertensive vascular disease (e.g., long term renal transplantation, end-stage renal, non-insulin dependent diabetes, hemo- and peritoneal dialysis, and liver disease) are all associated with deficiences in serum IMg^{2+} and intracellular deficits of Mg^{2+}, as assessed in red blood cells. The advent of ISE's for IMg^{2+} has opened up new vistas in studying and understanding the role of Mg^{2+} in the etiology and treatment of numerous vascular diseases.

13. Acknowledgments

We are grateful to The NIH (Research Grant AA-08674) for supporting some of these studies. We are most grateful to our collaborators, Michelle Barron, Joshua Burack, Eli A. Friedman, Raj K.Gupta, Hartmut Ising, Mariana S. Markell and Lawrence M. Resnick), without whom these clinical studies could not have been done.

14. References

1. Altura, B.T., Shirey, T.L., Young, C.C., Hiti, J., Dell'Orfano, K., Handwerker, S.M. and Altura, B.M. (1992): A new method for the rapid determination of ionized Mg^{2+} in whole blood, serum, and plasma. Methods Find. Clin. Exp. Pharmacol. 14:297-304.

2. Altura, B.T. and Altura, B.M. (1992): Measurement of ionized magnesium in whole blood, plasma and serum with a new novel ion-selective electrode in healthy and diseased human subjects. Magnesium Trace Elem. 10: 90-98.

3. Altura, B.T., Shirey, T.L., Young, C.C., Dell'Orfano, K. and Altura, B.M. (1992): Characterization and studies of a new ion selective electrode for free extracellular magnesium ions in whole blood, plasma and serum. In: Electrolytes, Blood Gases, and Other Critical Analytes: The Patient, the Measurement and the Government, ed. P.D'Orazio, M. Buritt and S.F. Sena, pp. 152-173. Madison: Omni Press.

4. Altura, B.T., Shirey, T.L., Young, C.C., Dell'Orfano, K., Hiti, J., Welsh, R., Yeh, Q., Barbour, R.L. and Altura, B.M. (1994): Characterization of a new ion selective electrode for ionized magnesium in whole blood, plasma, serum and aqueous samples. Scand. J. Clin. Lab. Invest. 54 (Suppl. 217):21-36.

5. Altura, B.M. and Altura, B.T. (1995): Magnesium in cardiovascular biology. Scientific Am. Science & Med. 2:28-37.

6. Altura, B.T., Bertschat, F., Jeremias, A. Ising, H. and Altura, B.M. (1994): Comparative findings on serum IMg^{2+} of normal and diseased human subjects with the NOVA and KONE ISE's for Mg^{2+}. Scand. J. Clin. Lab. Invest. 54 (Suppl. 217):77-82.

7. Altura, B.M. and Altura, B.T. (1996): Role of magnesium in patho-physiological processes and the clinical utility of magnesium ion selective electrodes. Scand. J. Clin. Lab. Invest. 56(suppl. 224): 211-234.

8. Altura, B.T., Memon, Z.I., Zhang, A., et al (1997): Low levels of serum ionized magnesium are found in patients early after stroke which result in rapid elevation in cytosolic free calcium and spasm in cerebral vascular muscle cells. Neuroscience Lett. 230: 37-40.

9. Gao, S.Z., Schroeder, J.S., Hunt, S. and Stinson, E.B. (1988): Retransplantation for severe accelerated coronary artery disease in heart transplant recipients. Am. J. Cardiol. 62:876-881.

10. Fassbinder, W., Challah,S. and Brynger, H. (1987): Long term results of renal transplantation in Europe. Trans. Proc. 19:3754.

11. Markell, M.S., Altura, B.T., Barbour, R.L. and Altura, B.M. (1993): Ionized and total magnesium levels in cyclosporin-treated renal transplant recipients: relationship with cholesterol and cyclosporin levels. Clin. Sci. 85:315-318.

12. Markell, M.S., Altura, B.T., Sarn, Y., Barbour, R.L., Friedman, E.A. and Altura, B.M. (1994): Relationship of ionized magnesium and cyclosporine level in renal transplant recipients. Ann. N.Y. Acad. Sci. 696:408-411.

13. Moller, D.E., ed. (1993); Insulin Resistance. New York: Wiley.

14. U.S. Renal Data System, USRDS 1992 Annual Data Report, The National Institutes of Health, NIDDK, Bethesda, MD, Aug. 1992.

15. Valporean, M.L. and VanStone, J.C. (1993): Dialysate magnesium. Seminars in Dialysis 6:46-51.

16. Markell, M.S., Altura, B.T., Sarn, Y., Delano, B.G., Ifudu, O., Friedman, E.A. and Altura, B.M. (1993): Deficiency of serum ionized magnesium in patients receiving hemodialysis or peritoneal dialysis. ASAIO J. 39:M801-M804.

17. Altura, B.M. and Altura, B.T. (1995): Magnesium metabolism and cardiovascular pathobiology. In: Magnesium and Physical Activity, ed, L. Vecchiet, pp. 37-70. Lancashire: Parthenon Press.

18. Resnick, L.M., Altura, B.T., Gupta, R.K., Alderman, M.H. and Altura, B.M. (1993): Intracellular and extracellular magnesium depletion in type 2 (non-insulin-dependent) diabetes mellitus. Diabetologia 36:767-770.

19. Banerji, M.A. and Lebovitz, H.E. (1989): Insulin-sensitive and insulin-resistant variants in NIDDM. Diabetes 38:784-792.

20. Chaiken, R.L., Banerji, M.A., Huey, H. and Lebovitz, H.E. (1993): Do Blacks with NIDDM have an insulin-resistance syndrome? Diabetes 42:444-449.

21. Barron, M. Altura, B.T., Banerji, M.A., Lebovitz, H.E. and Altura, B.M. (1995): Unpublished data.

INFLUENCE OF MAGNESIUM ON THE ESSENTIAL OIL PRODUCTION IN CHAMOMILE CULTURES

É. SZÕKE [1], E. MÁDAY [1], É. LEMBERKOVICS [1], A.S. KISS [2], ZS. MUSKÁTH [3]
[1] *Semmelweis University of Medicine, Institute of Pharmacognosy 1085 Budapest, Üllõi út 26.,* , [2] *Attila József University, 6726 Szeged,*
[3] *Semmelweis University of Medicine, Institute of PharmacyBudapest IX., Hõgyes Endre u. 9., Hungary*

1. Introduction

Magnesium plays a part on every cell of the living organism, almost in all life processes. As for its general biological importance we mention its role in the mitochondria and ribosomes.
Many authors have pointed out that the reduced energy supply and protein synthesis are caused by a lack of magnesium, which thus inhibits the growth of plants. The magnesium is indispensable for plant development. In connection with the utilization of natural magnesium sources in crop plants, it was pointed out there that a deeper knowledge of the effect of the magnesium supply provides an opportunity for enhancement of the effectivity of agricultural production. In order to achieve this goal, it is first absolutely necessary to know how plants absorb magnesium from the soil and what role this element plays in their life processes [1]. In order to get a thorougher knowledge of the role of magnesium we analysed the effect of $MgSO_4$ on organized chamomile cultures [2].
In the course of our experiment we followed the changes in the essential oil output and composition parallel with the biomass production, with special regard to the quantity of components considered important from a therapeutical poin of view [3,4].
Twenty per cent of the magnesium is stably incorporated in the plant cells, e.g. in the chlorophyll; the rest is to be found in the ribosomes, mitochondria, enzymes. It enhances the enzyme- and hormone activity, stimulates the synthesis of gibberellin [5] and auxin [6], activates the phosphorylation processes. Some 50 per cent of the magnesium is present in ionic form in the plant. It is of very great importance from more than one point of view, partly because of the high hydrating effect; it causes experimentally 10-15 per cent reduction of evaporation, whereby the drought tolerance of plants increases [7]. On the other hand, today the importance of high Mg content of medicinal plants, and tea blands or pharmaceutical products prepared from them is steadly increasing. Namely, in some parts of Hungary the drinkind water is Mg - deficient, which may lead to thedevelopment of various deseases, e.g. heart troubles and circulatory disturbances,

T. Theophanides and J. Anastassopoulou (eds.), Magnesium: Current Status and New Developments, 407-411.

high blood pressure, paradenthosis. Mg therapy is employed in many further fieldsof indication, e.g. physical disorders, allergy, menstrual complaints. Furthemore, it has been found that the maintance of pregnancy can be ensured by adding Mg ion, and it is also indispensable for the healthy development of the foetus.

We chose the chamomile as the object of our experiments because it is even today our most sought after medicinal plant on both the domestic and foreign market. A considerable proportion of the drug is now produced by field cultivation of the plant in the course of which it would be expedient to use $MgSO_4$ to increase the yield.

2. Materials and methods

The organized cultures of chamomile (*Matricaria recutita* L., syn.: *Matricaria chamomilla* L. Asteraceae) were produced from callus tissues of inflorescence origin on Murashige - Skoog [8] culture medium of mineral- and vitamine composition, containing 10% coconut milk. After transferred several times the cultures were raised on hormone - free MS medium (120 ml) at 26°C, in light (2500 Lux daily 10 - hour illumination in 500 ml Erlenmeyer flask).

In order to analyse the effect of $MgSO_4$ we added it to the culture medium in doses reduced to one - half (185 mg/l) and redoubled (740 mg/l), respectively, in addition to the 370 mg/l concentration prescribed by Murashige - Skoog [8]. The control cultures were kept on culture medium free from $MgSO_4$.

We characterized the growth activity of the tissues [10,11]. The fresh weight values characteristic of the different $MgSO_4$ concentrations were obtained by weighting the fresh herb and roots separately for each culture (8 weeks old), and evaluating the results statistically.

Hairy root cultures: transformed root cultures of *Matricaria recutita* were obtained by infection of sterile two - month - old orgabized cultures with *Agrobacterium rhizogenes* strains A 4, 15834 and R 1601. Different strains showed various abilities to induce hairy roots in this plant. The promoting effect of acetosyringone on the *Agrobacterium* mediated hairy roots initiation in chamomile has been investigated. The antibiotic carbenicillin (800 mg/l) and the combination of cefotaxime (250 mg/l) with ampicillin (1000 mg/l) were added for several subcultures to eliminate the bacteria. The hairy roots exempted from bacteria were cultivated on solid and then in liquid hormone - free B5 (B50) and MS (MS0) medium (3,4) at 24 ± 2°C. We have also studied the effect of $MgSO_4$ on genetically transformed hairy root cultures of chamomile.The established hairy root cultures were maintained on solid B50 medium and subcultured every 6 weeks.

Cultural conditions of the experiment: After 8 subcultures the hairy roots were transferred to B50 medium containing $MgSO_4$ in the following concentrations: 0, 125, 250 and 500 mg/litre (original Gamborg's B5 medium contains 250 mg/l $MgSO_4$). Four pieces of hairy root tip approx. 1,5 cm in length were inoculated into every Petri dish. The cultural conditions were as mentioned above (24 ± 2°C, in darkness). The rate of root growth, expressed in mm/day, was estimated by measuring root length every 2 days during the first 30 days of culturing.

Chemical analyses: for the purpose of chemical analyses the plants were refrigerated at -40°C. The samples were homogenized with biomix (1000 r/min), and the essential oil was produced by means of a modified Clavenger apparatus, with water - steam destillation. The total amount of essential oil wsa determined by gravimetric method (VI. Swiss Pharmacopoeia). Finally the essential oil was used in the subsequent analyses (GC, GC - MS).
The qualitative and quantitative composition of the essential oil was exemined in Fisons GC 800 type gaschromatograph.
GC parameters: column: 30 × 0,32 mm I. D., 0,25 μm; stacionary: DB - 1701; column temperature: 60 - 23°C, 8°C/min, 230°C isoterm 3 min; detector: FID, 240°C; carrier gas: N_2, p N_2: 50 kPa; V N_2: 6,8ml/min; injector temperature: 200°C; injection: splitless: 10 sec (1/10); evaluation: standard addition and retention factors.
For mass spectrometric measuring a Jeol JMS - D 100 and Finnigan GCQ type gas chromatograph - mass spectrometer was used.

3. Results

Besides attaining a larger drug yield, the Mg ion content of the drug would thereby be increased as well. It is particlarly important to obtain data about how the $MgSO_4$ influences the essential oil content of the organized chamomile cultures (herb, roots) and genetically transformed hairy roots beside the biomass production, with special regard to the therapeutically important components. In this section, the results of experiments with plant tissue cultures will be utilized to demonstrate, that in addition to influence primary metabolic processes, magnesium also has effects on the secondary or special matabolism.
As $MgSO_4$ has an exceptionally positive effect on rooting of *M. recutita* L. organized cultures, it is interesting to demonstrate its effect on the growth of isolated roots and genetically transformed hairy root cultures. We observed the biomass production and essential oil formation in the cultures with respect to the therapeutically active compounds. The cultures grew well on the Murashige - Skoog medium (MS) supplemented with $MgSO_4$. It was established that the fresh weight of the root is parallel to the concentration of $MgSO_4$ salt.
When the cultures were raised on a medium free from $MgSO_4$ they hardly grew, and their root formarion was greatly inhibited.
In the herb of cultures raised on a culture medium lacking or deficient (185 mg/l) in magnesium chloroplast destruction can be observed. The gradual disintegration of the thylakoid membrane system of the chloroplasts is supposed to be connected with the fact that in Mg - deficient cells the stabilizing effect of the Mg ions is missing [12].
With the concentration of $MgSO_4$ redoubled in the culture medium (740 mg/l) grew about twofold, rising from 6,80 g to 12,99 g (compared to the MS basic medium). This tendency is reflected in the growth values too, in terms of both fresh and dry weight. The leaves are bright green.
The root formation is considerable even under the influence of 185 mg/l $MgSO_4$ (1/2 dose), but it is most intensive in response to the redoubled $MgSO_4$ concentration. It must

be noted, however, that with an increase in the concentration (from 185 mg/l to 740 mg/l) the weight of the roots grew but slightly.
Analysing the essential oil content (%) and essential oil production (mg/culture) of herb in the cultures we found that they also showed a significant increase under the influence of $MgSO_4$. At a concentration of 740 mg/l $MgSO_4$ in the culture medium the essential oil production became six times as much as in the control.
An analysis of the essential oil composition points out that in comparison to cultures raised on culture media free from $MgSO_4$ the application of Mg results in a considerable increase in the quantity of the components (t-β-farnesene, α-farnesene, spatulenol, spiroethers), though they reach maximum at lower $MgSO_4$ concentrations.
Similar tendency was observed in the hairy root cultures genetically transformed with *Agrobacterium rhizogenes* strains A 4, 15834 and R 1601. The essential oil content increased by adding $MgSO_4$ in 740 mg/l concentration. Some of the important essential oil components (t-β-farnesene, α-farnesene, spatulenol, spiroethers) increased considerably.
Summarizing our results we can declare that the $MgSO_4$ positively influences both the biomass rate and the essential oil production also of isolated roots and genetically transformed roots. Furthemore it has a significant effect on its composition too.

4. Conclusion

During our experiments we studied the influence of Mg ions on the hairy root cultures *Agrobacterium rhizogenes* strains A 4, 15834 and R 1601. As $MgSO_4$ has an exceptionally positive effect on rooting of *Matricaria recutita* L. (Asteraceae) organized cultures, it is interesting to demonstrate its effect on the growth of isolated roots and genetically transformed hairy roots.
The biomass production and essential oil formation in a the cultures were studied with respect to the therapeutically important compounds. These were identified by GC and GC - MS methods.
The magnesium ions positively influenced both the biomass rate and the essential oil production. Furthemore it has a significant effect on its composition.

References

1. Kiss, A.S. (1983): *Magnéziumtrágyázás, magnézium biológiában. Mezõgazdasági Kiadó*, Budapest.
2. Szõke, É., Kiss, A.S., Kursinszki, L., and Petri, G. (1989): Effect of magnesium ions on the chamomile tissue cultures. *Magnesium Research* **2**, 227.
3. Cartoni, G., Goretti, G., Russo, M.V., and Zacchei, P. (1990): Microcapillary gas - chromatographic analysis of chamomile. *Ann. Chim.* **80**, 523 - 532.
4. Gilbertson, G. and Koenig, T. R. (1981): Essential oils and related products. *Anal. Chem.* **53**, 61 - 77.
5. Kiss, A.S. (1977): Magnesium effect on the synthesis of embrional gibberelline. Proc. 17th Hung *Ann. Meet. Biochem. Kecskemét*, 119 - 120.
6. Kiss, A.S. (1980): The effect of magnesium on the auxin - content of plants. *Proc. 20th Hung. Ann. Meet. Biochem.* Siófok, 239 - 240.

7. Kiss, A.S. and Baranyi, A. (1979): A növények vízháztartása. *Agrokémiai Tájékoztató, Budapest*, **4** (4) 60 - 62.
8. Murashige, T. and Skoog, F. (1962): A revised medium for rapid growth and bio - assays with tobacco tissue cultures. *Physiol. Plant.* **15**, 473 - 497.
9. Szõke, É., Kuzovkina, I.N., Verzár - Petri, G., and Szmirnov, A.M. (1979): Növekedést befolyásoló anyagok hatása a kamilla (*Matricaria chamomilla* L.) kallusztenyészetek biomassza képzésére.*Herba Hungarica* **18**, 41 - 57.
10. Szõke, É., Shavarda, A.L., and Kuzovkina, I.N. (1979): Effect of culturing conditions on essential oil formation in callus tissue of wild chamomile inflorescences.*Sowiet Plant Physiology* **25**, 579 - 584.
11. Szõke, É., Kiss, S., Kursinszki, L., and Petri, G. (1992): Inhibion of Al toxicity with Mg and citric acid in tobacco tissue culture. *Magnesium Research* **5**, 83 - 87.
12. Kiss, A.S. (1989): Effect of magnesium ions on the anion uptake of plants.*Acta Agron. Hung.* **38**, 23-29.

AUTHOR INDEX

SUBJECT INDEX